ERRATUM

to *Wide Components in Double and Multiple Stars*
Edited by J. Dommanget, E. L. van Dessel, and Z. Kopal.
This index was omitted from the reprint of *Astrophysics and Space Science*, Volume 142, 1988.

INDEX

of topics in direct relation with the colloquium; no terms of general notion are included. 'f' stands for: 'and following pages'.

Area scanner, 14, 125

Barium stars in wide binaries, 199f
Birthrate function, 271, 274
Bruce Proper Motion Survey, 17

Cataclysmic variables in triple systems, 94
CCDM, 23f, 31, 37
Census, 7, 30
Commission 26, 5, 8, 11, 172, 305
Common proper motion stars, 7–8, 11, 13, 17f, 51f, 67, 91, 111, 116, 118f, 123f, 139, 145, 151, 179, 223f, 238f, 289
Common proper motion stars, degenerate components, 18

Disruption (dissociation), 68, 73f, 103, 119, 259, 268f, 289, 299, 306
Distribution
- of binaries, see 'frequency'
- of difference of magnitude, 140f, 240
- of mass and mass ratios, 141–142, 239, 245f, 289
- of semi-axis major, 7, 137, 268f, 279, 284, 289
- of separations, 7, 51–52, 68f, 93, 112f, 131, 136, 224, 239f, 245f, 277

Evolution
- binary (orbital) evolution, 6–7, 170, 267–268, 274, 301–302
- galactic, see 'galactic dynamics'
- stellar, 103, 242

Formation (fission, fragmentation), 89, 103, 123, 173, 241, 278, 289, 300
Frequency of binaries, 49, 68f, 239, 245f, 278, 282

Galactic dynamics and evolution, 6, 169, 286, 289
Galactic formation and structure, 6, 170

Hierarchical systems, 69, 115
Hipparcos, 7, 13, 23f, 60f, 97, 161, 174, 208, 242, 255, 257, 259f, 287, 298
Hipparcos Input Catalogue, 23f, 31, 33, 37, 211–212, 258f

Initial mass function, 142–143, 247–248
Interferometric measurements, 64
Invisible companions (third companions), 13, 124, 211, 224

LDS, 17f, 56, 67, 72, 244
LHS, 50f, 77
Limiting separation (maximum-, cut-off-), 7, 112f, 224, 245–246, 263, 267, 274, 289, 299f
Linkage of reference systems, 61, 64

Mass loss, 7, 94, 170, 241, 290f
Mass-luminosity relation, 141, 219, 245, 249
Moving groups, 58, 74, 104, 111, 119, 174

Nearby stars, Section 2 – 47f
Nemesis, 137

Occultation technique, 12, 203f
Optical (random) pairs, 8, 17, 67, 77, 86, 92, 93, 103f, 118, 129, 237f, 256–257, 259
Orbits (orbital elements, orbital motion), 39f, 153f, 181f, 192, 194, 217f, 255f, 279, 296
Orbital poles (planes), 171f

Period-eccentricity relation, 169–170, 289, 294
Photocenter, 28, 62
Planetary objects, 12

Quadrant ambiguity, 101

Radiostars in wide binaries, 61f
Reference systems, see 'linkage'

Selection effects, 7, 21, 49, 94, 113, 131, 136, 139–140, 239, 271, 289, 301
Space Telescope, 13, 19, 23, 240
Speckle Interferometry, 14, 21, 93, 99f, 210
Substellar objects, 12, 15
Superclusters, 145

Trapezia systems, 11, 68f, 79f, 85f, 111, 115–116, 299
Tycho, 238

Velocity (space-), 57f, 103, 119, 143, 174, 239

WDS, 21, 77
W UMa systems in wide binaries, 94

WIDE COMPONENTS IN DOUBLE AND MULTIPLE STARS

WIDE COMPONENTS IN DOUBLE AND MULTIPLE STARS

Proceedings of the 97th Colloquium of the International Astronomical Union held in Brussels, Belgium, 8–13 June, 1987

Edited by

J. DOMMANGET
Department of Astrometry, Royal Observatory of Belgium, Brussels, Belgium

E. L. VAN DESSEL
Department of Astrophysics, Royal Observatory of Belgium, Brussels, Belgium

and

Z. KOPAL
Department of Astronomy, University of Manchester, England

Reprinted from Astrophysics and Space Science
Volume 142, 1988

Kluwer Academic Publishers
Dordrecht/Boston/London

Library of Congress Cataloging in Publication Data

CIP

CIP-data appear on seperate card.
ISBN 90-277-2737-6

Published by D. Reidel Publishing Company,
P.O. Box 17, 3300 AA Dordrecht, The Netherlands.

Sold and distributed in the U.S.A. and Canada
by Kluwer Academic Publishers,
101 Philip Drive, Norwell, MA 02061, U.S.A.

In all other countries, sold and distributed
by Kluwer Academic Publishers Group,
P.O. Box 322, 3300 AH Dordrecht, The Netherlands.

Kluwer Academic Publishers incorporates
the publishing programmes of
D. Reidel, Martinus Nijhoff, Dr W. Juk and MTP Press.

Printed in The Netherlands

WIDE COMPONENTS IN DOUBLE AND MULTIPLE STARS

Proceedings of the 97th Colloquium of the International Astronomical Union, held in Brussels, Belgium, 8–13 June, 1987

Edited by

J. DOMMANGET

Department of Astrometry, Royal Observatory of Belgium, Brussels, Belgium

E. L. VAN DESSEL

Department of Astrophysics, Royal Observatory of Belgium, Brussels, Belgium

and

Z. KOPAL

Department of Astronomy, University of Manchester, England

TABLE OF CONTENTS

OPENING SESSION

P. MELCHIOR / Welcome Address of the Director of the Royal Observatory, Belgium 3

J. DOMMANGET / General Introductory Lecture 5

FIRST SESSION
CATALOGUING WIDE PAIRS

M. G. FRACASTORO / The Concept of Star Multiplicity and its Historical Evolution 11

W. J. LUYTEN / Wide Double Stars with Common Proper Motion 17

C. E. WORLEY / Wide Pairs in the Washington Double Star Catalogue (WDS) 21

J. DOMMANGET / The Contribution of the Hipparcos Mission and of the CCDM Catalogue to the Study of Wide Systems 23

E. OBLAK / A Photometric Catalogue of Visual and Multiple Stellar Systems 31

YAN LIN-SHAN / A General Catalogue of Ephemerides and Apparent Orbits for Visual Binaries 39

SECOND SESSION
WIDE NEARBY BINARIES

W. GLIESE and H. JAHREISS / The Third Catalogue of Nearby Stars with Special Emphasis on Wide Binaries 49

A. R. UPGREN / The Search for Wide Pairs Among Nearby Stars 57
S. DÉBARBAT and F. CHOLLET / A Plea for Better Astrophysical and Astrometric Data in Case of Double Stars of Peculiar Interest 61

THIRD SESSION

WIDE COMPONENTS IN MULTIPLE SYSTEMS

A. POVEDA / Distant Companions in Double and Multiple Stars 67
G. N. SALUKVADZE and G. SH. JAVAKHISHVILI / Trapezium-Type Wide Systems 79
A. I. YATSENKO / The Kinematics of Trapezium-Type Systems 85
T. J. HERCZEG / A Survey of the Third Components in Spectroscopic and Eclipsing Systems 89
L. BORRIELLO / Imaging Approach to Multiple Star Recognition: The Algorithm, the Results 97
C. WAELKENS and P. LAMPENS / Physical Parameters of the Massive Multiple System Eta Orionis 99
J. P. ANOSOVA / A Search and Classification of Multiple Systems in the General Field 103

FOURTH SESSION

OBSERVATIONS OF WIDE PAIRS AND THEIR RESULTS

H. A. ABT / Visual Binary Separations as Functions of Primary Types, Ages and Locations 111
K. D. RAKOS and O. G. FRANZ / Photoelectric Observations of CPM Stars in the BVRI System 123
T. MAZEH and D. W. LATHAM / On the Wide Binaries in Selected Area 57 131
J.-L. HALBWACHS / Studies of a Sample of CPM Stars 139
O. J. EGGEN / Wide Binaries and Old Disk Population Groups 145
N. WIETH-KNUDSEN / Orbit Determination from a Short Arc Through the Coefficients of Power Series in Time 153
H. H. BERNSTEIN, R. HERING, and H. G. WALTER / Astrometric Parameters of Visual Double Stars Derived from Simulated Hipparcos Measurements 161

POSTER SESSION

J. DOMMANGET / Tentative Comments about the Meeting 169
J. DOMMANGET / The Space Arrangement of the Orbital Planes of the Visual Double Stars 171
L. O. LODEN / A Possible Connection between Wide Binaries and Open Clusters 177
A. A. KISSELEV and O. V. KIYAEVA / New Orbits of Four Visual Double Stars by the Method of Apparent Motion Parameters (AMP) 181

M. L. BOUGEARD / Are Double Star Observations The Most Suspicious Ones in the Paris Astrolabe Programme? 185
J. A. DOCOBO and J. F. LING / Visual Double Stars in Professional Spanish Astronomy: Past, present and Future Projects 191
J. A. DOCOBO, A. ELIPE, and A. ABAD / A Method for Calculating the Orbits of Interferometer/Spectroscopic Binaries from Combined data 195
R. B. CULVER and P. A. IANNA / The Frequency of Wide Binaries Among the Southern Barium Stars 199
M. FROESCHLE and C. MEYER / Observation et Rèsolution du Système Triple η Virginis par la Méthode des Occultations 203
R. PANNUNZIO and R. MORBIDELLI / The Researches at Torino on Wide Double Stars 207
S. WEICHINGER and K. D. RAKOS / Precise Radial Velocity Measurements of Visual Binary Stars 211
M. SCARDIA / Mesures Photographiques d'Étoiles Doubles Visuelles Larges 213
M. SCARDIA / Positions Précises d'Étoiles Doubles Visuelles Larges 215
S. SIREGAR / Software for Calculating Orbit and Mass of Visual Binaries 217
E. W. WEISS / VRI Photometry of Late Dwarf Common Proper Motion Stars 223
D. SINACHOPOULOS, H. NICKLAS, and M. GEFFERT / CCD Suitable for Double Star Astrometry 227
H. ZINNECKER, A. CHELLI, L. CARRASCO, I. CRUZZ-GONZALES, and C. PERRIER / Infrared Companions to Pre-main Sequence Stars 231
J. P. ANOSOVA and V. V. Orlov / Dynamical Evolution of Triple Systems (Scientific Film) 233

FIFTH SESSION

STATISTICAL STUDIES ON WIDE PAIRS

J. L. HALBWACHS / Statistical Studies on Wide Pairs 237
S. VERESHCHAGIN, Z. KRAICHEVA, E. POPOVA, A. TUTUKOV, and L. YUNGELSON / Statistical Study of Visual Binaries 245
P. BROSCHE and D. SINACHOPOULOS / The Optical Fraction of a Hipparcos Sample of Wide Binaries 255
H. HENSBERGE, A. JONGENEELEN, and R. LE POOLE / Wide Binaries in a Hipparcos Sample Centered Around $b = 0°$, $1° = 300°$ 259

SIXTH SESSION

ORIGIN AND EVOLUTION OF WIDE PAIRS

I. WASSERMAN / Theoretical Models for Wide Binary Evolution 267
M. D. WEINBERG / What Can We Learn from Studies of Wide Binary Statistics 277
D. VALLS-GABAUD / Evidence of Mass Loss in Visual Binary Stars 289
Z. KOPAL / Concluding Remarks 305

1. H. Hensberge
2. Mrs. Kopal
3. L. O. Loden
4. M. Froeschlé
5. C. Meyer
6. S. Débarbat
7. D. Valls-Gabaud
8. L. Yungelson
9. J.-L. Halbwachs
10. E. Van Albada - Van Dien
11. T. Herczeg
12. A. I. Yatsenko
13. Z. Kopal
14. W. Gliese
15. T.Mazeh
16. A. Upgren
17. H. H. Bernstein
18. M. D. Weinberg
19. W. J. Luyten
20. I. Wasserman
21. Mrs. Wieth-Knudsen
22. Mrs. Luyten
23. M. G. Lattanzi
24. N. P. Wieth-Knudsen
25. J. Delhaye
26. R. Morbidelli
27. G. N. Salukvadze
28. M. Scardia
29. C. E. Worley
30. S. Vereshchagin
21. D. Sinachopoulos
32. S. Siregar
33. R. B. Culver
34. Yan Lin-Shan
35. E. Oblak
36. K. Rakos
37. P. A. Ianna
38. J. Smak
39. A. Poveda
40. R. Dejaiffe
41. H. A. Abt
42. E. W. Weis
43. O. Nys
44. Mrs. Bacchus
45. L. H. Louys
46. Mrs. Dommanget
47. G. Roland
48. E. Van Dessel
49. T. Pauwels
50. J. Dommanget
51. P. Bacchus
52. C. Waelkens
53. P. L. Bernacca
54. L. Borriello

SCIENTIFIC ORGANIZING COMMITTEE

H. A. ABT, Kitt Peak National Observatory, Tucson, U.S.A.
J. DELHAYE, Observatoire de Paris, Paris, France
J. DOMMANGET, Observatoire Royal de Belgique, Bruxelles, Belgium, President
M. FRACASTORO, Osservatorio Astronomico di Torino, Pino Torinese, Italy
W. GLIESE, Astronomisches Rechen-Institut, Heidelberg, Germany
Z. KOPAL, Department of Astronomy, University of Manchester, England
K. RAKOS, Institut für Astronomie, Wien, Austria
J. SMAK, Copernicus Astronomical Center, Warszawa, Poland
A. UPGREN, Van Vleck Observatory, Wesleyan University, Middletown, U.S.A
E. VAN DESSEL, Koninklijke Sterrenwacht van Belgie, Brussel, Belgium
S. VERESHCHAGIN, Astronomical Council of the U.S.S.R. Academy of Sciences, Moscow, U.S.S.R.

LOCAL ORGANIZING COMMITTEE

J. DOMMANGET, Observatoire Royal de Belgique, Bruxelles, Belgium
J. HENRARD, Facultés Universitaires N.D. de la Paix, Department de Mathématiques, Namur, Belgium
P. MELCHIOR, Observatoire Royal de Belgique, Bruxelles, Belgium
T. PAUWELS, Koninklijke Sterrenwacht van Belgie, Brussel, Belgium
E. VAN DESSEL, Koninklijke Sterrenwacht van Belgie, Brussel, Belgium, President
C. WAELKENS, Astronomisch Instituut, Katholieke Universiteit Leuven, Heverlee, Belgium
– with the assistance of S. DEWINTER (Mrs), R. DEJAIFFE, G. ROLAND and O. NYS of the Royal Observatory, Belgium

PARTICIPANTS

P. BACCHUS, Observatoire de l'Université de Lille, France
P. L. BERNACCA, Osservatorio Astrofisico, Asiago, Italy
H. H. BERNSTEIN, Astronomisches Rechen-Institut, Heidelberg, Germany
L. BORIELLO, CSATA, Valenzano-Bari, Italy
M. BOUGEARD, Observatoire de Paris, France
R. B. CULVER, Department of Physics, Colorado State University, Fort Collins, U.S.A.
S. DEBARBAT, Observatoire de Paris, France
R. DEJAIFFE, Observatoire Royal de Belgique, Bruxelles, Belgique
M. FROESCHLE, CERGA, Grasse, France
J. L. HALBWACHS, Obervatoire de l'Université de Strasbourg, France
H. HENSBERGE, Sterrewacht Leiden, The Netherlands
T. HERCZEG, Department of Physics and Astronomy, Norman University of Oklahoma, U.S.A.

P. A. IANNA, Department of Astronomy, University of Virginia, Charlottesville, Virginia, U.S.A.
A. JONGELEENEN, Sterrewacht Leiden, The Netherlands
M. G. LATTANZI, Osservatorio Astronomico di Torino, Pino Torinese, Italy
J. FF. LING LING, Observatorio Astronomico Ramon Maria Aller, Santiago de Compostela, Spain
L. O. LODEN, European Southern Observatory, Garching-bei-München, Germany
L. H. LOUYS, Observatoire Royal de Belgique, Bruxelles, Belgium
W. J. LUYTEN, University of Minnesota, Minneapolis, Minnesota, U.S.A.
T. MAZEH, Wise Observatory, Tel Aviv University, Tel Aviv, Israel
C. MEYER, CERGA, Grasse, France
R. MORBIDELLI, Observatorio Astronomico di Torino, Pino Torinese, Italy
O. NYS, Koninklijke Sterrewacht van Belgie, Brussel, Belgium
E. OBLAK, Observatoire de Besançon, France
A. POVEDA, Instituto de Astronomia, Universidad Nacional Autonoma de Mexico, Mexico City, Mexico
G. ROLAND, Observatoire Royal de Belgique, Bruxelles, Belgium
G. N. SALUKVADZE, Mt. Kanobili Astrophysical Observatory, Abastumani, U.S.S.R.
M. SCARDIA, Osservatorio Astronomico di Brera, Merate, Italy
D. SINACHOPOULOS, Universitätssternwarte Bonn, Germany
S. SIREGAR, Bosscha Observatory, Lembang, Indonesia
D. VALLS-GABAUD, Institut d'Astrophysique de Paris, France
E. VAN ALBADA-VAN DIEN, Amsterdam, The Netherlands
I. WASSERMAN, Department of Astronomy, Cornell University, Ithaca, N.Y., U.S.A.
M. D. WEINBERG, Institute for Advanced Study, Princeton, N.J., U.S.A.
E. W. WEISS, Van Vleck Observatory, Wesleyan University, Middletown, Conn., U.S.A.
N. P. WIETH-KNUDSEN, Frederiksberg, Denmark
C. E. WORLEY, U.S. Naval Observatory, Washington, D.C., U.S.A.
YAN LIN-SHAN, Shanghai Observatory, Academia Sinica, Shanghai, China
A. I. YATSENKO, Main Astronomical Observatory, Ukrainian Academy of Sciences, Kiev, U.S.S.R.
L. YUNGELSON, Astronomical Council of the U.S.S.R. Academy of Sciences, Moskva, U.S.S.R.

DID NOT PARTICIPATE BUT HAVE PRESENTED A PAPER

A. ABAD, Santiago de Compostella (E); J. P. ANOSOVA, Leningrad (SU); P. BROSCHE, Bonn (D); L. CARRASCO, Mexico City (MEX); A. CHELLI, Mexico City (MEX); F. CHOLLET, Paris (F); I. CRUZ-GONZALES, Mexico City (MEX); J. A. DOCOBO-DURANTEZ, Santiago de Compostella (E); O. J. EG-

GEN, Tucson (USA); A. ELIPE, Santiago de Compostella (E); O. G. FRANZ, Flagstaff (U.S.A.); M. GEFFERT, Bonn (D); R. HERING, Heidelberg (D); H. JAHREISS, Heidelberg (D); G. Sh. JAVAKHISHVILI, Abastumani (SU); O. V. KIYAEVA, Leningrad (SU); Z. KRAICHEVA, Sofia (BG); D. LATHAM, Cambridge (USA); R. LE POOLE, Leiden (NL); H. NICKLAS, Heidelberg (D); V. V. ORLOV, Leningrad (SU); R. PANNUNZIO, Torino (I); C. PERRIER, Lyon (F); E. POPOVA, Sofia (BG); A. TUTUKOV, Moskva (SU); H. G. WALTER, Heidelberg (D); S. WEICHINGER, Wien (A) and H. ZINNECKER, Edinburgh (GB).

Members of the Scientific and of the Local Organizing Committees.
From left to right: C. Waelkens, Th. Pauwels, O. Nys, S. Dewinter, G. Roland, H. A. Abt, E. Van Dessel, J. Dommanget, W. J. Luyten, W. Gliese, A. Upgren, Z. Kopal, J. Delhaye and S. Vereshchagin.
(In the background: the equestrian statue of King Albert I)

W. J. LUYTEN

The colloquium was dedicated to W. J. Luyten.

OPENING SESSION

Word of welcome by the Chairman of the Scientific Organizing Committee, Dr. J. Dommanget

First of all I have the great pleasure to convey to all of you all wishes for a successful meeting from the General Secretary of the IAU, J. P. Swings and from the Assistant General Secretary, D. McNally.

I have also to fulfill a particularly agreeable duty in informing you that the Scientific Organizing Committee at its meeting of yesterday, has unanimously decited to dedicate the present colloquium to our always active colleague here present:

W. J. LUYTEN

who devoted so much of his scientific activity and researches to wide systems. Professor Luyten is so well known that I cannot recall here all what he did not only for double star astronomy, but also more generally for proper motion surveys and white dwarf discoveries.

May I thus immediately ask you, Professor Luyten, to accept this expression of our gratitude.

W. J. Luyten expresses his thanks as it follows:

Je suis profondément touché de l'honneur que vous me faites spécialement parce que pour moi découvrir des naines blanches et des étoiles doubles à mouvement propre commun était une occupation extrêmement agréable, et je vous remercie du fond de mon coeur.

Ik ben diep ontroerd door de eer die gij mij bewezen hebt speciaal juist omdat het ontdekken van witte dwergen en wijde dubbelsterren een van de aangenaamste werkzaamheden is die ik ooit gehad heb en ik zeg U daarom dank uit de grond van mijn hart.

I am profoundly touched by the honor you have conferred upon me especially because the discovery of white dwarfs and wide doubles has always been one of my most pleasant research work, and I therefore want to thank you from all my heart.

Astrophysics and Space Science **142** (1988) 1.

WELCOME ADDRESS

P. Melchior
Director of the Royal Observatory of Belgium

It is a privilege for me to address this Colloquium and a great satisfaction to express our gratitude to the Ministers of Education of Belgium, to our National Fund for Scientific Research and, the Unisys Company for the appreciable support given for the practical organization of this Symposium.

We interpret this support as a recognition that, despite the well know difficulties of our epoch, pure scientific research is appreciated by our national authorities.

I then wish to thank this audience of foreign astronomers coming from 17 different countries for their participation, which will let us benefit, during the next five days, from the invaluable contribution of their high level of scientific competence and professional experience in the study of binary stars.

I would like to say that we are proud indeed that, for the second time, the International Astronomical Union has chosen Brussels to have a Symposium organized on binary stars.

The first one was in 1966 with the title 'On the Evolution of Double Stars'. To-day the subject is on 'Wide Components in Double and Multiple Stars'.

In the introduction of the 1966 Symposium, my predecessor the late Professor Velghe reminded in some detail the historical place taken by binary stars studies in the programmes of activity of the Royal Observatory of Belgium.

This started almost exactly one hundred years ago with the measurements by Niesten and Stuyvaert. Most famous have then been the contributions of Van Biesbroeck in the beginning of this century and, since 1935, those of Arend, whom we were especially happy to meet here.

I consider the organization here of a second Colloquium on the same topic as a testimony of the good health of scientific activity in our Royal Observatory. These researches of binary stars have not simply been maintained just like a long tradition by Jean Dommanget, the head of Department of Astrometry and Celestial Mechanics, but which is the most essential, has been efficiently developed in new directions and improved in quality.

Your presence here, dear colleagues, is a priceless encouragement for the Royal Observatory to persevere and to maintain this activity which we all know is of considerable meaning for fundamental astronomy, celestial mechanics, stellar evolution and cosmology.

As it is essentially at the initiative of J. Dommanget that the objectives and the Programme of the present Colloquium have been defined, I will let him now introduce your sessions in his capacity of Chairman of the Scientific Organizing Committee.

But before doing that I wish to express also our thanks to his colleagues of the Local Organizing Committee for their efficient work of preparation.

The Colloquium is yours. I have no doubt that it will be very fruitful.

Astrophysics and Space Science **142** (1988) 3.

GENERAL INTRODUCTORY LECTURE

J. DOMMANGET
Royal Observatory, Brussels, Belgium

Ladies and Gentlemen,

I first would like to join my acknowledgements to those expressed by Prof. P. Melchior to the representatives of all official organizations which made possible the present colloquium. But I also would like to express to him personally, my sincere thanks for having immediately supported the idea of organizing this colloquium in Brussels. It has been for me a permanent encouragement and has given me the necessary enthousiasm to conduct this enterprise to its final stage. I also have to convey my thanks to the Scientific Organizing Committee and to the Local Committee which, under the presidentship of Dr E. van Dessel, has been very active and helpful especially during these last months.

Now before starting with the working sessions, I have the duty to introduce this colloquium and to explain on this occasion the origin and the reason for its organisation. May I say that I do it with a particular pleasure because it gives me the opportunity to recall personal ideas on some features of double star astronomy that should be more often present in our minds for an increasing contribution to the advancement of our science.

It may be surprising for an objective observer of the history of astronomical research to state how few are today the astronomers actively interested in double and multiple stars, when these celestial objects are so numerous compared to the single stars. This is especially the case for visual binaries. Actually, some 260 astronomers of the 6000 members of the International Astronomical Union, are members of Commission 42 which interest concerns 1000 to 2000 very close or interacting binaries, but only 65 are members of Commission 26 of which interest covers some 70000 known double and multiple visual systems. And in this category of small stellar systems, the main interest goes to the visual close pairs, leaving the wide and medium-wide systems as the 'neglected children' of all celestial bodies.

The astronomy of double stars has been one of the 'rosettes' of our science in the last century. At the start, wide and medium wide visual systems only were considered because of the ease of their observation. Later on, during the present century, the interest of the observers has been more and more directed to the very close visual pairs, the most sensitive to orbital motion and thus the most favourable for orbit computations and consequent mass determinations. Consequently also, new techniques were studied in view to reach closer and closer visual systems so that to-day, we fortunately have at our disposal an important panel of various observation techniques. At the same time, the

Astrophysics and Space Science **142** (1988) 5–8.

extraordinary development of astrophysics induced a high interest for real close systems (spectroscopic and photometric binaries) and produced many important results.

All these researches, highly desirable and logical, are to-day contributing to the construction of a bridge joining the fields of visual binaries and of spectroscopic ones. Unfortunately, if one considers the whole domain of double star astronomy an enormous disequilibrium appears to the prejudice of the wide pairs, which were progressively neglected.

The reasons for this situation are manyfold: no appreciable orbital motion, difficulty to be recognised as physical systems and, probably more crucial, the fact that many astronomers do not see any interest to consider them.

This is very regrettable; because one should remember that a binary star is not only an assembly of two single stars with their classical observable physical parameters but is also specifically characterized by unique particular orbital parameters that not only give information on the present dynamical state of the system, but may also teach very much about its origin and its evolution. It is a matter of fact that researches on stellar evolution and on galactic dynamics and structure generally do not consider sufficiently the contributions that double and multiple star studies may add to the understanding of the formation and evolutionary mechanisms. Are we sure, for instance, that a system of two single stars undergoing the classical stellar evolution process, does not suffer concomitant orbital evolution? And thus that its present orbit does not keep the traces of the physical evolution of its components? If this is so, any discovery on double star orbital evolution will have a by-product in the field of single star evolution knowledge. Of course, differences exist between the evolution of the components of very close interacting pairs and the evolution of a single star for evident reasons, but it is difficult to accept that such differences exist also in the case of medium wide and wide pairs, which are not concerned by such interactions.

The same situation seems to exist in galactic research. I think that too few astronomers are aware of the contribution that double and multiple star researches may bring to our knowledge on galactic formation and evolution. How many accept to introduce as a fundamental data for the elaboration of their theories, the extraordinary high frequency of these systems and consider the mechanism of their formation?

This is why, in my opinion, it should be our duty to show more to our colleagues, how much double and multiple stars may play a fundamental role in stellar evolution and in galactic researches. But, therefore, we need ourselve to have some precise opinion on how double stars interfere with these other fields of research. And this means that we should not consider with less interest any of the categories of double and multiple stars – from the very close systems to the very wide ones.

Even when the medium wide and wide systems do not show orbital motion, and do not permit orbit computation and mass determination, they are not less important than the orbital ones. Their existence as physical systems confers on them the particular feature that both components have the same origin and the same evolutionary time scale. The real distance between the components as well as their relative linear motion and relative radial velocity are fundamental observable parameters that advantageously replace the orbital parameters for interesting statistical researches.

Having these considerations in mind already twenty years ago, I have proposed to organize at Brussels – as Prof. P. Melchior recalled it – an IAU colloquium entitled 'On the Evolution of Double Stars'. More than 40 participants attended; 36 papers were presented and many important results concerning different categories of double and multiple stars came to our knowledge.

One of the sessions was devoted to *Internal Structure and Evolutionary Changes in Binaries*. It has been the start of a long series of important meetings which dealt with the astrophysical and dynamical evolution of a specific categorie of systems: the interacting binaries. But the 1966th colloquium did not lead to any equivalent sequence of meetings, researches and results on the origin and evolution of the medium wide and wide binaries. Another session entitled *Mass Loss and Variations of the Orbital Elements* was also devoted nearly exclusively to the very close systems. At that time only a few participants considered the visual medium wide and wide systems. W. J. Luyten showed how the frequency of the common proper motion pairs peters out for separations of the order of 3 to 4000 AU. I myself, in an introductory lecture on the 'definition' of a double star, came to the same cutt-off limit and W. Heintz showed interest in the distribution diagram of the double star components' true separation considering the wide pairs as well as the close ones. But no special results were really produced on the evolution and origin of the systems going from the classical visual ones to the wide ones, which actually I had in mind when proposing to organize that colloquium.

From the successive meetings organized later by the International Astronomical Union on Double Stars and even from recent researches, it is not sure to me that a marked progress has been made on this important question, while simultaneously very important results were obtained for close and interacting (spectroscopic and photometric) binaries.

My personal impression is that our neglect of wide systems will be remedied only when a better census of all binaries will be attained. From observations made during these last few years in the frame of the *Hipparcos Input Catalogue* preparation, it systematically appears that roughly half of the medium-wide pairs with separation between a few arcseconds and some thirty arcseconds are not registered as such in our double star catalogues. For wider systems, the situation is difficult to anticipate. It is thus sure that we are to-day far from having a sufficiently completed census of the binaries and that all our statistics are distorted by selection effects which are difficult to estimate for appropriate correction. Systematic research should be undertaken for medium wide and wide pairs by all possible observation techniques because the lack of sufficient and accurate information on wide pairs makes especially impossible the correct establishment of what should be considered as the most important diagram in double star astronomy: *the distribution diagram of the semi-axis major*. Fortunately, it seems that since a few years more astronomers are aware of this: it is encouraging to see that one of the most important international reviews in astronomy, *The Astronomical Journal*, has chosen this diagram to illustrate the cover page of one of its last issues (Vol. 92, No. 1569, 1986).

Of course, the first and main question will be: how can we distinguish between the wide pairs, the true physical ones from the optical? It is a very difficult problem but it is worth to make all efforts to solve it because the importance of the stake.

Till now, one successful technique has been mainly considered for systematic surveys. It is based on the unmarked orbital motion in wide systems and on the consequent identical proper motion of their components. Three names come immediately to our minds: W. S. Burnham, H. L. Giclas and our here present and always active colleague W. J. Luyten, who is carrying on his exceptionally extended survey of common proper motion pairs.

As a result of all these considerations, I thought that time was come to 'prick the chart' and – with the encouraging advices of many of you – to organize a meeting specifically devoted to wide systems only. The goal of this colloquium should be to clarify our ideas on their problems but also simultaneously to develop a new strong current of research inside Commission 26, which unfortunately is unrighteously considered by the Executive Committee of the IAU to be at death's door. The list of papers proposed for this meeting shows that the situation is less dramatic than one could imagine, and that wide double star astronomy will lead to some encouraging development.

I wish to all of you, a pleasant, stimulating and fruitful meeting.

FIRST SESSION

CATALOGUING WIDE PAIRS

Chairman: H. A. ABT

THE CONCEPT OF STAR MULTIPLICITY AND ITS HISTORICAL EVOLUTION

Introductory Lecture

M. G. FRACASTORO

Osservatorio Astronomico di Torino, Italia

Summary. Great technical improvements have been achieved during latest decades in binary star research (angular resolution, photometry, radial velocities). Consequently, there is now a significant overlapping among the traditional groups. Furthermore, we can tackle the study of couples where the companion is a submassive object.

More attention is now devoted to multiple systems, their structure and evolutionary story.

The scenario is now ready for a new insight into the problem of formation and evolution of stars.

Couples of stars had already been noticed in ancient times and labelled with a single name, plus the adjective δυπλός. It may be pointed out that even in early Catalogues a double star occupies a single entry.

With the use of the telescope, as a consequence of the increased number of observable objects and the improved resolving power, the number of these couples was rapidly increasing. However, the general feeling was that the proximity of two stars was mainly due to chance. Only when an orbital motion was eventually ascertained, did W. Herschel accept the idea of a physical link between the two stars.

As a matter of fact, the existence of a gravitational link has been always fundamental in every definition of a binary system. The first leap happened when IAU Commission 26 accepted in its field of research the so-called Trapezia – namely systems where the total mechanical energy is probably positive, rather than negative.

The next step was accomplished when Comm. 26 itself changed its name (the proposal came from Prof. Dommanget), including officially multiple systems with widely separated components. The problem of their origin and mechanical evolution became very important, beside the classical target of knowing their orbital elements.

More recently, another revolutionary step has been made by Comm. 26, adopting the common proper motion stars in its great 'family'.

As a matter of fact, those weak and widely separated components of multiple systems were rather neglected or poorly considered until recent times. Yet, it was known since a long time that couples are observed in excess with respect to those predicted by Poisson formula even for very large separations, as shown by this very old table:

Astrophysics and Space Science **142** (1988) 11–15.

TABLE I

Separation (arc sec)	0–5	5–10	10–15	15–20	20–25
Number of couples					
Observed	469	550	443	473	549
Expected	47	141	234	328	422
N_{obs}/N_{exp}	10.0	3.90	1.89	1.44	1.30

Much more recent are the results obtained from an analysis of the *Yale Catalogue of Bright Stars*, as shown by Table II, where the smoothed values of $\log N_{obs}/N_{exp}$ are reported in terms of the separation s, given in the ordinate, and the magnitude m_B of the companion, given as abscissa.

TABLE II

m_B ⟶ s	4–5	5–6	6–7	7–8	8–9	9–10	10–11	11–12	12–13
≤ 0″9	+5.7	+5.7	+5.2	+4.7	+3.2	+3.0	+2.2	+1.4	–
–	–	–	–	–	–	–	–	–	–
4.0–7.9	+3.0	+3.1	+2.8	+2.6	+2.2	+1.7	+1.2	+0.7	+0.4
8.0–15.9	+2.7	+2.6	+2.3	+2.0	+1.5	+1.1	+0.7	+0.3	−0.2
16–32	+2.1	+1.7	+1.5	+1.3	+0.9	+0.5	+0.1	−0.2	−0.5
32–64	+1.0	+1.2	+0.8	+0.7	+0.4	+0.1	−0.4	−0.9	−1.4

N_{exp} is deduced from the Poisson formula. The line inside the Table separates the areas where N_{obs} is larger than expected (prevalence of physical couples) from the areas where N_{obs} is smaller (incompleteness of data).

Let us now consider the lower limit for observing visual binaries. During recent decades, we had great improvements mainly due to interferometric techniques, which were usefully added to the astrometric ones. Furthermore, we may quote the technique of occultations by the lunar limb, which proves itself very useful for statistics of very close pairs.

Several astronomers worked very hard for decades with astrometric techniques, searching for substellar ($\mathfrak{M}_B \sim 0.01\ \mathfrak{M}_\odot$) or even planetary objects ($\mathfrak{M}_B \sim 0.001\ \mathfrak{M}_\odot$). Their work was extremely careful, however, this procedure remains dangerously subject to systematic errors, or to sudden changes in the instrumentation, not to speak of the reliability of the adopted reference frame. So their results are limited to a very few systems and furthermore they are still controversial. Instead, to ascertain the existence of these submassive components, possibly down to planetary masses, is a very important task. Not only does the astronomical community wait for an indisputable word on the presence and possibly the frequency in our Galaxy of other planetary systems. This is an absolutely vital argument for science as a whole.

To its solution a significant contribution can be given by modern spectrographic techniques. Accurate and assiduous measurements of radial velocities may eventually ascertain the existence of small periodic oscillations about the mass center of a system star + planet (for instance, the orbital velocity of our Sun around the mass center of the system Sun + Jupiter is 13 m s^{-1}). Unfortunately, the lapse of time required for an achievement of this kind cannot be shortened and it remains of the order of decades. We only may hope to 'see' these Jupiters directly with the Space Telescope, emerging out of the noise produced by their stellar mate. For instance, assuming $m_* - m_J = -22$, we may predict $m_J = +27$ for 61 Cygni and $m_J = +31$ for the Barnard star.

As for visual pairs, the group of eclipsing binaries appears once again extremely large. Up to $m = 2$, 6 EB are listed out of 128 stars (4.7%). Up to $m = 5$, the catalogued EB are 43 out of 4578 stars (0.9%). These percentages become extremely small as m increases, and this is obviously due to practical reasons: there is presumably one million of eclipsing binaries up to 14^m, but there will never be enough observers to discover all of them, and even less to study them.

This is not at all regrettable: in my opinion a hundred light curves, accurate enough and possibly repeated year after year when it is necessary (see for instance the case of RS CVn, the light curve of which changes with time with a period of about ten years) are of much more profit than one million of purely qualitative data. No need to say that periodical variations of the epoches of minima can reveal the presence of a third body, transferring the object from the group of doubles to that of multiple stars.

The accuracy of photoelectric photometry is nowadays more or less the same as 50 years ago. The number N of received photons and its fluctuations $N^{-1/2}$ remain as a natural limit. However, we may explore the image on a network of sensorial elements (CCD) and obtain a two-dimensional map of the image, which is consequently explored as an extended source.

A good prospect for future research is that the traditional barriers between different observing techniques are now in several cases broken down: a classic example is Capella, whose orbit is now one of the best known, astrometrically as well as interferometrically, and also it takes advantage of accurate measurements of radial velocities. Of course, Capella has also the benefit of being a very bright star.

In conclusion, the following patterns of research appear very promising:

(1) Systematic surveys of widely separated components, up to now rather overlooked, should be encouraged, trying – as a first target – to build up a satisfactory statistics of third bodies (Hipparcos, Dommanget).

(2) The consequences of adopting trapezia and common proper motion stars within the great family of objects pertaining to comm. 26 should be adequately evaluated. In particular, the mechanical study of multiple systems (see the Colloquium in Turku next week), should be emphasized in order to draw a model of their dynamical evolution.

Special addendum for comm. 42:

(3) To study photospheric phenomena of stars and their possible periodicity, in the frame of interactions between the two components, up to larger separations.

To attain these goals, it seems necessary to take fully profit from large telescopes and modern techniques, giving priority to quality (Area Scanners, Speckle, and other interferometric methods) rather than to quantity. Systematic measurements of radial velocities should be carried on for all kinds of binaries, whenever it is needed.

For statistical purposes, as far as they are really needed, it seems wiser to carry out exhaustive investigations on Selected Areas, or perhaps to analyze systematically all stars of Gliese's catalogue.

When the computer is used as a tool for calculating new orbits, one should keep in mind that it cannot create information where this one is not sufficient.

Some of these suggestions were expressed several decades ago by W. H. van den Bos, and even before by G. P. Kuiper. Unfortunately, both of them remained unheeded, in spite of their great authority.

In conclusion, to coordinate the current research on double and multiple stars might not be a sufficient target for comm. 26. The astronomical community needs precise guidelines, as perhaps a list of those objects which deserve thorough investigation with the most advanced techniques, leaving the other innumerable objects to merely statistical purposes.

References

Batten, A. H.: 1973, *Binary and Multiple Systems*, Pergamon Press, London.
Bos, W. H. van den: 1948, *Trans. IAU* **VII**, Zürich, 1961, **XI**, Berkeley.
Dommanget, J. and Nys, O.: 1967, *Trans. IAU* **XIII**, Praha, 1976, **XVI**, Grenoble.
Evans, D. S.: 1976, *Trans. IAU* **XVI**, Grenoble.
Fracastoro, M. G.: 1968, *Vistas Astron.* **XII**, 196.
Fracastoro, M. G.: 1969, *Mem. Soc. Astron. Italiana* **XL**, 309.
Fracastoro, M. G.: 1983, *Rend. Acc. Lincei* **73**, 226.
Heintz, W. D.: 1978, *Double Stars*, D. Reidel Publ. Co., Dordrecht, Holland.
Heintz, W. D.: 1976, *Monthly Notices Roy. Astron. Soc.* **175**, 533.
Van de Kamp, P.: 1964, *Trans. IAU* **XII**, Hamburg.
Kuiper, G. P.: 1935, *Trans. IAU* **V**, Paris.
Muller, P.: 1952, *Trans. IAU* **VIII**, Roma.

LODEN – May I give the following comment. In the early sixties I tried to publish a catalogue of wide binaries. It was refused, for according to the referee's comments it was 'astrophysical nonsense'. About five years later I made another attempt. It was also refused, this time because it was 'astrophysically uninteresting'. To-day we are attending a colloquium on that subject, so it is evident that the opinion has changed quite a lot during the last decade.

FRACASTORO – I agree and I may add that fifty years ago the measurement of stellar radial velocities was not considered an important item for comm. 26 (see *IAU Trans.* Vol. V).

ABT – I am surprised by the reactions to Dr Loden's paper on wide doubles because in the 1960s there was much interest in open clusters, and multiple systems are really small clusters. Such systems are just as useful as clusters for studies of stellar evolution – perhaps more useful because they are much more plentiful than clusters.

WORLEY – You mention the area scanner as an applicable new technique for astrometry of binary systems. The only substantial list of such double star astrometry contains 15–20% gross errors in the listed position angles and separations. Therefore, I do not think that the area scanner has proved itself as an acceptable astrometric instrument.

FRACASTORO – I agree that Area Scanners are not suitable for astrometric couples. However, I could not avoid to mention them in a survey of modern techniques. Whether they are accurate or not, it is a questionable matter and I am not willing to give judgement on this.

ABT – If a new technique gives inaccurate results initially, does that necessarily mean that the technique is hopeless? Are there not usually problems that must be solved before perfecting a new technique?

WORLEY – I was just making the point that the area scanner has not yet proved itself for astrometric observations of double stars.

HERCZEG – Is the accuracy of a few tenths of km s^{-1} a matter of very special equipment and very special research programs or already routine?

I think for most observers an accuracy of about 0.5 km s^{-1} and the quoted high accuracies are still in the future.

FRACASTORO – Yes, for most observers. However, the achievement of such an important goal as the discovery of planetary bodies deserves, in my opinion, the access to the most specialized instruments. These can give right now the required accuracy.

MAZEH – I am engaged in a radial-velocity survey to find sub-stellar companions to closeby late-type stars. The survey takes place in the Harvard–Smithsonian Center for Astrophysics together with Dr Latham, and we have reached a precision of about 0.5 km s^{-1}.

ABT – Although that accuracy is characteristic of many automatic radial-velocity systems and sharp-lined spectra, what is impressive of the Latham-Mazeh work is that it is being done to magnitudes as faint as 15th.

WIDE DOUBLE STARS WITH COMMON PROPER MOTION

W. J. LUYTEN

Space Science Center, University of Minnesota, Minneapolis, U.S.A.

Summary. More than 6000 cpm systems including 400 pairs having at least one degenerate component have been discovered to-day by the author.

The others are mainly Main-Sequence stars less luminous than the Sun precisely the kind which fits in the best populous part of luminosity function.

Ever since I began my blink surveys for proper motions I have been interested in, and published lists of wide double stars with common proper motion. The first such list appeared just sixty years ago in *Harvard Bulletin* No. 852 and gave data for sixteen such objects found on Harvard plates. A few more short lists appeared in the *Monthly Notices of the Royal Astronomical Society* but the first extensive catalogue of LDS stars was issued in 1941 when I had completed the Bruce Proper Motion Survey in the southern hemisphere (*Publ. Obs. Minn.* **3**, No. 3). It gave data for 832 such objects. After we had blinked some 300 pairs of Bruce plates in the northern hemisphere, we added another 238 objects to the LDS total. The great increase came when we started blinking the Palomar Schmidt plates and the total number of LDS stars has now reached 6124.

In the early lists we included some known doubles for the reason that subsequent statistical analysis might find it useful to know how many such objects had been found in any given area. Furthermore, it is quite probable that, especially among the wider pairs with smaller motions, a number will prove to be optical pairs but I hope and believe that this number will be small. Hence, I think it appears reasonable to assume that the total number of new physical binaries contained in all the LDS lists is at least 6000 – which is considerably more than have been found by any other double star observer.

At the Double Star Conference in Nice in 1969 I pointed out that these common proper motion pairs constitute the most common, and the most representative physical binaries in space – at least as of now.

We should remember that while of the 6000 stars which the average human eye could see in the entire sky, probably not more than thirty, or one half of one percent are less luminous than the Sun, that probably of the 700 odd stars nearer than ten parsecs at least 96% are less luminous than the Sun. There is not even ONE real yellow giant, such as Capella, Pollux or Arcturus nearer than ten parsec and only about four Main-Sequence A stars.

I selected ten pages at random in the *ADS Catalogue* (pages 37, 157, 199, 305, 453, 799, 875, 1059, 1153, and 1253) and counted 129 visual doubles listed on them – of which one is optical. Of the remaining 128 there are 50 for which it is not possible to estimate the luminosity but for 78 one can make at least a rough guess from the spectral class or the proper motion. Among the 78 primaries there are 75 more luminous, and

Astrophysics and Space Science **142** (1988) 17–19.

only 3 or 4% less luminous than the Sun. In my recent *Schmidt Survey Publication LXIV* a list of 827 LDS stars is given, with approximate colours for 823 of them. Among the 823 primaries there are 9 earlier than g, 54 have the colour g, 113 g–k, 254 k, 109 k–m, 264 m, and 20 degenerates. Assuming that half of the g objects are more luminous, and half are less luminous than the Sun, would give a total of 36 among 823 primaries more luminous than the Sun, or about 4%. Among the 830 secondaries (there are 7 triples) all are presumably less luminous than the Sun, hence the final count stands at 36 more luminous, and 1617 less luminous than the Sun. It is clear, therefore, that the double stars with common proper motion are considerably more representative of the stars in space than are the visual double stars.

One more point of interest: while among the stars nearer than ten parsecs there is not one yellow giant, or even an O or B star and only four Main-Sequence A stars, there are at least forty and possibly sixty degenerates. Among my LDS stars I have published 400 pairs with at least one probably degenerate component, and perhaps a dozen double degenerates. Since my colour estimates are very crude it is quite possible that a number of those objects I have designated as probably degenerate will prove to be high-velocity Main-Sequence stars or subdwarfs. The number of real degenerates among them should probably be at least two hundred.

When, in due course of time, accurate orbital motions are determined for these pairs, and these are compared to the orbital motions of pairs of Main-Sequence stars of similar apparent magnitude, proper motion, and separation, it will become possible to determine masses for these degenerate objects without invoking any theory except the law of gravitation.

Furthermore, because the components of these pairs are many, many stellar diameters apart, it will be possible to analyze the evolution of these objects without being bothered by the interactions prevalent in close pairs.

Visual Double Star observers have generally concentrated on apparently bright stars – brighter than 10 V – while spectroscopic observers have, naturally, favored stars of large mass – O and B stars – for their analysis of spectroscopic binaries, since these produce the largest orbital velocities. Perhaps these are the reasons that, in spite of my large number of publications, and the thousands of data published, the double stars with common proper motion have been universally ignored.

Even as recently as 1983 – at the Strasbourg Conference on Double Stars in the HR diagram, none of the eighteen papers presented even indicated that they were aware of the existence of the LDS stars.

Thus, Pedoussaut, Ginestet, and Carquillatt presented a paper on binaries outside the Main Sequence, but dealt exclusively with giants, supergiants, O stars, and cataclysmic variables but did not mention the enormously much more numerous degenerate stars. They complained about a flagrant lack of observations of binaries of spectral class M even main sequence M stars but ignored the fact that probably at least a quarter of my 6000 LDS stars are Main-Sequence M stars.

Halbwachs presented a paper on Main-Sequence objects in the *Bright Star Catalogue* but obviously this restricts the analysis to stars much more luminous than the Sun.

Mayor and Mermilliod entitled their paper 'Duplicity in the HR Diagram', but dealt mainly with the eccentricities in binary orbits in nearby open clusters, and their data include only 33 stars later than F5, hence, again very few stars less luminous than the Sun were considered.

We could go on almost endlessly – quoting from the tri-annual reports of commission 26, but we should realize that our basic star catalogues are no longer just the *Almagest* or the *Bright Star Catalogue*. When the Space Telescope begins operations, we shall probably be able to get down to magnitude 25, and that will mean that a star of magnitude 16 V, or 17 V is then a relatively bright star.

ABT – I completely agree with Dr Luyten about the usefulness of common-proper motion pairs because the components have common origins, ages, motions, etc. But if I may say, in selecting the Aitken catalog for comparison you selected the most conservative of the compilers. If you had used the Burnham catalog, you would have found many more catalogued wide pairs, as I think that you know.

LUYTEN – I used the ADS Catalogue simply because it was more recent and had many more entries than Burnham.

POVEDA – Have you estimated what is the typical separation, in AU, of the LDS stars observed in the Palomar plates.

LUYTEN – This is difficult to say because the discovery of these common proper motion pairs is a very subjective thing.

But if, as a guess you take a separation of 20 arc sec and a motion of 0.2 arc sec as fairly representative then it would be a few hundred to a thousand astronomical units.

WIDE PAIRS IN THE WASHINGTON DOUBLE STAR CATALOGUE (WDS)*

C. E. WORLEY

U.S. Naval Observatory, Washington, D.C., U.S.A.

Summary. The *Catalog of Observations*, maintained at the U.S. Naval Observatory as an integral part of the *Washington Double Star Catalog*, is fast becoming a complete data base with the continued accession of the older data. For the wide pairs contained therein, the principal interest is whether or not they are physical. This report describes the observational history of various series of older observations, and points out what needs to be done to provide a reliable judgement as to the physical relationship of such pairs.

WEINBERG – In order to use a catalog to test our theoretical understanding of wide binary formation and evolution, one needs to know the observational selection effects, such as limiting magnitude, etc. In this view is it possible to estimate the completeness of WDS, for example, in the way that Poveda and his collaborators did for IDS?

WORLEY – As there has been no uniformity in the way double stars have been discovered, the WDS continues to suffer the same problems as IDS. I believe that a filtering similar to that employed by Poveda would still yield comparable results, since the data base is still composed of similar pairs.

HERCZEG – How many (about what percentage) of those 75 000 pairs listed in the Data Base have not just the discovery observation, but also a few, 2 or 3, later observations? By comparison of these data we could have at least an idea if a certain pair is 'fixed' or not.

WORLEY – The majority of the pairs in the data base only one observation, and generally no other information, such as proper motion or spectral type. Since a single early observation may be some what erroneous, the combination of it with a subsequent measure may not reliably indicate a 'fixed' pair.

GLIESE – Does the *Visual Double Star Data Base* include speckle observations?

WORLEY – Yes, it does!

* An expanded version of this paper will be presented at the *IAU Colloquium* **100** 'Fundamentals of Astrometry' in Belgrade, 8–11 September, 1987.

Astrophysics and Space Science **142** (1988) 21.

THE CONTRIBUTION OF THE HIPPARCOS MISSION AND OF THE CCDM CATALOGUE TO THE STUDY OF WIDE SYSTEMS

J. DOMMANGET

Royal Observatory, Brussels, Belgium

Summary. The general concept of the *Hipparcos* astrometric mission is first recalled as well as the problems related to the observation technique of this satellite. The *Inca Data Base* and the *Hipparcos Input Catalogue* are described and the place of the double and multiple stars in the mission and in the *Input Catalogue* is discussed.

The need of a specific catalogue for these celestial *objects* (CCDM) is shown and its format is given. The contribution of the *Hipparcos* mission and of the CCDM to double star astronomy and more precisely to wide double and multiple systems is finally detailed.

1. Fundamental Characteristics of the Hipparcos Mission

It is well known that the Hipparcos mission consists of the astrometric observation of some 100 000 stars uniformly distributed on the sky in order to establish a highly accurate stellar reference system linked to an extragalactic reference frame (inertial).

Positions, parallaxes and proper motions are expected to be determined in the best conditions with an accuracy of $\pm 0\farcs002$. The Hipparcos satellite being unable to observe stars fainter than the 13th magnitude, linkage to the faint extragalactic radio-sources will be obtained by use of the Hubble Space Telescope and of the VLBI techniques. The satellite (payload) – which is presently build – has a weight of some 215 kg and approximately the size of a cubic meter (solar panels folded) (Figure 1). The observation technique consists in scanning the sky with a slowly rotating telescope observing simultaneously two fields following each other along a great circle at an angular distance of 58°. Both fields cover a surface of $0\fdg9 \times 0\fdg9$ and their images are optically superimposed in the focal plane of the 29 cm aperture Schmidt telescope. The pole of the scanned great circle is moving slowly so that finally in the 2.5 years interval of observation time, each star is registered from some 60 to some 90 times yearly as a mean.

The isolation of the programme-stars in each field is made by an Image Dissector Tube (IDT) of which the Instantaneous Field of View (IFOV) has a diameter of the order of 30″. This IFOV is continuously centered on the observed star and follows it when during its transit in one or another of the conjugated fields, it crosses a modulating grid placed in the focal plane of the telescope. Each of the 2688 slits of this grid being as large as $1\farcs2$, it appears finally that the position of each star should be known beforehand with an error less than 1″ in order to solve the steps ambiguity. On the other hand, to enable a correct allowance of observation time to each star as a function of its brightness, the magnitudes should be known with an error smaller than 0.3 m.

These two last conditions were the essential data to consider when establishing the list of 100 000 programme-stars called the *Hipparcos Input Catalogue* (HIC).

Astrophysics and Space Science **142** (1988) 23–30.

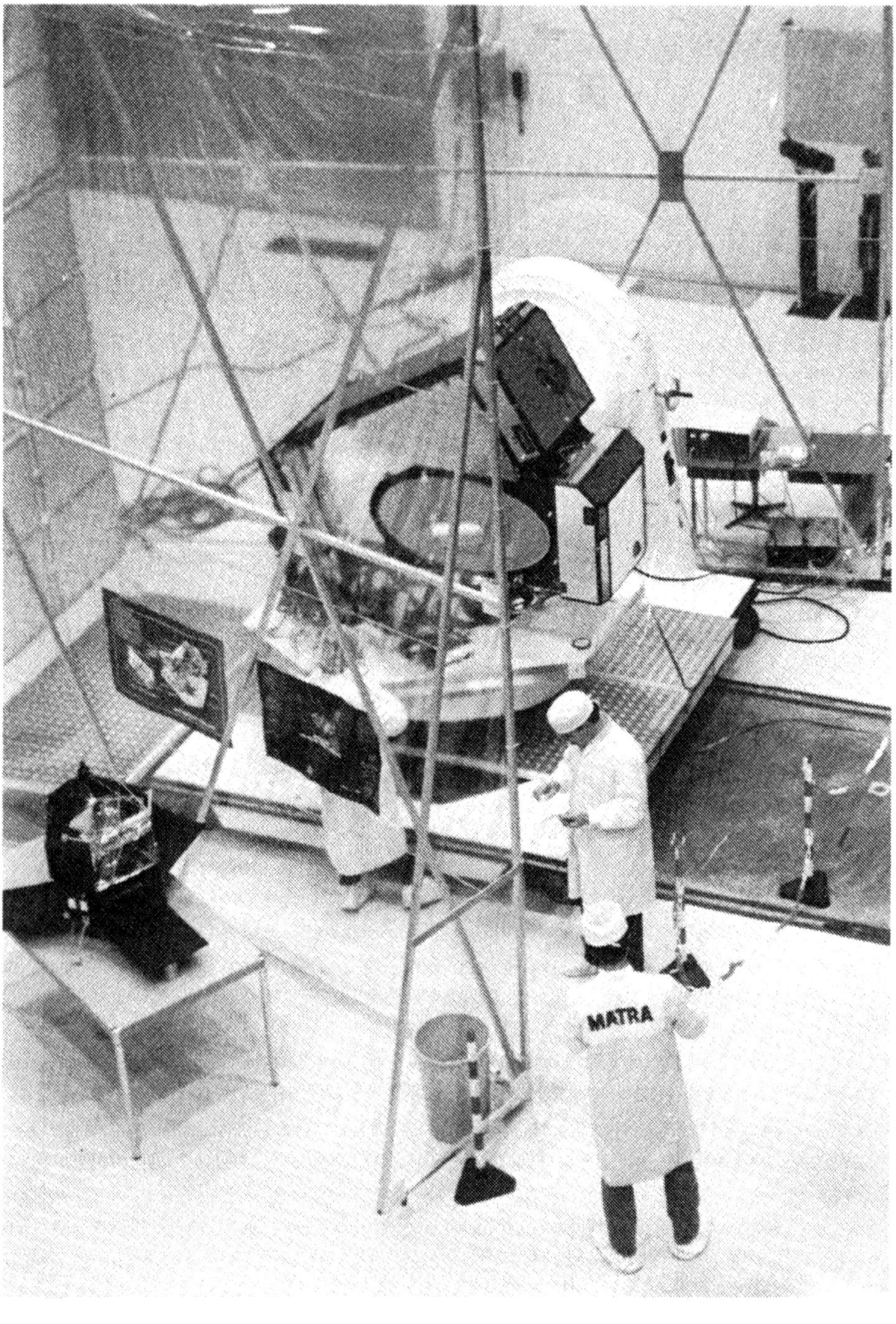

Photo: D. Morin, Observatoire de Meudon.

Fig. 1. Hipparcos payload in the clean room of IAL-Space at Liège (Belgium) in October 1986.

2. The Hipparcos Input Catalogue Preparation and the Double Star Problems

2.1. The HIC and the INput CAtalogue consortium (INCA)

It has been felt quite soon that the establishment of the HIC would need an international and interdisciplinary collaboration. This was realized by the appointment of a wide consortium of representatives of more than 20 institutions mainly European under the leadership of Mrs C. Turon of the Observatoire de Meudon. The work was distributed between task leaders and also coordinators of working groups in charge of the problems concerning specific celestial objects as: variable stars, double stars, asteroids, extra-galactic radio-sources, etc. Double and multiple stars were committed to the care of the author of the present paper.

At the start it has been decided that the HIC would contain at least all stars brighter than a given magnitude (say around 8 or 9), the Survey, and would be completed by fainter stars, chosen for specific programmes of research. An Announcement of Opportunity was thus made to the international astronomical community and a Selection Committee under the presidentship of Prof. A. Blaauw was settled to give advice to the INCA Consortium for realizing the HIC.

As an answer to this announcement, some 214 proposals were received, containing a total of 810 000 stars. Because of uncertainties on the identifications, unaccuracies and errors on the positions given by the proposals and in the reference catalogues, a few years were devoted to clean this list for redundancies. This tedious work conducted by Mrs A. Gomez and her collaborators at Meudon finally produced a list of 213 771* stars, called the *INCA Data Base*.

Taking into account the priorities established by the Selection Committee between the proposed list of stars as well as between categories of stars inside each of them, the conflict appearing between stars located in a same small area on the sky and the results of successive simulations of the observing strategy (with the condition that the stars should be uniformly distributed on the sky), a subset of 114 500 stars has been established which may be considered as a first version of the definite *Input Catalogue* (IC-1). This catalogue includes all stars brighter than a limiting magnitude decreasing from 7.6 to 8.7 approximately with increasing galactic latitude (the Survey).

The establishment of this *Input Catalogue* needed many new astrometric and photometric observations which are presently collected together with old existing data in the *INCA Date Base*. In the case of double and multiple stars, some specific problems appeared and will be discussed in the following.

2.2. Double and multiple star problems

It has been seen that to enable a valuable observation of any celestial object by the satellite, the IFOV should be precisely and continuously centered on it during its transit in the fields of the telescope. This is also needed to permit a correct analysis of the

* The figures concerning the *Inca Data Base* and the *Input Catalogue* and given in the present paper, are from a recent report by the INCA Consortium to the Selection Committee (Paris, 6–7 April, 1987).

	CCDM N° (1900)	Comp.	Name	Δα (1900)	Δδ (1900)	ref. n° cat	ref. n° star	rel. position t	θ	ρ	n	m	Sp	μ_α	μ_δ	Identifications DM	Identifications others	ADS	Index N° (1900)	Hipparcos INCA DATA BASE
Double stars	00000+8208	A		− 292	+ 257	29	000514					104		*W0026	S0024	81 842.0	11.8	43	00000N8208	
	00000+8208	B	STF3059					1833	331	23	7	119						43	00000N8208	
	00000+2946	A		+ 111	+ 220	19	000517					86	F0	*W0028	S0031	29 5059.0	+30 3.6	47	00000N2946	+29 5059 2A 3 77
	00000+2946	B	STF3058	+ 183	+ 303	19	000519	1831	051	127	12	101		*W0048	S0017		+30 4.6	47	00000N2946	
	00001+0131	A										100							00001N0131	
	00001+0131	B	J 1801					1945	064	45	2	110							00001N0131	
	00013+5140	A										96				51 3782.0		66	00013N5140	
	00013+5140	B	A 1252					1906	190	6	4	120						66	00013N5140	
	00013+0121	A		+ 249	+ 625	19	000758					105		*E0023	S0067	1 4830.0	+01 6.6	67	00013N0121	
	00013+0121	B	HJ 1000	+ 228	+ 554	19	000757	1908	206	73	5	106		*E0040	S0073		+01 5.6	67	00013N0121	
	00032+2418	A										127							00032N2418	
	00032+2418	B	POU 3					1898	285	95	1	140							00032N2418	
	00036+3656	A										121							00036N3656	
	00036+3656	B	ALI 242					1929	144	107	2	126							00036N3656	
	02356+4859	A		− 275	+ 179	19	031367					92	F5	*E0029	S0010	48 737.0	+49 311.6	2051	02356N4859	+48 737 0AB3 85
	02356+4859	B	HU 539					1959	047	3	12	94						2051	02356N4859	
	12180−5138	A										94	G5	W0074	N0103	−51 6641.2	239906.8		12180S5138	
	12180−5138	B	R 198					1880	051	32	7	101							12180S5138	
Triple stars	02125+5934	A		− 131	+ 58	19	026861					69	A2	*W0008	S0007	59 467.0	+59 256.6		02125N5934	+59 467 1A 3 69
	02125+5934	B	STT 26	− 414	− 539	19	026850	1875	200	633	10	74	G5	*W0012	S0017	59 466.0	+59 255.6	N	02125N5934	+59 466 0B 3 72
	02125+5934	C	STT 26					1911	133	722	1	110							02125N5934	
	02127+5421	A										100				54 518.0		1773	02127N5421	
	02127+5421	B	ES 950					1910	099	47	2	107						1773	02127N5421	
	02127+5421	C	ES 950					1910	287	159	1	124						1773	02127N5421	
	02142−3008	A		− 251	+ 62	21	026919					88	F5	*W0018	N0015	−30 815.2	167693.8		02142S3008	−30 815 21A 3 86
	02142−3008	B	B 34					1944	028	15	7	100							02142S3008	−30 815 22B 3 86
	02142−3008	C	B 34	+ 16	− 13	21	026926	1931	102	347	0	108		*E0016	N0006		167694.8		02142S3008	
	17160+0137	A										94					.9	10486	17160N0137	
	17160+0137	B	STF2150					1941	204	98	14	103					.9	10486	17160N0137	+01 3422 1B 3 80
	17160+0137	C	STF2150	+1049	+1427	19	224492	1905	057	1637	5	79	F8	*E0001	N0011	1 3422.0	+01 1959.6	10486	17160N0137	
	17160−1536	A										105				−15 4525.0		10489	17160S1536	
	17160−1536	B	FOX					1908	269	94	1	107						10489	17160S1536	
	17160−1536	C	FOX					1908	064	109	1	118						10489	17160S1536	
	17169+3236	A		+ 146	− 657	19	224511					54	G0	*E0126	S1043	32 2896.0	+32 1511.6	10488	17169N3236	+32 2896 1A 3 54
	17169+3236	B	DOR 544					1923	335	2300	8	97					.9	10488N	17169N3236	+32 2896 2B 3 97
	17169+3236	BC	DOR 544					1900	216	87	2	129					.9	10488	17169N3236	+32 2896 3C 3 129

Fig. 2. Present format of the CCDM illustrated by some examples taken in the lists of the double and triple systems. The CCDM number in the first column, constructed as the Index number on the basis of the star position (to $\pm 0^{m}.1$ and to $\pm 1'$) will later be referred to equinox 2000. A letter placed before that of the component means that the position (θ, ρ) is given relative to the component corresponding to that letter. The balances $\Delta\alpha$ (in $0^{s}.01$) and $\Delta\delta$ (in $0''.1$) have to be added, respectively, to the right ascension and the declination given by the CCDM number to obtain the accurate position of the concerned component. The relative positions are here provisionally that of the Index (1976.5) and may thus not be entirely in agreement with the absolute positions of the components.

photometric signal. In case of a double star, one should know first how to center the IFOV: on the photocenter, at mid-distance of both components or on one of its components? Whatever the wishes of the Data Reduction Consortia may be, it appeared indispensable to know for each double and multiple system, accurate positions and magnitudes for all individual components.

Unfortunately, when the INCA consortium was created, the only general catalogue existing on double stars was the Index (1976,5) which appeared unpractical for our purpose because of: (a) insufficient accuracy of the positions given by the Index number ($\pm 1'$), (b) incompleteness of the DM identifications ensuring connections with fundamental stellar catalogues, and (c) difficulties in recognising the degree of multiplicity of the systems by automatic computer technique. It has thus been decided to establish on the basis of this Index (1976.5) a specific *Catalogue of the Components of Double and Multiple stars* (CCDM) that should function as a fundamental double and multiple star data base at the disposal of the Inca Consortium but also as a new original tool in double star astronomy (Dommanget, 1983).

3. The CCDM Catalogue – Past, Present, and Future

One of the fundamental differences between the CCDM and the Index lies in an important format modification that consists in allocating one full record to each double or multiple system component. Another difference lies in the astrometric characteristic of the CCDM obtained by introducing in two specific columns, the balances in right ascension and in declination between the position given by the Index number (to $\pm 1'$) and the accurate position (to $\pm 0''.1$) of each component (Figure 2).

Systematic investigations have been made for completion of the components identification by DM, AGK2/3 and SAO numbers as well as for systematic detection of different types of errors.

Another important information that may be found in the CCDM consists in the identification of the systems and their components in the INCA Data Base.

Presently, the CCDM contains:

56 400 double systems;
4 400 triple systems;
1 000 quadruple systems;
400 systems having a multiplicity higher than 4; and
100 systems with incoherent data in the Index (1976.5),

which leads to a total of 62.300 systems that should be compared to the 70 300 entries existing in the Index (1976.5), from which they are deduced. Some 20 000 accurate positions ($\pm 0''.1$) of components or of systems obtained from existing stellar catalogues have already been introduced in the CCDM; a few thousands other newly determinated accurate positions are being added.

To enable accurate positioning of the IFOV for each system, coordinates are needed for:

individual components when $\rho > 3''$

and

photocenter only when $\rho < 3''$,

whatever the decision may be concerning the point in the system on which the IFOV should be centered. Therefore, special effort is now made to urgently complete the catalogue in this way for all *Hipparcos* stars before the launch of the satellite (presently foreseen for april 1989). Relative positions of the components in case of $\rho < 3''$ may be collected by ground based observations during the 2.5 years operation time of the satellite, these information being needed only for the reduction of the observations.

Concerning the magnitudes, one should make reference to a photometric catalogue for double and multiple stars in preparation by E. Oblak (see next paper in this volume).

Later, in addition to the present work, various researches will be continued: (1) on identification of systems in stellar catalogues as well as in the *INCA Data Base*, (2) on the completion of the CCDM by accurate positions, and (3) on updating the CCDM with the most recent version of the Index at Washington.

4. Double and Multiple Stars in the HIC

The identification of double and multiple stars in the *INCA Data Base* (213 771 stars) on the basis of the CCDM (in its present state), has given the following results:

double systems:	16 500 of which	{8.000 with $\rho < 3''$; 8.500 with $\rho > 3''$}
triple systems	2 500	
quadruple systems:	700	
more than 4 components:	300	
Total	20 000	

For the double stars only, the situation in the *Input Catalogue* (IC-1 = 114 500 stars) is as follows:

Angular separation	$\Delta m < 3$ m		$\Delta m > 3$ m
$\rho < 1''$	2466	3.950 (for $\rho < 1''$ to $1'' < \rho < 3''$)	3711
$1'' < \rho < 3''$	1484		
$3'' < \rho < 5''$	565	2.714 (for $3'' < \rho < 5''$ to $40'' < \rho$)	
$5'' < \rho < 7''$	373		
$7'' < \rho < 10''$	354		
$10'' < \rho < 15''$	328		
$15'' < \rho < 20''$	247		
$20'' < \rho < 30''$	361		
$30'' < \rho < 40''$	131		
$40'' < \rho$	355		
Total	6.664		3711

These figures concern only known visual and astrometric systems to which some thousands spectroscopic and eclipsing systems have to be added. Their presence in the *Inca Data Base* is to be found in their belonging, for the brightest ones to the 'survey' and for the faintest ones, to the 23 proposals received by the European Space Agency for such objects. One should regret here that only five of their authors are members of our IAU commission 26! Fortunately, in spite of this, it may be mentioned that all orbital visual pairs of an up-dated Uccle private catalogue, all spectroscopic pairs of Batten's seventh catalogue and all occultation binaries of Evan's catalogue have been proposed by us and are thus included in the *INCA Data Base*. As a result, 94, 93, and 64%, respectively, of the stars of these catalogues are in IC-1!

Another origin for double and multiple systems in the *INCA Data Base* has to be found in proposals made for stars which by identification in the CCDM, have turned out to be wide components of such systems: the components ignored by the proposers have then been added in this base.

To complete our survey on the presence of double and multiple stars in the *INCA Data Base*, we also should remember that there inevitably exist in this base – and thus also in the HIC – many unknown double stars that will be discovered by the satellite itself through a systematic detection of anomalies that will appear in the images of such systems, for angular separations between $0\overset{''}{.}020$ and say $0\overset{''}{.}200$ or $0\overset{''}{.}300$ (astrometric pairs) (Dommanget, 1985).

This is the basic situation of double star astronomy in the *Hipparcos* mission, situation from which we may expect very much.

5. Contribution of the Hipparcos Mission and the CCDM to Wide Systems

As can be seen from the discussion above, an important fraction of the *Hipparcos* observation programme will bear on wide systems. Accurate parallax and proper motion determinations for the components of the double stars having $\rho > 3''$ may lead to the recognition of physical and optical pairs and thus to an interesting sample for statistical studies on the boundary between both physical and optical domains.

On the other hand, some 15700 possible companions of nearby stars (at less than 25 parsecs) have been proposed by us to ESA and are included in the *INCA Data Base*. Of these stars, 4652 are kept in the *Input Catalogue* (IC-1) and it may be expected that a few percents of them will show parallaxes and proper motions in accordance with the dynamical characteristics of a physical connection, leading to the discovery of new physical nearby wide systems.

We must also mention that in view to complete the CCDM with accurate positions for all components of the double and multiple systems that may appear in the HIC, the collaboration of many colleagues has been obtained and a Working Group 'Double Stars' has been appointed inside INCA, of which we are in charge as coordinators. From the experience of all of the observers of astrographic positions, it systematically appeared that on each plate taken especially for our purpose, approximately twice as much wide systems as those already recorded in the Index, were found. This should be considered for a systematic ground based survey to be undertaken if we want to complete to some extend, the census of double and multiple stars.

Concerning the contribution of the CCDM to double star astronomy and to wide systems more specifically, it is difficult to separate it from that of the satellite. Both are fundamentally interconnected: the CCDM would probably not have been undertaken at least under its present format without the Hipparcos project: the *Hipparcos Input Catalogue* would probably not contain so many binaries, especially wide ones, without having had the CCDM at our disposal.

Finally, we also may not forget to mention how much the attraction for double and multiple stars has been increasing during these last years not only inside the *Input Catalogue Consortium* (INCA) but also inside both reduction consortia (NDAC and FAST) who will have to deal with specific, difficult but very interesting problems. This new interest of many colleagues who originally were not personally concerned with double star astronomy, may be considered as another very important contribution of the Hipparcos satellite (even before its launch) and the CCDM catalogue.

Acknowledgements

It is a pleasure to give credit to my colleague O. Nys of the Royal Observatory, Belgium for his constant and efficient contribution to the realisation of the CCDM and all problems related to the identification of the double and multiple stars in the *INCA Data Base* as well as to all foreign collaborators to our project for their effective help and various contributions.

We also want to acknowledge the 'Fonds de la Recherche Fondamentale Collective' (Belgium) which under project No. 2.9009.79 has brought a substantial support to the participation of the Department of Astrometry and Celestial Mechanics of the Royal Observatory to the work of the INCA Consortium.

References

Dommanget, J.: 1983, 'Un catalogue des composantes d'étoiles doubles et multiples (CCDM)', Bulletin d'Information du CDS, No. 24, pp. 83–80.

Dommanget, J.: 1985, *Astrophys. Space Sci.* **110**, 47.

UPGREN – Could you review the precision of Hipparcos? In the specifications circulated a couple of years ago, an error in parallax of $\pm 0''.002$ was claimed for stars brighter than 8 or 9, rising up to about $\pm 0''.005$ at magnitude 12 or 13. The error in proper motion was not given, but may depend on the length of the mission. Could you discuss whether these intial parallax error estimates are still valid, and what the errors in position and in proper motion are expected to be?

DOMMANGET – As far as I know, a couple of years ago, the errors expected in position, parallax and proper motion were of the same order ($\pm 0''.002$ to $\pm 0''.005$). I have not been informed about any changes in that respect. Prof. Bernacca here present may perhaps give more up-dated information.

BERNACCA – Folowing extensive simulations of the mission, we expect that the r.m.s. mean error (sky average) on the astrometric parameters of stars of blue magnitude about $B = 9$ will be close to: $d\beta = 1.3$ mas, $\cos\beta\, d\lambda = 1.6$ mas, $d\mu_\beta = 1.6$ mas yr^{-1}, $\cos\beta\, d\mu_\lambda = 2.0$ mas yr^{-1}, $d\pi = 1.6$ mas.

A PHOTOMETRIC CATALOGUE OF VISUAL AND MULTIPLE STELLAR SYSTEMS

E. OBLAK*

Observatoire de Besançon, France

Summary. The catalogue gives the parameters of the *UBV*, *uvby*, and Geneva photometries for more than 2000 stars.

It gives also the first statistics with respect to the different caracteristics of visual systems: separation, difference of the magnitude of components, ... etc.

For 11713 components of stellar systems the catalogue, in its actual form, gives the parameters of the *UBV*, *uvbyβ* and Geneva photometric systems.

A concise statistics presents the distribution of the stars in the different photometric systems and the relations between the difference in the magnitudes and separation of the components.

1. Introduction

The literature gives a lot of very different photoelectric photometric systems. Some of them (*uvbyβ*, *UBV*) contain actually 50000 or more stars.

It is interesting now to examine these different systems with regard to the visual stellar systems (generally double stars with separation greater than 10 arc sec.

This examination would be very useful to take into account the photometry of the double stars in the *Hipparcos Input Catalogue* for which an extensive work has already been done at Bruxelles (CCDM) and at Geneva.

A compilation work concerning the photometric measures of the components of wide double stars was undertaken in 1987 and continued using the most recent versions of photometric catalogues.

2. The Photometric Catalogue

We want to integrate all photometric systems concerning the visual stellar systems.

We start from three extensive photometric systems:

UVB: Mermilliod (1987) and Wallenquist (1981).

uvbyβ: Hauck and Mermilliod (1985).

Geneva: Rufener (1987).

Two of them: *UBV* (for averages) and a new version of the Geneva Catalogue are not yet published and has been kindly communicated to us by Mermilliod and Rufener.

We also take into account for the *UBV* photometry the catalogue of Wallenquist for visual double and multiple systems (1064 systems).

Table I presents the content of four catalogues and concerns the photometric measures of one of more components, or joint photometry for the system.

* CNRS UA 040389.

Astrophysics and Space Science **142** (1988) 31–37.

TABLE I

Components of stellar systems for the three photometric systems

Catalogue	Total	*N* components
UBV (1987)	87 263	6423 (4358)
Wallenquist (1981)	2 275	2275
uvbyβ (1985)	40 482	2983 (2738)
Geneva (1987)	29 212	2447 (2158)

With the redundances 11 853 components or systems are concerned.

The parenthesis indicate the number of components with HD identifications.

Table II gives the details of the stellar systems, components and joint photometry for the three mentioned photometric systems.

TABLE II

Photometry of visual stellar systems for three photometric systems

Catalogue systems	*UBV*		*uvby*		Genève	
	(1987)	Total components	(1985)	Total components	(1987)	Total components
Doubles	709		423		137	
Triples	56		28		4	
Quadruples	13		5			
		1643		956		286
Quintuples	1					
Sextuples			1			
Systems	779		457		141	
One component	446	2089	105	1061	85	371
Joint photometry		4334		1922		2076
Total		6423		2983		2447

These measures have been integrated in the catalogue already obtained from the previous versions of photometric catalogues (Oblak, 1982).

The actual preliminary catalogue gives for each component or system the:
HD, DM, or ADS number,
the equatorial coordiantes,
the galactique longitude and latitude,
the separation of the components,
the HD spectral type or Geneva spectral type,
the magnitude and the parameters of the *UBV*,
uvbyβ and Geneva photometries.

The catalogue contains actually 11763 components or systems; 1313 systems have been measured at least in one of the three photometric systems.

For 396 components the difference in magnitudes between two photometric systems is greater than $0\overset{m}{.}2$.

These cases have been examined manually.

The last line of the Table III gives the number of data obtained by Grenon (1987) from about 75 photometric systems.

Among these, 7665 are common with the *Hipparcos Input Catalogue*, of which about a thousand components or systems have some identification or photometric problems and have also been examined manually.

They will be incorporated in the Catalogue.

The intersection between the different photometries gives the following results:

uvby	Geneva:	1093 components ,
uvbyβ	Geneva:	879 components ,
uvbyβ	*UBV*:	1114 components ,
Geneva	*UBV*:	1089 components ,
uvby Geneva	*UBV*:	515 components .

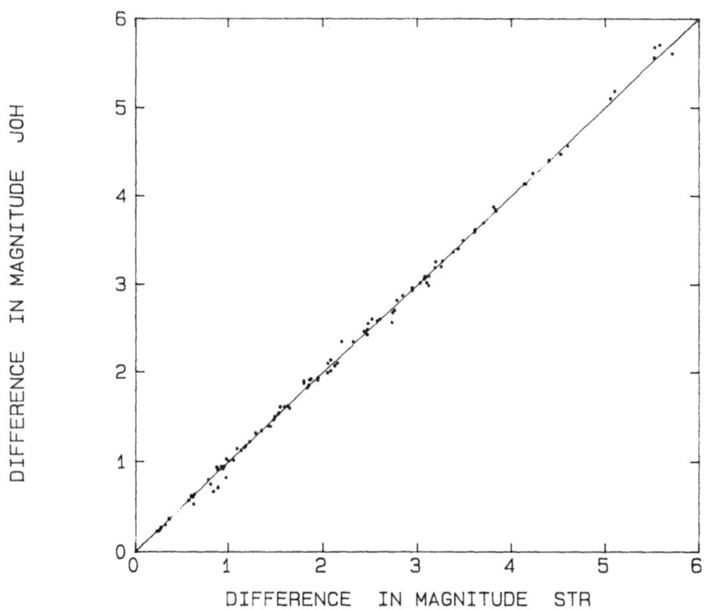

Fig. 1. Differences in magnitudes between the components in the *UBV* (JOH) and *uvby* (STR) photometric systems.

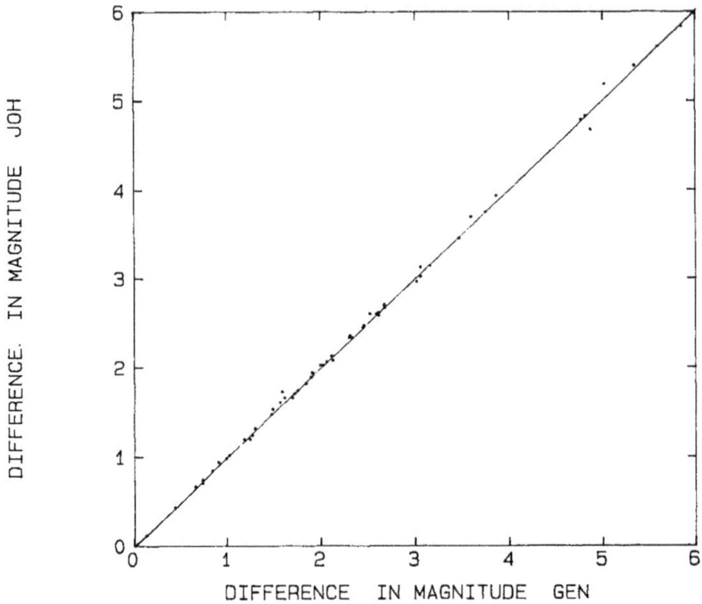

Fig. 2. Differences in magnitudes between the components in the *UBV* (JOH) and Geneva photometric systems.

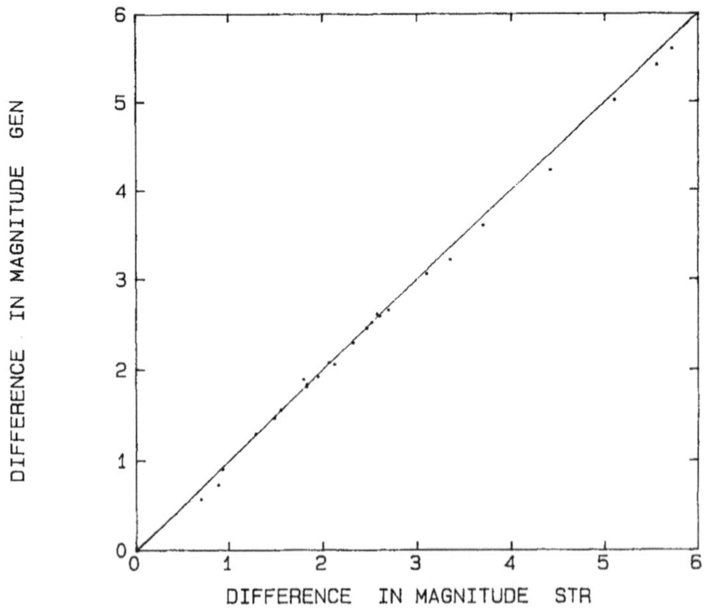

Fig. 3. Differences in magnitudes of the components in the Geneva and *uvby* (STR) photometric systems.

The comparison between the differences in magnitudes of the components for the three photometries shows a good agreement (Figures 1 to 3).

$$dm\ (\text{Strömgren} - \text{Geneva}) :\ +0\overset{\text{m}}{.}048 \pm 0\overset{\text{m}}{.}099\ (N = 25)\,,$$

$$dm\ (\text{Strömgren} - \text{Johnson}) :\ +0\overset{\text{m}}{.}003 \pm 0\overset{\text{m}}{.}003\ (N = 106)\,,$$

$$dm\ (\text{Geneva} - \text{Johnson}) :\ -0\overset{\text{m}}{.}015 \pm 0\overset{\text{m}}{.}002\ (N = 59)\,.$$

A small systematic effect persists between the Geneva and the Johnson system, especially for large differences in magnitudes.

3. Short Statistics of the Catalogue

The contents of the Catalogue is given in the Table III.

TABLE III

Photometric catalogue of visual double stars

Systems	N	Total of components
Doubles	1158	
Triples	128	
Quadruples	20	
Quintuples	7	
		2815
	1313	
Only one component	3101	
Joint photometry (D)	5847	
Total (3 photometries)		11763
All photometries		16766

We have also compared the differences in magnitudes between our catalogue and Wierzbinski's catalogue:

$$\text{For his list 1 } (\rho < 10''):\ -0\overset{\text{m}}{.}044 \pm 0\overset{\text{m}}{.}14\ (N = 55)\,.$$

$$\text{For his list 3 } (\rho > 10''):\ -0\overset{\text{m}}{.}034 \pm 0\overset{\text{m}}{.}34\ (N = 72)\,.$$

A systematic effect appears: $0\overset{\text{m}}{.}03$ to $0\overset{\text{m}}{.}04$.

The study of the differences in magnitudes *dm* of the components shows an absence of doubles with large *dm* for the small separations ($\rho > 20''$) (Figure 4).

For these small separations there appears a large dispersion also in the differences of dm between the different photometries than for large separation (Figure 5).

A complete statistical study is meant to be undertaken when the catalogue is finalized.

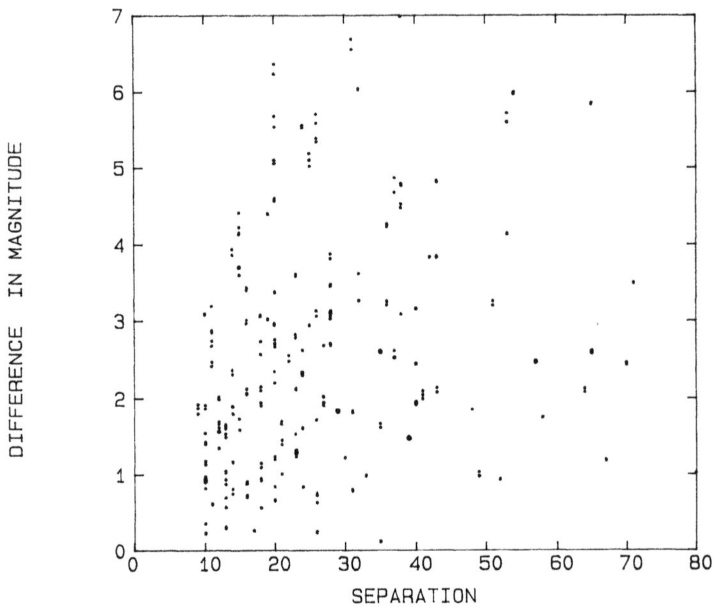

Fig. 4. Differences in magnitudes *dm* of the components as a function of the separation.

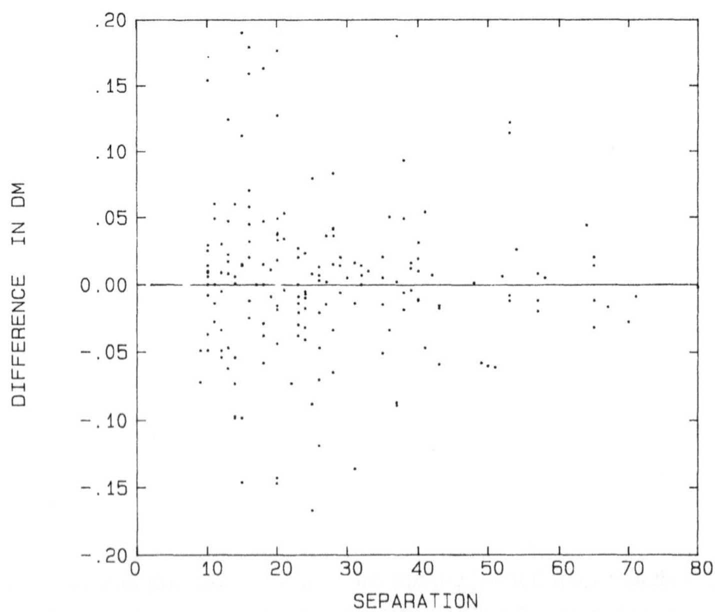

Fig. 5. Differences of the *dm* of the components as a function of the separation.

4. Future Work

With M. Mermilliod and M. Grenon, from the Geneva Observatory, we have examined all double stars presenting problems. This work is actually necessary for the *Hipparcos Input Catalogue*. The results will be then incorporated in the photometric catalogue of visual and multiple systems.

The constituted and analyzed final catalogue will allow to introduce the homogeneous magnitudes in the CCDM Catalogue.

We think to use also the catalogue of Wierzbinski of the differences in magnitude to obtain the magnitudes of the components when the joint photometry is given.

Acknowledgements

I would like to thank J. Dommanget, F. Rufener, M. Mermilliod, J. C. Mermilliod, and the staff of the Centre de Données Stellaires for communicating the catalogues of stellar data, some of them before publication. I also thank Prof. Golay and M. Grenon for their hospitality at the Geneva Observatory.

References

Grenon, M.: 1987, private communication.
Gronbech, B. and Olsen, E. H.: 1976, *Astron. Astrophys. Suppl* **25**, 213.
Gronbech, B. and Olsen, E. H.: 1977, *Astron. Astrophys. Suppl.* **27**, 443.
Hauck, B. and Mermilliod, M.: 1975, *Astron. Astrophys. Suppl.* **22**, 235.
Hauck, B. and Mermilliod, M.: 1985, *Astron. Astrophys. Suppl.* **60**, 61.
Lindroos, K.P.: 1981, *Stockholms Observatorium Report*, No. 18.
Mermilliod, J. C.: 1986, *Astron. Astrophys. Suppl.* **71**, 413.
Mermilliod, J. C.: 1987, private communication.
Nicolet, B.: 1978, *Astron. Astrophys. Suppl.* **34**, 1.
Oblak, E.: 1982, *Bull. Inf. CDS*, No. 22, 41.
Olsen, E. H.: 1982, *Astron. Astrophys. Suppl.* **48**, 165.
Rufener, F.: 1976, *Astron. Astrophys. Suppl.* **26**, 275.
Rufener, F.: 1987, private communication.
Wallenquist, A.: 1981, *Uppsala Universitets Astronomiska Observatorium*, Report No. 22.

ABT – Dr. Oblak is to be congratulated on producing a very useful catalog. Will it be published in *Astronomy and Astrophysics Supplements?*

OBLAK – Yes, it will be as soon as possible.

GLIESE – Does your catalogue give only directly observed data? Not combined magnitudes later on splitted into magnitudes colours for component *A* and for *B* (?) on the base of estimated magnitude difference?

OBLAK – The published catalogue will give only directly the observed data.

YUNGELSON – Did you check if all pairs included in your catalogue are really physical ones?

OBLAK – We weren't interested especially in this problem. We had simply collected the data with photometry. But the physical connection at pairs may be checked, e.g., by computing the distances to components.

A GENERAL CATALOGUE OF EPHEMERIDES AND APPARENT ORBITS FOR VISUAL BINARIES

YAN LIN-SHAN

Shanghai Observatory, Academia Sinica, Shanghai, China

Summary. Informations on 736 pairs of visual binaries are given in the form of the Tables. The General Catalogue gives ephemerides (t, $\theta°$, ρ'') in 20 years (1984–2003) for each pair with the figures of its apparent elliptical orbit where the positions of secondary component relative to the primary one at different epochs are indicated. *The General Catalogue* contains four parts: Part one – Source and Grading of Orbit; Part two – Ephemérides and Atlas of Apparent Orbits; Part three – Classifications of Visual Binary Stars; Part four – Indexes of Visual Binary Stars. The principle of calculation and the statistical data are presented in this paper. There are seven statistical tables, giving the elements distribution of true and apparent orbits, grading distribution of orbits, number distribution with different physical property of component of binary star. The number of binary stars in anyone constellation, the number of binary stars brighter than $6^m.5$. The number of binary stars nearer than 25 parsec.

1. Introduction

The ephemerides of binary stars are the important informations for astronomical observation and research. After 1950, many ephemerides had been compiled by Müller (1954), Müller and Mayer (1964, 1969), Müller and Couteau (1979), and Wepner (1967), and so on. In recent years, orbital elements of several hundred pairs have been re-computed and many new binaries were discovered. It becomes necessary to set up perfect ephemerides. Our General Catalogue is based on the Finsen and Worley (1970) third catalogue of orbits of visual binary stars and on our many new data and new results collected. Informations on 736 pairs of visual binaries are given in the form of the Tables. The General Catalogue gives ephemerides (t, $\theta°$, ρ'') in 20 years (1984–2003) for each pair with the figures of its apparent elliptical orbit where the positions of the secondary component relative to the primary one at different epoch are indicated. Apparent elliptical elements are given below each atlas. At this point, our General Catalogue is only one. It is different from all other ephemerides.

2. Principle of Establishing Apparent Orbit

In the case of the orbital elements (a, e, i, Ω, ω, p, T) of visual binaries have been known, we could calculate the orbital motion on tangential plane on the celestial sphere ephemerides.

First at all, we set up two coordinate systems to find the elements of apparent ellipse. Let a and e represent the size and form of the ellipse, θ-position angle of major axis and (x_0', y_0'-coordinates of apparent ellipse center.

Take rectangular coordinate system $S - xyz$. The primary S is the origin of coordiantes, xy-plane with reference to true orbit plane. The a-axis towards the

Astrophysics and Space Science **142** (1988) 39–46.

periastron, and y-, z-axes both follow the right-hand rule. Since the primary S is at a focus of the true ellipse orbit, the equation takes the form

$$\frac{x^2}{a^2} + \frac{y^2}{a^2(1 - e^2)} = 1\,, \tag{1}$$

As for $S - x'y'z'$ system, the origin is the primary star S, apparent orbital plane is $x'y'$-plane, but x'-axis towards the north, y' towards the east, z' is perpendicular to $x'y'$-plane and follows the right-hand rule.

2.1. Coefficients of the equation of the apparent ellipse

In the $S - x'y'z'$ system, the apparent orbit equation takes form

$$Ax'^2 + 2Bx'y' + Cy'^2 + 2Dx' + 2Ey' + F = 0\,. \tag{2}$$

These coefficients will be determined. Relate the two coordinate systems by Eulerian angles $i\omega\Omega$, we have

$$\begin{pmatrix} x' \\ y' \\ z' \end{pmatrix} = Rz(-\Omega)Rx(-i)Rz(-\omega)\begin{pmatrix} x \\ y \\ z \end{pmatrix}, \tag{3}$$

Rz, Rx show the orthogonal rotation matrix, defined by

$$Rx(-i) = \begin{pmatrix} 1 & 0 & 0 \\ 0 & \cos(-i) & \sin(-i) \\ 0 & \sin(-i) & \cos(-i) \end{pmatrix}, \qquad Rz(-\omega) = \begin{pmatrix} \cos(-\omega) & \sin(-\omega) & 0 \\ -\sin(-\omega) & \cos(-\omega) & 0 \\ 0 & 0 & 1 \end{pmatrix}.$$

The direction of ω is positive when we rotate any point counter-clockwise in the true ellipse, its coordinate $z = 0$. From Equation (3), x', y' are linear function of x, y. Substitute these x, y-values into Equation (1), we may obtain these coefficients A, $B, \ldots, F$, which are the functions of the elements of a, e, i, ω, Ω, respectively.

2.2. The position of the apparent ellipse

The coordinates of true ellipse center in the $S - xyz$ system are $x_0 = -ae$, $y_0 = 0$. Substitute these coordinate values into Equation (3), we find (x_0', y_0'), but the major axis position angle θ of apparent ellipse is found from the equation

$$\cot 2\theta = \frac{A - C}{2B}\,.$$

2.3. The size and form of the apparent ellipse

Take the coordinate system $O - XYZ$: O is the origin and denotes the center of apparent ellipse, its major axis is the X-axis, so that

$$\begin{pmatrix} X \\ Y \end{pmatrix} = \begin{pmatrix} \cos\theta & \sin\theta \\ -\sin\theta & \cos\theta \end{pmatrix} \begin{pmatrix} x' - x_0' \\ y' - y_0' \end{pmatrix} . \tag{4}$$

If we substitute Equation (4) into Equation (2), we obtain

$$mX^2 + nY^2 + k = 0 .$$

Hence, a', e' are derived from two equations

$$a'^2 = -\frac{k}{m} , \qquad a'^2(1 - e'^2) = -\frac{k}{n} .$$

2.4. Figures of the apparent elliptical orbit

The sizes of major axes of 736 pairs are not the same. In order to harmonizing, all the major axes have to be the same, and $2a' = 10$ cm. Hence, the scale of figures is different and shows

$$M = 50\ mm/a' .$$

2.5. Computation of ephemerides

When the elements are known, the apparent position angle θ and the angular distance ρ for the time t are derived from the following equations

$$E - e\sin E = \frac{2\pi}{P}(t - T) , \qquad x = \cos E - e , \qquad y = \sqrt{1 - e^2}\sin E ;$$

the symbols E, T, and e represent the eccentric anomaly, epoch of periastron passage and eccentricity, respectively. We substitute each set of x, y-values successively into Equation (3), then get

$$x_i' = \rho_i \sin\theta_i , \qquad y_i' = \rho_i \cos\theta_i ,$$

and, $\theta°$, ρ''-values for any time are derived.

3. Statistical Results of Orbits

3.1. The elements of true orbit

Our General Catalogue gives the elements of true orbit of 736 pairs of visual binary stars up to 1982, the most part (94%) of the orbital elements of these pairs were computed

TABLE I

Orbits distribution by years of announcement

y	1900–1949	1950–1959	1960–1969	1970–1979	1980–1982
N	9.1	149	314	134	9.8
%	6	20	43	18	13

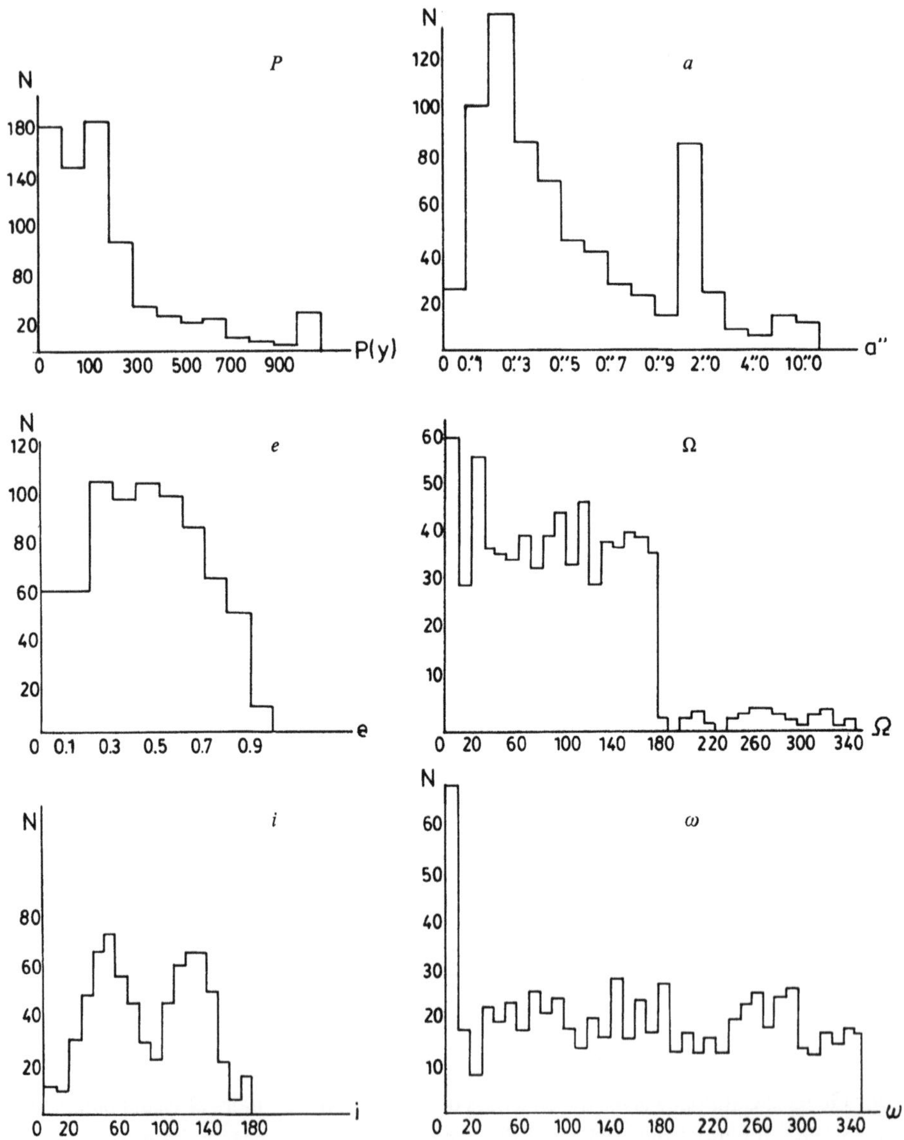

Fig. 1. Distribution of numbers of the elements of true orbits of 736 pairs.

after 1950, and the 13% were done after 1980. The Catalogue, therefore, appears an entirely new aspect. The calculation of orbits were distributed by years (Table I).

The numbers distribution of orbital elements – *P*, *a*, *e*, Ω, *i*, ω – of the 736 pairs are all given in Figure 1.

P – period of revolution (in year). There are 515 pairs (70%) with a *P* shorter than 200 yr, 22 pairs (3%) with a period longer than one thousand year, 31 pairs (4%) with a period shorter than 10 yr.

a – semi-major axis of true orbit (in second of are). The frequency maximum of a'' appears between 0.″2–0.″3 for $a'' < 1''$, and between 1″–2″ for $a'' > 1''$. The maximum a''-value is 43.″9, the minimum is 0.″018.

e – eccentricity of true orbit. From *e* in Figure 1 we can find the normal distribution of eccentricity. The maximum stands near 0.″5. It appears a symmetric distribution in two wings. There are 32 pairs with circular orbits, i.e., $e = 0$, and 12 paris for $e > 0.9$.

Ω – position angle of the ascending node (in degree). We can easily find out 2 groups in Ω in Figure 1, one is 0°–180°, the another is 180°–360°. The average pair numbers of the former is about 40 and only 3 in the later. Both groups fluctuate around the average pair number but smaller is the later.

i – inclination of orbital plane (in degree). Two maximums of *i* distribution are shown in *i* of Figure 1 and with similar *N*. The average *N* is 65. One maximum is between 50°–60°, the other between 120°–140°. Two wings drop down symmetrically and sharply.

ω – longitude of periastron relative to ascending node (in degree). ω of Figure 1 shows the distribution according to ω in our General Catalogue. When $\omega < 10°$, the pairs are very concentrated (68 pairs in total). While ω is between 10°–360°, its distribution is a constant fundamentally, the average is 19 and the fluctuation is very small.

3.2. Grading of Orbits

The reliability of ephemerides are dependent upon the characterof the orbits. Because the visual binaries are far away from us and their motions are very minute as well as the revolution periods are very long, the grading of their orbits are different from that of some bodies' orbits of solar system. The five grading standards defined by Finsen and Worley are:

Grade 1 – Definitive: Several revolutions well observed;

Grade 2 – Reliable: At least one revolution well observed;

Grade 3 – Preliminary: Elements especially *P* and a not likely to be grossly in error. In general, the observations define at least, half of the orbit;

Grade 4 – Premature: Individual elements entitled to little weight, but a^3/P^2 may be accepted with some confidence. In general, less than half of the orbit is defined by the observation.

Grade 5 – Indeterminate: Observations are very short with small curvature.

Table II gives the distribution of numbers in grading.

In brief, grades 1 and 2 are in the minority, about 26%. Grades 3 and 4 are in the majority, about 63%. Grade 5 is also in the minority but much less than 7%.

TABLE II

Distribution of numbers in grading

Grading	I	II	III	IV	V
N	62	130	239	223	51
%	8	18	33	30	7

3.3. The elements of apparent orbits

The elements of apparent orbits are given below the atlas for each pair. The numbers of distribution in semi-major axis a' and eccentricity e' are shown in Figure 2.

A comparison to the semi-major axis of the true orbit with that of the apparent orbit, the similarity can be found between two figures. As between e with e', a large discrepancy there exists.

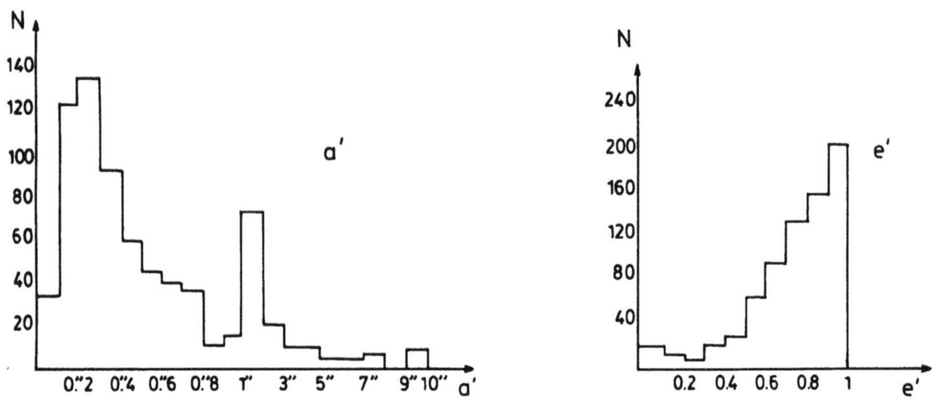

Fig. 2.

4. Survey of the Classification and Index

For the research of physical properties, the pairs with special type components in this General Catalogue were classified. The kinds and numbers are shown in Table III.

TABLE III

Type	Var.	Spectra.	W. dwarf	Giant	X-ray	Radio stars	Double dwarf	Double gaint	Gaint-dwarf
N	14	26	2	5	5	4	120	5	7

There are four kinds Index: i.e., constillation, magnitude, parallax, and Chinese star name. Their numbers of distribution of VBS are given as follow.

The numbers of binary star in each constellation for 88 constellations on whole sky are given in Table IV. There is no General Catalogue binary star in 8 constellations, 13 constellations contain over 20 pairs (314 pairs in total).

TABLE IV

Numbers of distribution for VBS in constellation

Constel.	*N*	Constel.	*N*	Constel.	*N*	Constel.	*N*
And	22	Col	2	Lac	7	Psc	17
Ant	3	Com	8	Leo	18	Pup	7
Aps	1	CrA	3	Lep	6	Pyx	5
Aql	8	CrB	4	Lib	7	Ret	0
Aqr	15	Crt	5	LMi	1	Scl	4
Ara	1	Cru	0	Lup	5	Sco	8
Ari	10	Crv	2	Lyn	10	Sct	1
Aur	6	CVn	6	Lyr	11	Ser	11
Boo	25	Cyg	20	Men	0	Sex	3
Cae	1	Del	11	Mic	2	Sge	1
Cam	18	Dor	0	Mon	8	Sgr	11
Cap	3	Dra	22	Mus	1	Tau	24
Car	3	Equ	4	Nor	2	Tel	2
Cas	20	Eri	8	Oct	1	TrA	0
Cen	13	For	6	Oph	25	Tri	3
Cep	19	Gem	15	Ori	21	Tuc	4
Cet	28	Gru	1	Pav	4	UMa	21
Cha	0	Her	32	Peg	33	UMi	2
Cir	2	Hor	3	Per	21	Vel	8
CMa	7	Hya	16	Phe	4	Vir	13
CMi	4	Hyi	0	Pic	2	Vol	0
Cnc	14	Ind	4	PsA	1	Vul	6

Table V shows the numbers of the pairs in magnitude $< 6\overset{m}{.}5$, parallax $> 0\overset{''}{.}04$ and the binary stars with Chinese names.

TABLE V

Distributions of numbers in Mag. π and name

Type	Mag $< 6\overset{''}{.}5$	$\pi > 0\overset{''}{.}04$	Chin. name star
N	24.7	4.0	13.4
%	33.6	5.4	18.2

References

Müller, P.: 1954, *J. Observateurs* **37**.
Finsen, W. S. and Worley, C. E.: 1970, *Republic Obs. Johannesburg Circ.* **7**, No. 129.
Müller, P. and Couteau, P.: 1979, Publ. l'Obs. de Paris.
Müller, P. and Mayer, Cl.: 1964, *Notes et Informations*, Publication de l'Obs. de Paris.
Müller, P. and Mayer, Cl.: 1969, Publication de l'Obs. de Paris.
Wepner, W.: 1976, Treugesell-Verlag.

SECOND SESSION

WIDE NEARBY BINARIES

Chairman: J. DOMMANGET

THE THIRD CATALOGUE OF NEARBY STARS WITH SPECIAL EMPHASIS ON WIDE BINARIES

Introductory Lecture

W. GLIESE and H. JAHREISS
Astronomisches Rechen-Institut, Heidelberg, F.R.G.

Summary. Thirty-six percent of the 1641 visual components nearer than 22.5 parsecs listed in 1969 were members of systems. Meanwhile for an extension to 25 pc of the Sun about 2000 additional objects are found, however, only 22% of them are known today as members of systems. Angular separations are observed from a few tenths of an arcsec to more than one degree. The widest linear separations at the sphere seem to be larger than 10^5 astronomical units. The observed frequency distribution of the projected separations for all nearby stars is given and compared with the frequency distribution of more complete subsamples. The 32 presumable systems with the widest separations are discussed with respect to their astrometric and spectroscopic properties.

1. Introduction

At present, we are compiling stars and their astrometric, spectroscopic, and photometric data for a 'Third Catalogue of Nearby Stars' which will be extended to 25 parsecs of the Sun. The predecessor (Gliese, 1969) had a distance limit of 22.5 pc; the *Catalogue of Stars within Twenty-Five Parsecs of the Sun* (Woolley *et al.*, 1970) was restricted mainly to objects with known trigonometric parallaxes.

Meanwhile, van Altena (1985) generously has made available to us a preliminary version of the forthcoming *Yale Catalogue of Trigonometric Stellar Parallaxes*. Furthermore, spectroscopic data and numerous photometric series became known to us. Spectral type-luminosity relations and colour-luminosity relations, partly derived by the observers, partly calibrated at Heidelberg allowed the detection of many new nearby members. Particularly the number of faint red dwarf stars has been increased significantly. In spite of this fact one has to bear in mind that still severe selection effects are present already within 20 pc; and an extension of the distance limit from 20 to 25 pc is equivalent to a doubling of the volume of space to be considered.

2. The Relative Binary Frequency

In 1969 the relative binary frequency, i.e., the number of components in double or multiple systems relative to the number of all stars considered was

- by 61 stars within 0–5 pc 59%,
- by 221 stars within 5–10 pc 51%,
- by 1071 stars within 10–20 pc 41%.

* Presented by W. Gliese.

Astrophysics and Space Science **142** (1988) 49–56.

In these figures spectroscopic and probable astrometric binaries were included. The corresponding percentages are at present the following

- by 72 stars within 0–5 pc 63%,
- by 258 stars within 5–10 pc 50%.

We hesitate to give a guess for the stars beyond 10 pc, since our compilation is not yet completely finished with respect to radial velocities and other spectroscopic data. Furthermore, the results of several programmes investigating the percentage of spectroscopic binaries among the nearby K and M dwarf are not yet available. In a preliminary report Halbwachs (1985) expects an increase from 4 to 15% detected spectroscopic binaries among the nearby K dwarfs. For comparison the corresponding relative binary frequency for a 'complete' and well investigated subsample of the LHS (372 stars with $\mu \geq 1''$ per year, $\delta \geq 30°$ and $|b| \geq +10°$) amounts to only 30%. But, one has to recall that on the scale of the Palomar plates separations of less than 10″ may cause detection problems (Dawson, 1986).

In 1969 we listed 1328 singles and systems with 1641 visual components; in 1979 these numbers were supplemented to altogether 1594 objects with 1972 components. Then 36% of the 1641 stars were members of binary or multiple systems, however, only 22% of the about 2000 objects added since 1969 are known as members of systems. This is a selection effect as the already known nearest neighbours of the Sun were preferred on observational programmes, particularly programmes for separating very close binaries. Angular distances of a few tenths of an arc sec to more than one degree are detected.

3. Wide Binaries

Distances between the components can be expressed in astronomical units (AU) if the parallax of a system is known. Of course, in many systems the components are as close as planets to our Sun; the semi-major axis of the Pluto orbit is 40 AU. But detected were also separations of 100, 1000, or even 10000 AU between components.

Such values are measured as tangential distances at the sphere but we should like to know the real space separation. For the estimation of the radial distance between the components we need the knowledge of the parallaxes of the objects supposed to be gravitationally bound. The problem hereby is the reliability of parallax results.

One parsec is equal to 206265 AU. At the catalogue limit a parallax error of 4% corresponds to a linear difference of one parsec and this accuracy is attainable today only in a few exceptional cases. As an example with favourable conditions we have our nearest neighbours, the triple system Alpha Centauri A, B and Proxima Cen, obviously a wide multiple system about 1.3 pc away from the Sun. The tangential separation of more than 2° is slightly more than 10000 AU. The difference between the two trigonometric parallaxes is $0''.0218 \pm 0''.0062$ which corresponds to a separation of 7800 AU in the radial direction. In space, Proxima is nearly 13000 AU apart from α Cen A and B, an estimate with a standard error of $\pm$ 2200 AU. In all other cases the uncertainties in the trigonometric measurements do not allow a reliable determination of separations in the radial direction.

Therefore, search for wide binaries means a search for common proper motion (c.p.m.) objects. Luyten's NLTT volumes (1979, 1980) listing the stars with yearly proper motions ≥ 0″.18 proved to be one of the best sources for such objects in the solar neighbourhood.

Let us insert here a few words of gratitude to W. J. Luyten for his immense number of proper motions measurements which include the large majority of known stars within 25 parsecs of the Sun; 0″.18 per year corresponds to 21 km s^{-1} at the limit of our catalogue. Today photometric programmes for NLTT stars are going on yielding data by which many more faint nearby objects are detected.

Of course, c.p.m. alone does not decide on gravitational bounding of two stars. Large differences between the parallax data are in contradiction to such a connection. Equal tangential velocities at different distances of the Sun require smaller proper motions with decreasing parallaxes.

How to define a lower limit of separations for the term 'wide'? From column 2 of Table I one can deduce a nearly monotonic decrease of the observed frequency distribution of separations with increasing astronomical units. Already Dommanget (1964) had mentioned that the distances between the components are uniformly distributed on a logarithmic scale between a few AU and a few thousand AU. Therefore, there is no distinct limit from which onwards systems should be called 'wide' as a separated class.

4. The Distribution of the Projected Separations

In addition to the frequency distribution of the projected separation for all known visual binaries within 25 pc in column 2, Table I gives the corresponding values for the stars

TABLE I

Frequency of nearby visual binaries with respect to different separations (in AU) between their components

AU	25 pc	5 pc	10 pc	LHS sel
	8	1	4	3
2	23	2	5	3
5	29	2	4	2
10	48	2	8	7
20	89	4	16	3
50	76	2	9	6
100	70	1	7	3
200	90	1	7	10
500	59		2	2
1000	49		3	5
2000	33		1	3
5000	10			1
10000	9	1	1	
20000	8		2	
50000	2			
100000	3 (?)			
Sum	608	16	69	48

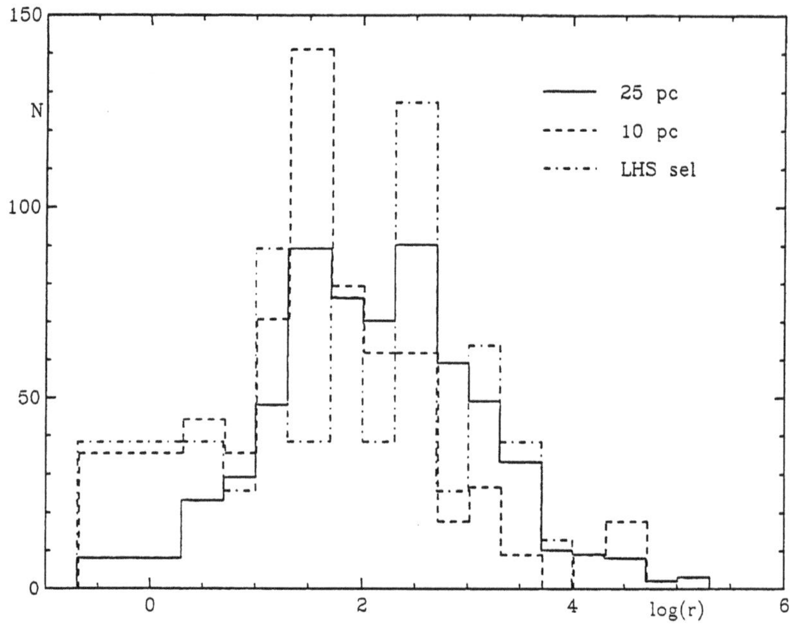

Fig. 1. The distribution of projected separations of nearby visual binaries versus log r (AU). Stars within 25 pc (full line); normalized to the latter, stars within 10 pc (dashed line); LHS selected sample dashed-dotted line).

within 5 pc, 10 pc and for the stars in the LHS subset in columns 3, 4, and 5, respectively. Likewise, Figure 1 shows the distribution of the 25 pc sample together with the normalized distributions of the 10 pc sample and the selected LHS sample. Compared with the other frequency distribution the 25 pc sample reveals a strong deficit of separations less than 20 AU and two peaks at 20–50 AU and 200–500 AU seem to be indicated. Whereas the 5 pc and 10 pc sample which should be more complete show only the peak at 20–50 AU. The latter is in accordance with Heintz (1969) (the omitted correction $\log a = \log r + 0.1 - \log \pi$ would only slightly change our distribution function) and other finding (see for example, Retterer and King, 1982). Yet, the second maximum is also indicated in the selected LHS sample, i.e., high-velocity stars. So, one may speculate on the presence of a bimodal frequency distribution for the separation of nearby binaries composed by two unimodal functions for low-velocity and high-velocity stars.

However, taking into consideration the numbers of objects with different separations in this sample the observed fluctuations along the frequency function must not be overemphasized.

5. The Widest Pairs

Our nearby-star material contains 32 systems with projected tangential separations exceeding 5000 AU between their components. They are listed in Table II in order of

TABLE II

Wide binary and multiple systems in the solar neighbourhood

, GJ ame, DM	Separation angular	linear	N	Trig. parallax	s.e.	Proper motion		RV	V	$B-V$	M_V
03	16°	1 150 000	2	0″.0535	± 0″.0025	1″.89	127°		14.18	1.48	12.82
34				0.0489	± 0.0052	1.84	112		15.05	1.86	13.50
31 A(SB), B, C	6°16′	1 300 000	6	0.0740	± 0.0082	0.504	241	+ 12	3.14	0.19	2.49
32 A, B				0.0720	± 0.0046	0.505	240	+ 27	4.15	0.37	3.44
48	5°10′	170 000	3	0.1152	± 0.0044	1.748	98	+ 6	10.04	1.46	10.34
22 A, B				0.1011	± 0.0022	1.783	101	+ 10	10.42	1.50	10.42
55 A, B, C	1°14′	100 000	4	0.0421	± 0.0022	0.645	29		7.7	0.85	5.8
9476				0.0457	± 0.0095	0.694	35	− 30	7.81	0.69	6.11
49 A, B	1°28′	90 000	3	0.0717	± 0.0119	0.471	211	− 11	4.06	0.50	3.34
49 C				0.0273	± 0.0205	0.543	214		12.0		9.2
95, 394 (RV v)	1° 7′	50 000	4	0.0789	± 0.0035	0.180	256	+ 10	4.84	0.52	4.33
57°1266				0.0203	± 0.0170	0.202	258	+ 13	8.22	0.92	4.76
81	1°58′	50 000	2	0.1547	± 0.0057	0.368	117	+ 6	1.15	0.09	2.10
79				0.1294	± 0.0090	0.363	117	+ 10	6.47	1.10	7.03
75°510	0°33′	45 000	3	0.0421	± 0.0136	0.437	271	− 35	10.24	1.29	8.36
75°511 A, B				0.0437	± 0.0119	0.445	267	− 3	10.40	1.27	8.60
03	1°18′	40 000	3	0.1085	± 0.0053	0.429	142	+ 1	8.81	1.43	8.99
99 A, B				0.1240	± 0.0055	0.430	142	− 4	10.98	1.58	11.45
54.2 (SB)	24′.4	33 000	3	0.0449	± 0.0054	0.32	76	+ 51	3.85	1.13	2.11
D−65°272				0.0434	± 0.0119	0.36	75	+ 19	7.85	0.65	6.04
29.2 A	19′.5	22 600	2	0.0519	± 0.0096	0.704	193	− 163	9.62	0.74	8.20
29.2 B						0.700	193		13.86	1.44	12.44
36 A, B	20′.2	21 000	3	0.0629	± 0.0048	0.370	168	+ 4	6.90	0.81	5.90
36 C				0.0487	± 0.0050	0.318	172		15.40	1.84	13.84
38 A	22′.5	21 000	2	0.0641	± 0.0086	1.688	350	+ 103	5.36	0.58	4.39
38 B						1.688	350		16.59	1.20	15.62
57.1 A, B	13′.5	17 400	4	0.0464	± 0.0025	0.445	178	+ 4	4.99	0.47	3.32
55.4 A, B				0.0452	± 0.0039	0.431	178	+ 6	8.39	0.98	6.67
6.1 A, B	12′.3	16 900	3	0.0439	± 0.0042	0.202	224	− 6	3.56	0.06	1.78
6.1 C						0.215	220		10.15	1.36	8.37
Ma A(SB), B(SB)	11′.8	16 200	6	0.0437	± 0.0061	0.125	105	− 9	2.27	0.02	0.47SB
UMa SB				0.0442	± 0.0075	0.121	101	− 9	4.00	0.17	2.23SB
51.1	12′.4	15 100	3	0.0492	± 0.0136	0.260	140	+ 25	5.52	0.54	3.98
50.1 A, B				0.0299	± 0.0119	0.275	143	+ 26	7.3	0.84	4.7
15.2 A(SB), B	11′	14 600	4	0.0453	± 0.0054	0.287	254	− 12	5.69	0.50	3.97SB
15.2 C						0.305	254		12.30	1.40	10.58
0 A, B	7′.7	11 500	3	0.0400	± 0.0085	0.674	71		8.66	1.15	6.87
0 C						0.668	66		12		10
2	12′	10 900	2	0.0664	± 0.0090	0.244	92	+ 36	5.00	0.53	4.11
1				0.0623	± 0.0068	0.264	90	+ 37	7.97	1.09	6.94
4 A, B	12′.3	10 600	3	0.0699	± 0.0058	0.573	153	− 13	5.33	0.56	4.56
5				0.0727	± 0.0028	0.573	153	− 11	9.97	1.47	9.28
9 A, B	2°11′	10 400	3	0.7501	± 0.0047	3.664	281	− 22	0.01	0.64	4.39
1				0.7719	± 0.0041	3.809	282	− 16	11.12	1.90	15.56
4 A, B	12′.0	9 500	4	0.0757	± 0.0042	0.430	169	+ 30	0.71	0.80	0.10
5 A, B				0.0768	± 0.0028	0.430	169	+ 36	10.20	1.48	9.62
6 A, B	18′.8	8 900	3	0.1266	± 0.0047	0.468	220	− 10	3.58	0.47	4.09
1						0.633	205		15		

Table II (continued)

Gl, GJ Name, DM	Separation angular	linear	N	Trig. parallax	s.e.	Proper motion		RV	V	$B-V$	M_V
838.1 A	7!2	8700	2	0"0499	± 0"0078	0"548	92°	−11	8.65	1.04	7.14
838.1 B				0.0397	± 0.0039	0.504	91		14.89	1.80	12.69
848.1 A	6!8	8400	3	0.0487	± 0.0119	0.36	112		10.50	1.21	8.94
848.1 B, C						0.36	112		12.51	1.51	10.95
515	8!3	8000	2	0.0618	± 0.0027	1.188	248		12.35	0.07	11.30
514.1				0.0684	± 0.0049	1.202	246		14.25	1.65	13.42
579.2 A	5!0	7400	2	0.0407	± 0.0030	3.681	196	+292	9.04	0.78	7.08
579.2 B						3.681	196	+306	9.42	0.86	7.46
ξ Sco A(SB), B, C	4!7	6600	6	0.0425	± 0.0036	0.07	254	−29	4.77	0.47	2.91SB
ADS 9910 A, B				0.0438	± 0.0045	0.08	247	−34	7.46	0.74	5.67
55.3 A, B	5!3	6200	4	0.0517	± 0.0088	0.40	74	+9	4.99	0.46	3.56
55.1 A, B				0.0298	± 0.0102	0.40	76		7.75	0.95	5.12
+10°2347	3!9	5900	2	0.0391	± 0.0205	0.340	289		7.54	1.06	5.50
LP 493–64						0.340	289		14.8:		12.8:
620.1 A	5!8	5100	2	0.0673	± 0.0077	0.08	96	+12	5.39	0.64	4.53
620.1 B				0.0626	± 0.0116	0.08	90	+43	11.00	−0.14	9.98

decreasing linear separations which were computed from the angular separation and the parallax of the primary. These objects were cited in the literatur as pairs with common (or nearly common) proper motion. In some of these cases, however, it is very questionable whether or not such two stars are really gravitationally bound. Each of these apparent systems needs a particular examination.

The first entry, the two red dwarfs GJ 1003/1034 separated by more than 16° certainly do not form a physical system. In other cases the parallax values do not agree: Gl 549 AB–C, Gl 586 AB–C, Gl 395/394/+57°1266, and perhaps Gl 161.1/160.1, or the differences in the observed radial velocities (RV) are remarkably large: Gl 331/332, +75°510/75°511, Gl 154.1/−65°272.

Figure 2 shows a colour-magnitude-diagram of the widest pairs. From Table II only objects with reliable absolute magnitudes M_V (s.e. $(M_V) \leq 0\overset{m}{.}30$) are plotted. Each system is represented by two dots: the primary and the wide companion connected by a straight line (a broken line means a common proper motion pair which probably is not gravitationally bound). Small dots mark single components. However, many of the primaries and a few of the distant companions form themselves a close double or even multiple system. Such objects are shown as big dots (●).

Column 4 in Table II gives the total number of objects in the wide systems including also known spectroscopic binary (SB) components. We learn from Figure 2 that the majority of the widest systems is composed of more than two stars favouring the region of the Main Sequence from A to G, mostly as primaries.

In addition to objects of luminosity class V we see a pair of yellow giants in the upper part of our diagram: Capella A and B accompanied by a pair of red dwarfs. Below the

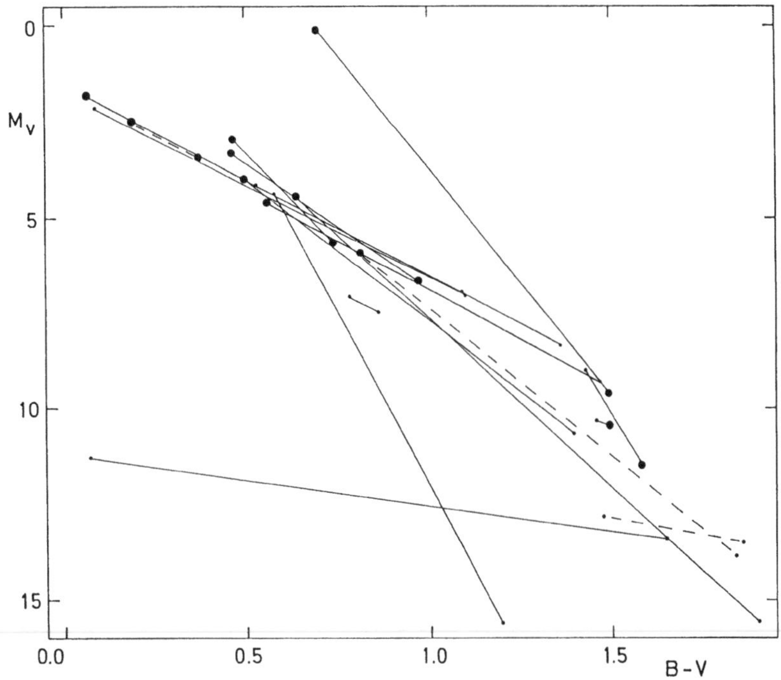

Fig. 2. Colour-luminosity diagram of the widest pairs in the solar neighbourhood. Explanation: see text.

Main Sequence there is the wide pair of K subdwarfs $-15°4042/4041$ (Gl 579.2) with large radial velocities of about $+300$ km s^{-1}. Van Biesbroeck 3 (Gl 288B) was recently detected to be a low luminosity white dwarf star (Kunkel *et al.*, 1984). The diagram contains one further white dwarf $-7°3632$ with a slightly fainter red companion Ross 476 (Gl 515, 514.1).

A systematic search among the selected LHS sample for up to now unknown c.p.m. pairs yielded one possible candidate: LHS 228 and LHS 231 with an angular separation of 252.8 arc min have almost identical proper motions 1″.123 in 122°.0 to 1″.104 in 122°.8. For LHS 228 exists photometry ($V = 15.47$, $B - V = 1.73$) and a trigonometric parallax $\pi_t = 0''.0181 \pm 0''.0034$ from USNO, whereas we have not such data for LHS 231 ($m_R = 15.4$, $m_{pg} = 17.6$, colour class *m*). If the combined parallax 0″.022 (from π_t and $B - V$) for LHS 228 should also be valid for LHS 231 a projected separation of 630000 AU would be obtained.

6. Conclusions

The 32 widest (larger than 5000 AU) common proper motion systems known today within 25 pc of the Sun are listed in Table II. Separations between 2000 and 5000 AU are observed for 33 additional systems and linear separations at the sphere between 1000 and 2000 AU are found for 49 systems. It would be of interest to consider also the colour-magnitude-diagram of these objects. Probably the percentage of multiple systems with more than two components will decrease with decreasing separations. Among these objects there will be many pairs with uncertain parallax data. Special investigations of each pair may show whether spectroscopic and/or photometric determinations of the luminosity can improve the measured distances, particularly for wide components with apparently large differences between their trigonometric parallax data.

Systematic search through the NLTT volumes will yield new c.p.m. objects. Also Luyten's LDS lists should be examined. Together with programmes for broad-band photometry of such stars this work will extend remarkably our knowledge of wide systems in the solar neighbourhood.

References

Dawson, P. C.: 1986, *Astrophys. J.* **311**, 984.
Dommanget, J.: 1964, *Comm. Obs. Roy Belg. Série B*, No. 56.
Gliese, W.: 1969, *Veröffentl. Astron. Rechen-Inst.*, No. 22.
Halbwachs, J. L.: 1985, *Bull. Inf. CDS* **28**, 25.
Kunkel, W. E., Liebert, J., and Boroson, T. A.: 1984, *Publ. Astron. Soc. Pacific* **96**, 891.
Luyten, W.: 1979, *NLTT Catalogue*, Vols. 1 and 2, University of Minnesota, Minneapolis.
Luyten, W.: 1980, *NLTT Catalogue*, Vols. 3 and 4, University of Minnesota, Minneapolis.
Heintz, W. D.: 1969, *J. Roy. Astron. Soc. Can.* **63**, 275.
Retterer, J. M. and King, I. R.: 1982, *Astrophys. J.* **254**, 214.
van Altena, W.: 1985, private communication.
Woolley, R., Epps, E. A., Penston, M. J., and Pocock, S. B.: 1970, *Roy. Observ. Ann.* **5**.

WEINBERG – In your systematic search for wide binaries, what was the largest angular separation that you considered?

GLIESE – The largest angular separation which I included in my list is somewhat larger than 16° – however, I am in doubt whether this pair is gravitationally bound. The next object in this list is a system of 4 components separated from a double system with common proper motion and parallax, separation 6°16′.

LUYTEN – Isn't the separation of Alpha and Proxima Centauri 14000 AU.

GLIESE – With the parallax values given in the new Yale catalogue we find about 13000 AU ± 2200 AU.

POVEDA – How many binaries do you have with separations larger than 5000 AU?

GLIESE – We have listed 32 such systems, however some of them, probably are not gravitationally bound.

ABT – You found that when you proceed to larger distances, the numbers of companions decreases significantly. Is it possible to extrapolate to very near stars to estimate the frequency of companions among the nearest stars?

GLIESE – We did not estimate in this way. From 1969 to 1987 the percentage of the known relation binary frequency (number of components in system relative to the number of all stars) for the stars nearer than five parsecs of the Sun increased from 59 to 63%.

HERCZEG – What is the role of photometric parallaxes in constructing the new *Catalogue of Nearby Stars*. I know you consider them for the 'final' value of π, but one can make a case that at or over 25 pc distance, good photometric parallaxes (spectral types or colours) might well be more reliable than average trigonometric parallaxes.

GLIESE – For many stars only photometric parallaxes are known and will be used. If a trigonometric parallax with a small error (s.e. ≤ 10%) was measured it will be used without any change. However, for many stars, mainly at distances of 20 ... 25 pc, trigonometric and photometric (+ spectroscopic) parallaxes will be combined to a weighted mean parallax ('resulting parallax').

THE SEARCH FOR WIDE PAIRS AMONG NEARBY STARS

A. R. UPGREN

Van Vleck Observatory, Wesleyan University, Middletown, Conn., U.S.A.

Summary. The detectability and confirmation of wide pairs among nearby stars is difficult due to uncertainties in their motions arising from errors in the observational data. Statistical tests for wide pairs reveal the increasing unevenness in space motions with decreasing age. They cannot so easily distinguish between the supposed finer structure of wide pairs and triples, and larger clumps commonly associated with small moving groups and star streams. Improved radial velocities and parallaxes which are now possible, will decrease the error in space motion to well under one km s^{-1}, and this will permit such distinction to be made.

Space velocities have been determined for a group of about 200 dwarf K and M stars with parallaxes, proper motions and photometry determined in the course of the Van Vleck astrometric program and radial velocities from several sources. Space velocities have been determined for these stars and also for the Main-Sequence stars within 22 pc of the Sun and earlier than G0 according to the lists of Gliese (1969) and Gliese and Jahreiss (1979). These brighter stars are divided into four groups by $B - V$, or by spectral class for the very few without photometry. The five groups are defined in Table I.

TABLE I

Stars with space velocities

Group	$B - V$	N
1	< 0.29	44
2	0.29 to 0.44	52
3	0.45 to 0.53	69
4	0.54 to 0.61	95
5	0.95 to 1.55	196

The groups of stars, separately and together, were divided into intervals, or volume elements, of velocity space, oriented along the U, V, and W axes of the galactic coordinate system. Since all groups of stars reveal motions which are at least approximately normally distributed along any axis (as revealed by the velocity ellipsoid model of stellar motions) the volume elements must be of unequal size, increasing in length along an axis with distance from the centroid in velocity space in order to have equal probability. This is explained in more detail by Upgren and Chabotte (1983) in a preliminary study of this problem. That paper has examined the frequency of wide pairs (pairs of such separation that their connection is revealed only by a parallel space

Astrophysics and Space Science **142** (1988) 57–60.

motion). The possible existence of such wide pairs and multiples and the references for such work is given in that paper and need not be repeated here. In brief, the only evidence for such widely separated pairs comes from the paper by Vandervoort (1968) who discovered that about half of the 20 closest A-type stars (specifically those with $B - V$ less than about 0.2) formed small groups with mean separations in relative speed of 1 to 5 km s^{-1} and separations of 10 to 30 pc. The evidence for pairs is marginal and no other positive evidence has been found for older stars. The question of pairs as simply a fine-structure phenomenon of larger moving groups is still an open one.

The tests for the incidence of wide pairs above the number expected in a random distribution cited in that paper were repeated here for the late-type dwarfs (group 5 of Table I) and extended to the other four groups of brighter stars. The divisions along the Main Sequence represent different average ages of stars. The A-type stars are the youngest, and groups 3 and 4 may have an average age as old as the later dwarfs.

The motions of the stars in all groups are normally distributed along any axis. This is revealed by the Kolmogorov–Smirnov test as well as by cumulative probability plots, or probit plots along the U, V, and W axes. The probit plots for the Van Vleck stars (group 5) are shown and discussed by Upgren and Lu (1986). However, as indicated above for these stars, no evidence is present for the existence of pairs above that predicted by a random distribution. The results for the first four groups, which will be discussed in detail elsewhere, are identical to those for this last group.

The precision necessary for the detection of wide pairs depends on their separation in velocity space in comparison to the error along each direction. For the majority of the stars in question here, this latter figure can be approximated quite easily, since the stars under discussion here can all be expected to be uniformly distributed within the sphere bounded by a radius of 22 pc from the Sun, the limiting distance of the nearby star catalogues. The average parallax of a star in each group is given by $1/[22(0.5)^{1/3}]$ or $0''.057$ with half of the stars falling within $0''.007$ of this value. The expression for the variance in transverse velocity can be given by

$$\sigma_\tau^2 = \left(\frac{4.74}{\pi}\right)^2 \sigma_\mu^2 + \left(\frac{4.74\mu}{\pi^2}\right)^2 \sigma_\pi^2 . \tag{1}$$

The first term, containing the variance in proper motion is quite negligible when compared to the second term. Thus,

$$\sigma_\tau = \left(\frac{4.74\mu}{\pi^2}\right) \sigma_\pi . \tag{2}$$

The standard error in the parallax can be taken to be very close to $0''.010$ for most of the stars. This is because the majority of stars brighter than the sixth magnitude have parallaxes determined at the Allegheny Observatory. Virtually all of the stars in the first four groups are brighter than this magnitude limit. A minority of the stars have a parallax determination from one or two other observatories, but they are of lower precision and would not greatly affect the overall error.

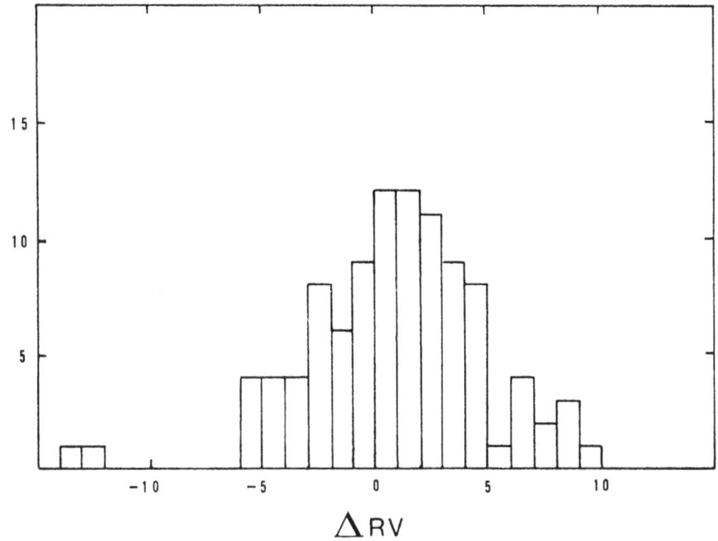

Fig. 1. Frequency distribution of differences in radial velocity for 96 stars.

The stars in the last group have nearly the same uncertainty in parallax. Three determinations have been made for the standard error in the Van Vleck parallax data. They are listed and discussed by Upgren *et al.* (1987) and range between 0".008 and 0".010. All are in agreement that despite the range in the individual published errors for these stars, they can be satisfactorily represented by a single value of this size.

The error in transverse velocity given by (2) reduces to 14.6 times the proper motion for a dispersion of 0".010, thus the uncertainty is related to the proper motion as expected. For the majority of stars having transverse velocities between 20 and 30 km s^{-1}, this amounts to between 3.5 and 5.2 km s^{-1}.

The errors in the radial velocities of these stars have been less certain, but new evidence has become available which allows such an evaluation to be made with confidence. The primary source for radial velocities of the dK–M stars of the last group has been the work of Wilson (1967) with two other studies being also deserving of mention; those of Dyer (1954) and Joy (1947). Recently acquired radial velocity measures of stars in the Van Vleck sample have been obtained in the course of the program of the Center for Astrophysics measured on their 1.5 m reflector. These data have errors ranging between about $\frac{1}{2}$ to 1 km s^{-1}. Of the stars measured so far, an even 100 overlap with the list of Wilson and a much smaller number are in common with the other two sources. The distribution of the differences in radial velocity between Wilson and the CfA data is shown in Figure 1 for 96 of the stars. Four of the stars showed variations of more than 20 km s^{-1} and thus must be considered to be variable in radial velocity, and the two stars with differences between -10 and -15 km s^{-1} may also be the result of real variation. The remaining stars clearly reveal a normal distribution.

When these residuals are fit to a Gaussian distribution, they show a mean at $+1.0$ and a dispersion of ± 4.0 km s^{-1} in the sense Wilson minus the CfA data. The resuts for 13 stars in common with Dyer show $+1.1 \pm 4.7$ km s^{-1} and for 20 stars in common with Joy, the mean difference is $+4.2 \pm 5.2$ km s^{-1}. Although the differences for these two groups are close to that for the stars in common with Wilson, the small numbers of stars involved does not lead to very certain results, and the distributions cannot be as surely said to be Gaussian in form.

The feature to be noted here is that the radial and the transverse components of the motions of the stars are of about the same size, being generally of the order of about 5 km s^{-1}. This represents the error in each of the velocity components in the galactic coordinate system, regardless of the direction cosines of a particular star. This uncertainty is of such a magnitude that the failure of most investigators to detect wide pairs with closely parallel space motions may be as much due to the uncertainty in the data as to the true nonexistence of such a fine-structure feature in the motions of the nearby common stars.

Although the available data may be insufficient to settle the wide-pair question at the present time, new techniques promise much better data in the next few years. The modern radial velocity measuring devices are well known with proven results. Less well known are the improvements in parallax now coming into widespread usage. A large improvement in the precision of parallaxes determined from conventional photographic plates taken with long-focus refractors has just recently been demonstrated. Upgren *et al.* (1987) find parallax errors to be reduced by one half from the use of automatic measuring machines (specifically the PDS microdensitometer) over the customary hand measuring techniques. Thus about three fourths of the total variance is due just to the measuring process. The use of fine-grain emulsions and hypersensitization processes results in a further increase of precision. New methods employing ccd's and other devices which replace the photographic plate and the results of Hipparcos should lead to better precision in the distances to more stars. The combination of proven methods in both the radial and the transverse directions extended to large numbers of nearby stars must and will allow the finer structure typified by the wide pair phenomenon to be studied and resolved.

References

Dyer, E. R.: 1954, *Astron. J.* **59**, 218.

Gliese, W.: 1969, Veröffentl. Astron. Rechen-Inst. Heidelberg **22**.

Gliese, W. and Jahreiss, H.: 1979, *Astron. Astrophys. Suppl.* **38**, 423.

Joy, A. H.: 1947, *Astrophys. J.* **105**, 96.

Upgren, A. R. and Chabotte, S. C.: 1983, in *Proceedings of the Symposium on Statistical Methods in Astronomy*, Strasbourg, France, p. 231.

Upgren, A. R. and Lu, P. K.: 1986, *Astron. J.* **92**, 903.

Upgren, A. R., Weis, E. W., Fu, H.-H., and Lee, J. T.: 1987, 'Mapping the Sky – Past Heritage and Future Directions', in S. Debarbat, J. A. Eddy, H. K. Eichhorn, and A. R. Upgren (eds.), *IAU Symp.* **133**.

Vandervoort, P. O.: 1968, *Astrophys. J.* **152**, 895.

Wilson, O. C.: 1967, *Astron. J.* **72**, 905.

SIREGAR – What kind of statistical tests are in this work?

UPGREN – The test for the distributions being represented by a Gaussian distribution included the Kolmogorov–Smirnov test as well as the more robust probit plots which I showed.

A PLEA FOR BETTER ASTROPHYSICAL AND ASTROMETRIC DATA IN CASE OF DOUBLE STARS OF PECULIAR INTEREST

S. DÉBARBAT and F. CHOLLET*

Observatoire de Paris, France

Summary. The *IAU Symposium* No. 61 (Perth, 1973, 'New Problems in Astrometry') makes clear that radiosources will play an important role for astrometry in view of the linkage of reference systems.

The purposes developed later were to link the various systems from the earth and the solar system to the most inertial possible one.

Extragalactic radiosources are the most reliable objects but they are faint. So, intermediate objects are needed. Among them the radiostars, optical objects having radioemission at a certain level and for given frequencies have a fundamental and key position.

Most of such radiostars are double or multiple 'binaries' as it can be seen from the list of reference stars issued by the Working Group of Commission 24, Working Group created in 1978. The last issue of the core list of radiostars was presented by Ch. de Vegt (Chairman of the WG) on the occasion of the last General Assembly of the IAU (November 1985).

A recent study of the system α Scorpii (Antares) has shown that the data concerning the magnitude and the spectral type are not accurate enough, according to the accuracy obtained for the astrometric position ($\pm 0^{s}.003$ for an astrolabe campaign in right ascension).

There is also a need for better knowledge about the orbits. In the case of the astrolabe observations, if the distance between the components is $<3''$ the astrometric position concerns the photocenter and accurate physical properties are needed for comparison with the radio position.

Better astrophysical properties are strongly needed for all radiostars to be used as references, during the time the astrometrists have to improve the accuracy of their measurements both in optical and radio fields.

The *IAU Symposium* No. 61 (Perth, 1973, 'New Problems in Astrometry') makes clear that radiosources will play an important role for astrometry in view of the linkage of reference systems. The purposes developed later were to link the various systems, from the Earth and the Solar System to the inertial reference frame. Extragalactic sources are the most reliable objects, but they are faint. So, intermediate objects are needed. Among them the radiostars, objects of our Galaxy having radioemission, at a certain level and for given frequencies, have fundamental and key positions.

Most of such radiostars are double or multiple 'binaries' with wide components in some cases, according to the value given to the distance between them. The last issue of the core list presented by Ch. de Vegt (then chairman of the IAU commission 24 Radio/Optical Reference Frame), during the General Assembly of the IAU (November 1985), contains double stars with known orbital elements and also spectroscopic or photometric 'binaries'. The knowledge of their orbits is of great importance, as well as their proper motions. From the catalogue by Heintz it can be seen that only a few stars have their orbit accurately calculated.

Some of the radiostars of the list of 52 objects can be separated by the meridian instruments or will be seen as binaries by the astrometric satellite Hipparcos. But other

* Presented by S. Débarbat.

Astrophysics and Space Science **142** (1988) 61–64.

instruments of high accuracy can contribute to the improvement of the astrometric position. Such are the various astrolabes still in service in the world. In this case, if the binary is not with very wide components, the astrometric position concerns the photocenter; the distance being small, the components are not separated in the field of view. On the other hand, and in any case, the companion cannot be seen if its magnitude is more than 6. But a mean-square error of $0''.03$ can be obtained for one campaign.

An example of such a double star is α Scorpii, for which in Valinhos (Brazil) a mean square error (Clauzet *et al.*, 1985, 1986a) in right ascension (the declination was not attainable) of $\pm 0^s.003$ was obtained. This accurate value is of particular interest according to the fact that radioastrometric observations from the Very Large Array (NRAO, USA) have a mean square error (Florkowski *et al.*, 1985) which is of the same order of magnitude: $\pm 0^s.001$, $\pm 0^s.002$. In this case the A and B components are separated by the VLA, and the given position is for component A. In the case of the astrolabe results, A and B (which separation is less than $3''$ and which magnitudes are about 1.3 and 5.5) are mixed and the photocenter is observed.

In a recent study of the results (Clauzet *et al.*, 1986b) a correction for the difference $\alpha(A) - \alpha(cg)$ has been given; *cg* is the gravity center which α and δ are given in the APFS, and the difference is $+ 0^s.029$. The difference $\alpha(cg) - \alpha$(photocenter) was also evaluated according to the method previously used in the case of the planets (Chollet, 1984) and the value given was $+ 0^s.028$. That is to say that what was observed was a point near to the component A. But the difference $\alpha(A) - \alpha$(photocenter) was given with a mean square error of $\pm 0^s.008$; this very large quantity arise from the uncertainties of the physical data. They were the following, given on lines 1 and 2 in Table I. According to

TABLE I

	Component A	Component B
1 Magnitude	0.88–1.08	5.2–6.5
2 Spectral type	M0–M3	A3–B2
3 Magnitude	1.0	5.5
4 Spectral type	M2	B2

various catalogues of measures, the visual magnitude for α Scorpii (which is a variable star) was found with values as different as 0.88, 0.91, 0.92, 0.96, 1.07, 1.08 allowing us to take 1.0(line 3) for the calculation of the correction. For the component B, the chosen value was 5.5 (line 3). For the spectral type of components, those given by catalogues were mostly M1 and M1.5 (component A) and B0, B2 for component B. With the chosen values (given on lines 3 and 4) the calculated correction applied to the measurements made with the astrolabe conducted to the following value for the right ascension of the observed object: $16^h26^m20^s.206 \pm 0.009$ s. This value was compared with the radiomeasurements (VLA): $16^h26^m20^s.199 \pm 0.001$ s. According to the errors the agreement was good, allowing us to consider that the calculations were appropriately made.

But the error, ± 0.009 s for the optical result, which is very large, is a combination of the error on the astrometric position (± 0.003 s) and an evaluation of the error on the correction (± 0.008 s) related to the astrophysical quantities (magnitude and spectral type). These last ones degradate the accuracy of the astrometric measurements. The example of α Scorpii shows that the effects of luminosity and colour must be taken into account, as well as the orbit, but it also demonstrates that better astrophysical quantities are strongly needed to avoid such a degradation of the astrometrics results. It also makes clear that it will be useful to pursue this experiment on various couples of binaries for which magnitude and spectral type are different from those of the example given by α Scorpii and also as different as possible. It would also be of interest to have various distances between the components.

From the list established by the IAU commission 24 Working Group 'On the Identification of Optical/Radio Astrometric Sources' Table II has been formed; it gives the differences in magnitude for the main component, together with the spectral type extracted from a selection of catalogues for astrophysical quantities from the *Centre de Données Stellaires* in Strasbourg. This list is limited to stars brighter than magnitude 6 which is the approximate limit for visual astrolabes.

Most of the radiostars are variable stars being binaries for which the orbits need to be known with the best possible accuracy, at the same time that magnitude and spectral type are better defined. But as it can be seen from the example of α Scorpii it appears not to be sufficient to have a more or less good mean evaluation of these quantities. To avoid the degradation of the astrometric measurements it would be necessary to have simultaneous astrophysical determination (magnitude, spectral type or colour indice) with appropriate accuracy, the variability of these double stars being sometimes very fast. Ephemerides of these quantities can give a good approximation but simultaneous measurements would be better.

TABLE II

Objects from WG Com. 24 List	Magnitudes	Spectral types
UU PSC	5.79–6.64	F0
ζ And	4.06–6.05	K0–K1 §
39 Cet	5.40–5.54	G0–G5
Algol	1.51–3.40	B8 §
HR 1099	5.70–5.71	G0–G9
b Per	4.59–4.98	A2 §
χ 1 Ori	4.39–4.43	F8–G0 §
α 1 Ori	0.42–1.30	M1–M2
KQ Pup	4.88–5.17	M2 §
σ Gem	4.22–4.29	K0–K1
HR 5110	4.96–4.98	F0–F2
σ CRB A	5.64–5.66	G0
α Sco A	0.88–1.08	M0–M1–M2 §
ε UMi	4.22–4.28	G5
9 Sgr	5.96–5.97	O5–O8
β Lyr	2.96–4.20	A8–B7 §
λ And	3.78–4.90	K0–G8 §

Double stars are very often 'Trouble Fête' when they are not considered for their own properties. They need to be studied very carefully by specialists for the purposes of the astrometric satellite Hipparcos. On the other hand, in the case of their use for linkage of reference frames through their optical and radio positions they will have influence in relation with three different well known points of view: they can be double stars with orbital elements, double stars with variations in magnitude and varying double stars for spectral type. Interferometric measurements are of particular interest for such objects: nine radiostars of the list of the WG of Commission 24 have been measured in this way, and so they are given by McAlister and Hartkopf in their *Catalogue of Interferometric Measurements of Binary Stars*. Their separation is not larger than 0".030 and eight of them are included in Table II, that is to say accessible to astrolabes (they are marked with §). It has also to be noted that this list includes 17 radioastars among the 52 which are in the core list of the Working Group. A few more will be attainable by the photoelectric astrolabes; the rest will be completely in hands of meridian circles. But one can assume that for all these instruments the physical properties will be of particular importance for comparison with the radio positions.

Better astrophysical properties for luminosity and colour are strongly needed for all radiostars to be used as references, during the time astrometrists have to improve the accuracy of their measurements, both for positions, proper motions, orbits, and both in optical and radio fields. For the purpose of the linkage of the different reference systems, it is suggested that a high priority should be given to objects of Table II. Being able to be observed by different types of instruments, they will serve as test-stars for the future in both cases of wide or not so wide components of double stars.

References

Chollet, F.: 1984, *Astron. Astrophys.* **139**, 215.

Clauzet, L. B. F., Attala, R., and Débarbat, S.: 1985, *Astron. Astrophys. Suppl. Ser.* **61**, 211.

Clauzet, L. B. F., Attala, R., and Débarbat, S.: 1986a, *Astron. Astrophys. Suppl. Ser.* **63**, 204.

Clauzet, L. B. F., Débarbat, S., and Chollet, F.: 1986b, *Astron. Astrophys.* **167**, 387.

Florkowski, D. R., Johnston, K. J., Wade, C. M., and de Vegt, Ch.: 1985, *Astron. J.* **90**, 2381.

Heintz, V. W.: 1960, *Veröffentlichungen der Sternwarte München*, Band 5, No. 10.

McAlister, H., Hartkopf, W. I.: 1984, *Catalog of Interferometric Measurements of Binary Stars*, CHARA Contributions, No. 1.

Catalogue du Centre de Données Stellaires de Strasbourg, SIMBAD.

LODEN – The two values of magnitude (and spectral type) you show for Antares, and its component, do they represent the real amplitude of the variability?

DEBARBAT – No! They are just two extreme values obtained from different sources in literature, so you see exactly the need I am speaking about.

THIRD SESSION

WIDE COMPONENTS IN MULTIPLE SYSTEMS

Chairman: J. DELHAYE

DISTANT COMPONENTS IN DOUBLE AND MULTIPLE STARS

Introductory Lecture

A. POVEDA

Instituto de Astronomia, Universidad Nacional Autonoma de Mexico, Mexico City, DF, Mexico

Summary. A discussion is presented of the IDS, LDS, and Gliese's catalogues, inasmuch as they are relevant to the multiplicity of field stars and the distribution of separations of the companions in binaries and triples.

The effect of dissociation of binaries with time is exhibited for a 'quasi-unbiased' sample of 106 nearby binaries taken from Gliese's catalogue. There are no binaries as old or older than the Sun with separations larger than 4000 AU, while binaries younger than the Sun may have separations as large as 180000 AU. The rate of this disruption seems too large to be produced by field stars; probably molecular clouds and spiral arms are the most effective agents.

It appears that the size of the inner commet cloud cannot be much larger than 2000–4000 AU.

1. Introduction

The IDS, the LDS, and Gliese's catalogue are the fundamental databases for visual and common proper motion double and multiple stars; unfortunately, the IDS contains a very large number of spurious and optical components, which make many statistical analyses based on it very unreliable. On the other hand, Luyten's double star catalogues, which are based on the similitary of the proper motions in a binary, are essentially free of optical components; however, since LDS stars are discovered by their proper motions as determined in photographic plates, very close pairs are unresolved and get lost; furthermore, LDS stars are very faint and consequently there are no spectral types and no adequate photometry available for most of them. Thus the IDS and LDS catalogues in several ways complement each other.

In order to eliminate from the IDS spurious and optical components, an effort was made (Poveda *et al.*, 1982) to model the process by which visual observers register a double star; as a result of this study we devised a 'one percent filter' which, when applied to the 1976 tape version of the IDS, resulted in the rejection of more than 18 000 pairs and multiples. The new filtered catalogue, FIDS for short, contained 51 201 entries of which 44 282 were pairs and 3086 triples and multiples. It appears that, in spite of the large number of stars rejected by the filter, there still remain a substantial number of opticals of intermediate and faint magnitudes (Abt, 1986). This is probably due to the fact that the application of the filter rests upon the apparent magnitude of the faint companion(s) of a system; if the listed magnitude of a secondary happens to be brighter than its true value, then the 1% filter might retain it when it should have been rejected. The magnitudes of the stars in the IDS are reasonably correct for $V < 8$, but between $V = 8$ and $V = 8.9$ they show a scatter of ± 0.75 mag when compared with the magnitudes of those stars for which Eggen has made photoelectric observations; for fainter magnitudes, there is, in addition, a systematic error in the sense that the IDS

Astrophysics and Space Science **142** (1988) 67–78.

magnitudes tend to be brighter than the photoelectric ones by half a magnitude or more (Abt, 1986). The errors (accidental and systematic) in the magnitudes of the IDS may explain the high proportion of opticals (about 30%) found by Abt when he observed spectroscopically a sample of trapezia taken from the catalogue of Allen *et al.* (1977). It seems necessary to filter again the IDS allowing for the errors in the magnitudes of the secondaries.

present colloquium, the following questions should be addressed:

(a) The frequency of duplicity and multiplicity among field stars.

(b) The distribution of separations.

(c) The dissociation of binaries with time.

In what follows I shall briefly discuss these points; I shall also refer to some unpublished work I have been doing with Christine Allen and which is relevant to the above problems.

2. The Frequency of Duplicity and Multiplicity among Field Stars

When one considers the process by which visual observers discover a double star it is possible to construct a model which describes the probability of discovery and cataloguing a binary whose primary has a given apparent magnitude m_1 and whose secondary has a separation s and an apparent magnitude m_2. We have constructed such a model and applied it to the filtered IDS (Poveda *et al.*, 1982). One of the results that emerged from the above study was that 100 field stars correspond to 10 single stars (with no visual companions), 63 visual binaries, 22 triples, and 5 quadruples; most of these secondaries are too faint and remain unrecognized either because they fall below the detection limit of the observers or because they drown in the glare of the primary when they are too close and faint.

Most of the visual pairs ($P > 100$ years) belong to the class of independent condensation binaries, which follow van Rhijn's luminosity function (Abt and Levy, 1976; Abt, 1983). However, Abt (1979) has found that each field star and each component

TABLE I

Multiplicity of field stars

$n \backslash p$	0.5	0.6	0.7	0.8	0.9	1.0
1	5.0	4.0	3.0	2.0	1.0	0.0
2	20.8	16.1	12.7	10.5	9.6	10.0
3	34.3	31.7	27.1	20.3	11.4	0
4	24.3	29.1	35.1	42.4	51.6	63.0
5	9.5	10.3	10.1	8.6	5.4	0
6	4.6	6.5	8.9	12.0	16.3	22.0
7	1.3	1.7	2.1	2.1	1.5	0
8	0.3	0.7	1.2	2.1	3.3	5.0
No.	333	355	377	400	422	444

of a visual system has a probability of about $\frac{1}{2}$ to $\frac{2}{3}$ of having a close companion ($P < 100$ years).

When one combines the two types of binaries one finds that the typical field star is, in fact, a multiple. We have calculated the distribution of the number of components expected for a sample of one hundred field stars for various values of the probability p of having a close companion; the results are shown in Table I. For the values $p = 0.5$ and $p = 0.6$ the most frequent multiplicity falls between 3 and 4. In other words, if one were able to observe without the limitations of present techniques all components, even the closest and faintest, one would find that more than half of the field stars are, in fact, triple or quadruple.

The existence of an appreciable number of optical components in the FIDS will reduce somewhat our estimate of visual duplicity and multiplicity; but since our determination depends mostly on the brightest part of the distribution of apparent magnitudes of the observed doubles and multiples, and since these are not seriously in error, we do not expect a large correction to our results.

3. The Distribution of Separations in FIDS

The multiple systems in FIDS can be divided in two groups: (a) trapezia and (b) hierarchical systems; since we are interested here in the distant companions we concentrate our attention on the hierarchical multiples, particularly on the triples.

In a classical study Ambartsumian (1937) investigated whether the visual double stars were in statistical equilibrium with field stars and, hence, whether the distribution of separations s was consistent with the hypothesis that the distribution of their binding energies $f(E)$ should follow $f(E) = \text{constant}$.

Let us suppose with Ambartsumian that all the primaries are concentrated in one point; then the secondaries form a cloud (like a globular cluster). If we assume $f(E) = \text{constant}$, as corresponds approximately to a situation of statistical equilibrium, then the projected density of secondaries should be $\rho(s) = A\ \mathrm{s}^{-1/2}$, with $A = \text{constant}$. Now, the angular separations of the double stars in Aitken's catalogue, converted into surface densities of secondaries, follow the relation $\rho(s) = A\ \mathrm{s}^{-2}$ instead of that of statistical equilibrium; the s^{-2} density law had been found years before by Öpik (1924). Ambartsumian's result shows very clearly that the system of wide double stars has not been perturbed significantly by passing stars, and that, therefore, the distribution of separations is, in a sense, a relic of the conditions of formation.

The situation is very different for triple or multiple stars, because their mutual internal interactions will modify their energies and their elements even in the absence of external interactions. For instance, it is well known that the fate of a trapezium system (Allen and Poveda, 1974) is to evolve into a hierarchical configuration or even to disolve into a binary. It has even been suggested (Poveda, 1977) that all multiple stars are born as trapezia, which evolve into hierarchical systems. In view of the above circumstances it is important to investigate the distribution of separations among the components of triple stars. For this purpose, we have elaborated a catalogue of hierarchical triple stars extracted from the FIDS. The surface densities were calculated by counting the number of companions in each of the geometrically growing intervals of separation: 0″–0.25″, 0.25″–0.5″, 0.5″–1″, 1″–2″, 2″–4″, etc. This was done both for the separations of the primary to the close companion and to the distant companion; the projected densities follow by simple division of the number of companions in each ring by the area of the corresponding ring. Note that if the surface density follows Öpik's law then the number of secondaries in the successive, geometrically growing, rings should be constant.

We have fitted by least-squares density laws of the form $\rho(s) = As^{-x}$ to the distribution of surface densities of binaries and of close and distant companions of triples, all taken from the FIDS; we found the following relations:

binaries in FIDS	$\rho(s) = 2.9\ s^{-2.04}$,	$s \leq 32''$,	$m_1 < 10$,
triples: close companions	$\rho(s) = 2.5\ s^{-2.07}$,	$s \leq 16''$,	$m_1 < 10$,
triples: distant companions	$\rho(s) = 12.1\ s^{-0.68}$,	$s \leq 16''$,	$m_1 < 10$;

where $\rho(s)$ is the surface density of companions normalized to 100. These fits are shown together with the observed densities in Figures 1, 2, and 3.

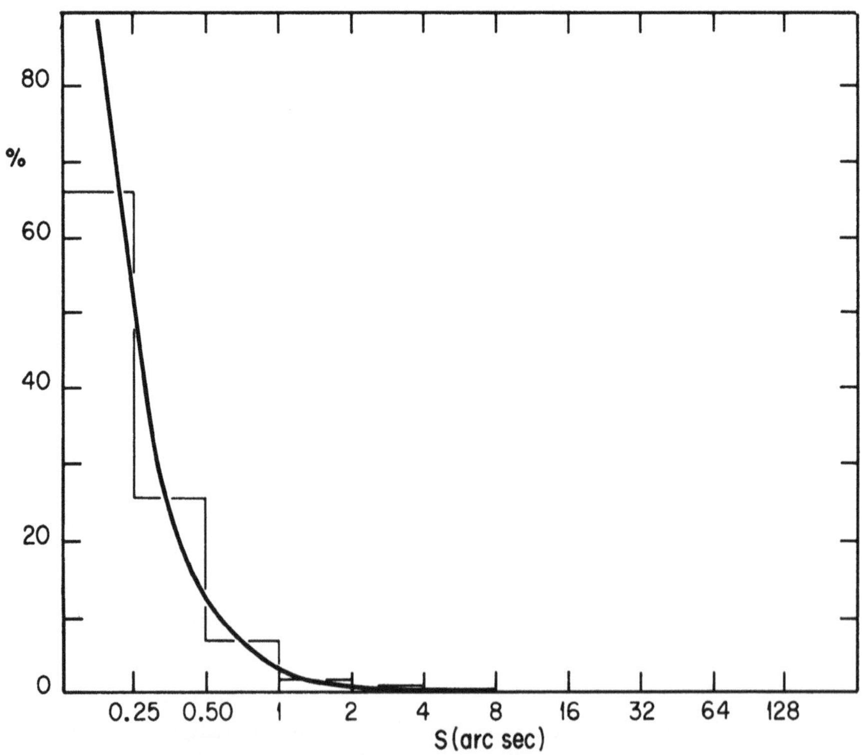

Fig. 1. Distribution of separations for binaries from the FIDS ($m_1 < 10$, $\Delta m_{1,2} \leq 1$). The histogram corresponds to surface densities of secondaries (normalized to 100) versus separation in sec of arc. The continuous line is the best fit $\rho(s) = 2.9\ s^{-2.04}$, $s \leq 32''$.

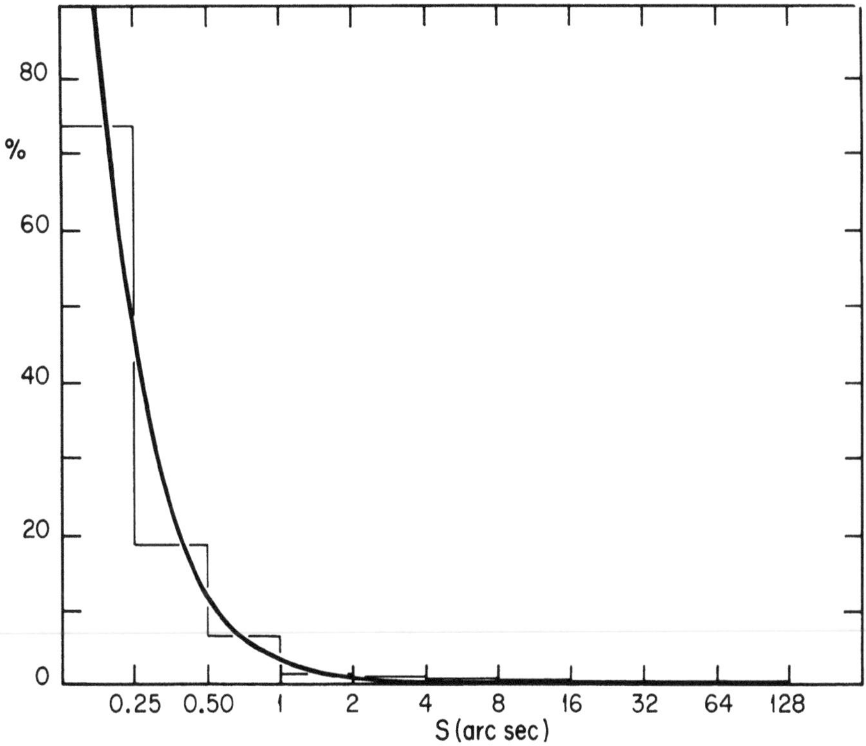

Fig. 2. Distribution of separations of close companions in triples from the FIDS ($m_1 < 10$). The ordinates are surface densities of close companions, normalized to 100. The continuous line is the best fit $\rho(s) = 2.5\ s^{-2.09},\ s \leq 16''$.

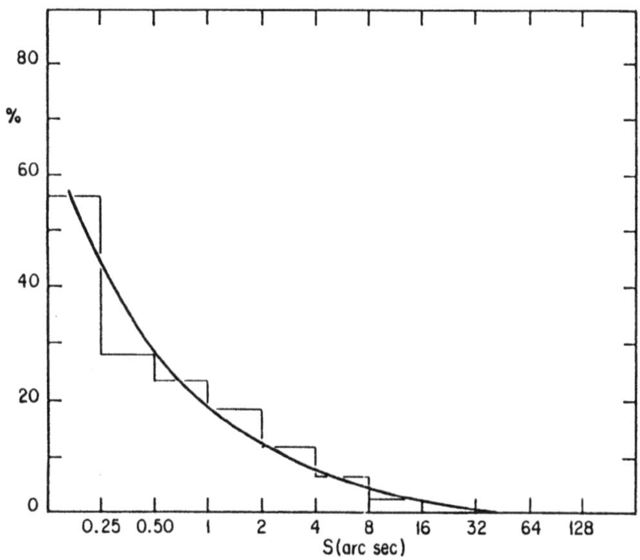

Fig. 3. Distribution of separations of distant companions in triples from the FIDS ($m_1 < 10$). The ordinates are surface densities of close companions, normalized to 100. The continuous line is the best fit $\rho(s) = 12.1\ s^{-0.68},\ s \leq 16''$.

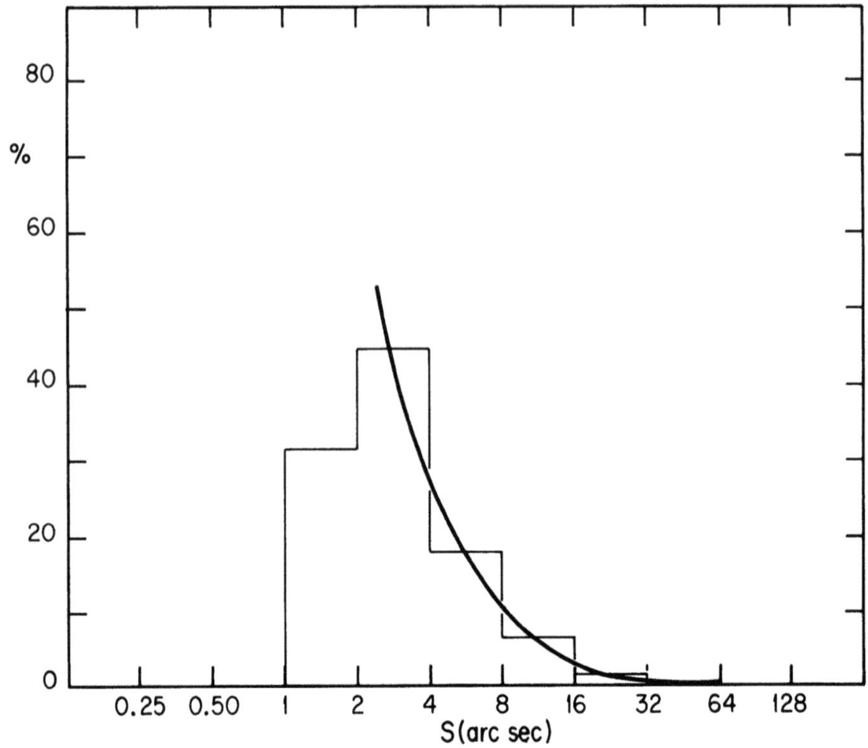

Fig. 4. Distribution of separations for a sample of binaries from the LDS catalogue. The ordinates are surface densities of companions, normalized to 100. The continuous line is the best fit $\rho(s) = 544.8\ s^{-2.05}$, $s \leq 128''$.

Note that the close companions follow rather closely the s^{-2} law, while the distant companions resemble the law for statistical equilibrium. For comparison, we also computed the fit corresponding to the binaries in FIDS which, for the sake of completeness, were restricted to $m_1 < 10$ and $\Delta m_{1,2} \leq 1$. We find that this larger sample conforms well to the s^{-2} law.

From the above results, we conclude that in the case of hierarchical triples the distant companion, which usually is weakly bound compared to the close pair, has experienced energy exchanges of a magnitude comparable to its binding energy, leading to its approach to statistical equilibrium. For the close companions, the same energy exchanges would not lead to statistical equilibrium because of their stronger binding energy.

The Luyten (1969) *Double Star Catalogue* is an alternative source of binaries that differ from those that prevail in the IDS because they are mostly intrinsically faint and, more important, because the physical reality of the pairs is assured by their common proper motion. For the purpose of studying the distribution of separations, we have selected the 562 pairs found by Luyten in the Palomar plates in the cap north of $+60°$ declination, which Luyten recommends for statistical studies.

For this sample of the LDS catalogue we fitted the normalized surface densities of companions; we found that Öpik's law is again satisfied by this completely different set of binaries. The resulting fit (see Figure 4) is

$$\rho(s) = 544.8\ \mathrm{s}^{-2.05}\ , \quad s \leq 128''\ .$$

The distributions of separations shown in the various histograms of Figures 1 to 4 do not exhibit any significant cutoff as one might expect from the effect of dissociation; we shall see below how this smooth decay to zero can be understood as the result of a mixture of binaries of different ages.

4. The Dissociation of Binaries with Time

It is well known that the random gravitational pulls that a star experiments while moving in the Galaxy change its energy, gradually leading to its relaxation. In the case of binaries, one expects that the differential pulls felt by the components of a pair will lead to the gradual increase of the binding energy and to its eventual dissolution. In fact, in his classical paper Ambartsumian also calculates the time of dissolution of a binary subject to random perturbations by passing stars. The importance of this phenomenon has led various authors (Chandrasekhar, Oort, Cruz-Gonzalez and Poveda, King, Heggie, Weissman, Clube and Napier, Fernandez, Hills, and others; see Weissman (1986) for a discussion on this subject) to investigate this process with greater precision. The effect of the perturbations caused by molecular clouds has been taken into account in modern calculations; although the uncertainties involved are very great, it appears that their importance in disrupting binaries is greater than that of individual stars.

Perhaps the earliest attempt to exhibit explicitly the effect of binary dissociation with time was done by Poveda and Allen (Allen, 1974; Poveda and Allen, 1975) using several samples of wide binaries, particularly the sample of 60 binaries in the solar vicinity compiled by Kuiper (1942). We found that stars could be classified as young or old (relative to the Sun), depending on their spectra, their flare activity or their membership in young moving clusters; furthermore, we divided the binaries in two groups, those with separations closer than 1000 AU and the rest. The surprising result was that among the group of close binaries we found two old binaries for each young one, while among those with separations larger than 1000 AU the ratio dropped to one; this indicated that the effect of dissociation is present all the way down to 1000 AU for stars older than the Sun. The theory of binary dissociation based on interactions with individual stars cannot explain such a short dissociation time.

More recently, Abt (1986) has studied the correlation between spectral type of the earliest star in a sample of multiple systems and the maximum separation in the system. The relation found by Abt: maximum separation (AU) = $3.9 \times 10^6\,(\mathrm{age})^{-0.3}$ also indicates dissolution times shorter than those expected from the theory of stellar perturbation, but still longer than the times we derived from Kuiper's sample.

In order to throw more light on this important problem we have repeated our analysis of 1974 (Poveda and Allen, 1987), but now using the binaries contained in Gliese's catalogue (1969). According to Gliese *et al.* (1986), the catalogue is fairly complete for stars down to absolute magnitude 13 within a sphere of radius 13 parsecs. Our own analysis shows that there is some incompleteness, particularly in the southern hemisphere, the younger stars being underrepresented because of the kinematical bias. In order to find the effect of dissociation as a function of time and separation, we need to work with a sample that is not biased, for instance, against young stars; in practice we have to strike a compromise between this requirement and the size of the sample. Therefore, we considered all the binaries and multiples in the catalogue (including the common proper motion companions) with resultant parallaxes $\geq 0.08''$. We separated the system in two age groups: (i) younger than the Sun (young) and (ii) as old or older than the Sun (old), according to the following criteria: a system was considered young if at least one of its stars (a) is a main sequence earlier than F7, (b) is a flare star, (c) has H-alpha emission, (d) is a member of a moving group younger than the Sun.

We found 106 binaries closer than 12.5 pc (nearly double the number we used before); if we take the ratio of old to young ones for separations up to 2000 AU we find that for each young star there are two old ones (see below); assuming a constant rate of star formation, we find that prior to the Sun's birth there must have elapsed somewhat less than two times the age of the Sun, i.e., 7.5 billion years, or a total age for the system of stars of approximately 12 billion years, which seems to be consistent with other evidence. We use this agreement as an independent check on the age ($3.5–4 \times 10^9$ years) of the dividing line between the two groups of stars.

TABLE II
Distribution of separations of young and old binaries in Gliese's catalogue ($\pi \geqq 0.08''$)

Separation (AU)	Number of binaries	Number of young ones	Number of old ones	Old/young	Expected number of old ones	Observed number of old ones/ expected number of old ones
≦500	83	29	54	1.9		
>500	23	10	13	1.3	20	0.7
≦1000	88	30	58	1.9		
>1000	18	9	9	1.0	18	0.5
≦2000	96	32	64	2.0		
>2000	10	7	3	0.4	14	0.2
≦2500	97	32	65	2.1		
>2500	9	7	2	0.3	14	0.1
≦4000	101	34	67	2.0		
>4000	5	5	0	0	10	0.0

In Table II we have divided the 106 binaries by their ages and increasing separations; the first column lists the separations s in astronomical units, the second column gives the numbers of binaries with separations less or equal to a given value s and those with larger separations. The third and fourth columns give the number of young and old binaries for the corresponding intervals of separation; the fifth column lists the ratio of the number of old binaries to young ones, and the sixth column the expected number of old binaries for separations larger than a given value of s. The number was calculated by multiplying the actual number of young binaries with separations larger than s by the expected ratio of old ones to young ones; we take this ratio to be that found for separations smaller than 2000 AU. This ratio is found to be fairly constant and equal to 2 for intervals of separations s smaller than 2000 AU. We assume this ratio to be representative of the intrinsic (undissociated) proportion of old to young binaries. When we find, as we do in Table II, that this ratio is significantly smaller than 2 for separations larger than 2000 AU, we consider this as evidence of the dissociation of the older binaries. Finally, column 7 gives the fraction of the actual number of old binaries that have survived dissociation.

Inspecting in Table II, column 2, the numbers of binaries (old plus young) as the separations increase, we see that the numbers decrease slowly, a result consistent with the observed behavior of the separations of the secondaries to binaries and triples in the FIDS and the LDS; but if we focus our attention on the distribution of separations of the *old* binaries in Table II (column 4), we notice a more dramatic decay, to the point that there are only 2 old binaries with separations larger than 2000 AU (we expect about

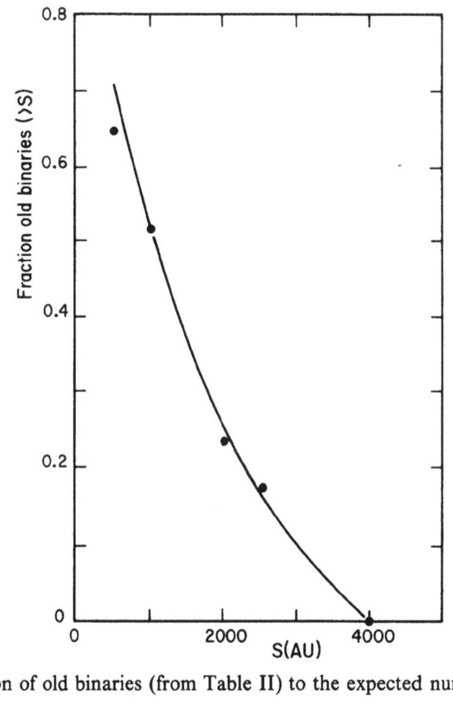

Fig. 5. The actual fraction of old binaries (from Table II) to the expected numbers for separations > s.

14), and *none* with separations larger than 4000 AU (we expect about 10). Taken at face value, this result would indicate that there is a cutoff for old binaries at separations of about 4000 AU. Actually, the cutoff is not really sharp and it begins to be quite noticeable already at 2000 AU (see Figure 5).

The fact that we have no old binaries with separations larger than 4000 AU is also a function of the volume of the sample; if we take a larger volume, for instance, a sphere with $r < 35$ pc, we find a remarkable old binary, LDS 519, which has a separation (in projection) of about 10000 AU. This extreme velocity pair has a galactocentric velocity of 465 km s^{-1} which allows it to reach distances of about 115 kpc from the galactic center (Allen *et al.*, 1987); the spectra of the stars in this binary show the metal deficiency characteristic of halo objects. We interpret the survival of this very old wide binary, whose separation is larger than our limit of 4000 AU, as the result of its galactic orbit, which keeps it far away from the galactic disk during most of the time; of course, it is in the disk where the dissociative encounters occur; when LDS 519 does cross the galactic disk, it does so at such large velocities that the encounters are less effective in dissociating it than is the case for the more normal disk stars.

The choice of a sphere $\pi \geq 0.08''$ around the Sun, as mentioned above, gives us a sample that is somewhat incomplete in slow moving objects, i.e., it is biased in favour of old stars; hence, if we were able to complete the sample we should expect to increase the proportion of young stars and to sharpen the relative decrease of old stars beyond 2000 AU. When we deal with a sample of binaries of all ages the young ones (which may have large separations because there has not been enough time for dissociation to occur) will populate the part of the distribution beyond 2000 AU, giving the impression of a smoother decrease, for wide binaries in general.

5. Conclusions

By combining the incidence of duplicity and multiplicity of visual binaries with that of spectroscopic ones, one finds that a typical field star should be a triple or quadruple. Very few field stars, of the order of five percent, are single.

The study of binaries and multiples in the FIDS and the LDS confirms Öpik's distribution $\rho(s) = A\,s^{-2}$ for pairs and close companions in triples. Meanwhile, the distant companions exhibit a different distribution, indicative of an approach to statistical equilibrium. The distribution of separations of the binaries in these catalogues does not show a sharp cutoff.

A quasi-unbiased sample of field stars taken from Gliese's catalogue shows clearly the effect of dissociation of binaries with time. Binaries younger than the Sun may have separations as large as 180000 AU, while there are no older ones with separations larger than 4000 AU; the effect of dissociation is noticeable already at 2000 AU. The interactions with field stars are insufficient to explain the observed rate of dissociation; probably molecular clouds and spiral arms are important perturbers. The observed cutoff for old binaries indicates that the radius of the inner comet cloud can not be much larger than 2000 to 4000 AU; clearly, the survival of a putative solar companion like Nemesis with $a \simeq 90000$ AU would be extremely unlikely.

Acknowledgements

I am grateful to W. Gliese for making available prior to publication an updated list of wide binaries, to H. Abt for interesting discussions on the incidence of opticals in the FIDS, and to C. Allen for some unpublished results.

References

Abt, H. A.: 1979, *Astron. J.* **84**, 1591.
Abt, H. A.: 1983, *Ann. Rev. Astron. Astrophys.* **21**, 243.
Abt, H. A.: 1986, *Astron. J.* **304**, 688.
Abt, H. A. and Levy, S. G.: 1976, *Astrophys. J. Suppl.* **30**, 273.
Allen, C.: 1974, Thesis, UNAM.
Allen, C. and Poveda, A.: 1974, 'The Stability of the Solar System and Small Stellar Systems', in Y. Kozai (ed.), *IAU Symp.* **62**, 239.
Allen, C., Tapia, M., and Parrao, L.: 1977, *Rev. Mexicana Astron. Astrof.* **3**, 119.
Allen, C., Martos, M. A., and Poveda, A.: 1987, *Rev. Mexicana Astron. Astrof.* **14**, 213.
Ambartsumian, V. A.: 1937, *Soviet Astron. J.* **14**, 207.
Gliese, W.: 1969, *Veröffentl. Astron. Rechen-Inst. Heidelberg* **22**.
Gliese, W., Jahreiss, H., and Upgren, A. R.: 1986, in R. Smoluchowski, J. N. Bahcall and M. S. Matthews (eds.), *The Galaxy and The Solar System*, The University of Arizona Press, Tucson, p. 13.
Jeffers, H. M., van den Bos, W. H., and Greeby, F. M.: 1963, *Index Catalogue of Visual Double Stars*, Lick Observatory, Mt. Hamilton.
Kuiper, G. P.: 1942, *Astrophys. J.* **95**, 201.
Luyten, W.: 1969, *Proper Motion Survey with the 48-Inch Schmidt Telescope*, XXI Double Stars with Common Proper Motion, University of Minnesota, Minneapolis.
Öpik, E.: 1924, *Publ. Tartu Univ. Obs.* **25**, No. 6.
Poveda, A.: 1977, *Rev. Mexicana Astron. Astrof.* **3**, 189.
Poveda, A. and Allen, C.: 1975, *Astrophys. J.* **200**, 42.
Poveda, A. and Allen, C.: 1987, in preparation.
Poveda, A., Allen, C., and Parrao, L.: 1982, *Astrophys. J.* **258**, 589.
Weissman, P. R.: 1986, in R. Smoluchowski, J. N. Bahcall, and M. S. Matthews (eds.), *The Galaxy and The Solar System*, The University of Arizona Press, Tucson, p. 204.

GLIESE – Today the Nearby Star Catalogue is not complete until a certain limit. Recently a few LHS stars were detected nearer than 10 or even 5 parsecs of the Sun. Of course, these objects have faint luminosities (about $M_r \sim 16$ or so).

POVEDA – We tried to strike a good compromise: on the one hand one needs to take a large volume of space in order to have a reasonable number of binaries, on the other hand we should limit the sample to a size such that the kinematical bias (for instance) is kept to a minimum; by taking the stars closer than 12.5 pc ($\pi > 0.08''$) we take a sample which does not seem to be very incomplete.

WORLEY – Please remember that IDS and WDS are *double* star catalogs, not *binary* star catalogs. They, therefore, necessarily include both optical and physical doubles. Many of the optical pairs are the result of including the astrographic doubles. These also cause magnitude inconsistencies because their magnitudes are photographic, not visual. Only pairs which have at least one astrometric differential measure are included in this data base, hence objects identified by Bahcall and Soneira are not included.

POVEDA – I think it is important to realize that the IDS and the WDS Catalogues contain such a large number of optical and spurious components.

WASSERMAN – Simulations by M. D. Weinberg, S. L. Shapiro, and I. Wasserman (*Astrophys. J.* **312**, **367**, 1987), although restricted to $a \gtrsim 10^{-2}$ pc, are not inconsistent with binary dissolution at $a \sim 10^{-2}$ pc. The half-lives computed are $\sim 10^{10}$ yr at $a \sim 10^{-2}$ pc. Dissolution is almost certainly due to disruptive encounters with subclumps within giant molecular clouds during penetrations. Although disruption during a single penetration is unlikely, a binary with $a \sim 10^{-2}$ pc will penetrate GMC's repeatedly. Encoutners with passing stars have only a negligible effect at such 'small' values of the semi-major axis. Non-destructive penetrations of GMC's would tend to result in slow increase of semi-major axes, making disruption upon subsequent penetrations more likely.

ABT – As I understand your model, the Trapezia systems evolve into hierarchical systems by having the closer stars move toward each other and distant stars move farther away. But if I understand comments made privately by Wasserman, wide pairs will never become closer together. Could Dr Wasserman explain?

POVEDA – There are two modes of evolution for the components in a multiple system:

(a) As the result of *internal* energy exchanges, a star may be thrown to an extended orbit (or ejected altogether); by energy conservation the remaining pair (in the case of a triple) will become more closely bound.

(b) *External* energy exchanges can make weakly bound members of system to be ejected altogether.

YUNGELSON – Our independent crude estimates of the effect of encounters with field stars and giant molecular clouds also had shown that the latter are most effective tool of disruption of binaries with $\sim M_\odot$ mass.

TRAPEZIUM-TYPE WIDE SYSTEMS

G. N. SALUKVADZE and G. SH. JAVAKHISHVILI*
Abastumani Astrophysical Observatory, Georgia, U.S.S.R.

Summary. On the basis of the Smithsonian Observatory Catalogues 15 Trapezium-type systems are found. Photometric distances are determined for the Trapezium components. A conclusion is drawn on physical relations of the components in the systems.

The fact of existence of Trapezium-type large stellar systems confirms the existence of Trapezia with positive total energy.

The idea on Trapezium-type multiple systems was introduced by Prof. V. A. Ambartsumian as a particular class of multiple stars in which the space distances between the components are of the same order of magnitude (Ambartsumian, 1954). The first list of Trapezia was compiled by Ambartsumian (1954) and a more complete list by one of the present authors (Salukvadze, 1978). These two ists contain such Trapezia, the distances between whose components are 0.02–0.20 pc (4×10^3–4×10^4 AU) – i.e., only close systems. All the basic conclusions on the Trapezia properties were drawn by Prof. V. A. Ambartsumian in terms of general theory assumptions just after this interesting class of multiple systems was separated.

At present an exceedingly small age of Trapezia is taken for granted. Its study enables a better understanding of star formation process in the Galaxy, in particular its group character. So far Prof. V. A. Ambartsumian's important suggestion on a positive total energy of Trapezium-type multiple systems is expected to be corroborated. Recently, however, on the basis of a thorough treatment of observational material for 15 Trapezia from the catalogue of Salukvadze (1978) available at present, a conclusion was drawn on the expansion of a certain number of Trapezia (Salukvadze, 1985).

From the standpoint of a dynamical evolution of Trapezium-type systems it is of interest to search and study such objects which are at earlier or later stages of evolution than those listed in the above-mentioned catalogues. Considering that Trapezium type systems are expanding – i.e., their average sizes increase with evolution – it is natural to believe that there are very close and wide systems accordingly.

Theoretical estimations show that with positive total energy the lifetime of a Trapezium is no more than 10^6 yr. If 5 km s^{-1} is adopted to be the expansion velocity, then the Trapezia keep their configuration until their sizes become of the order of 5 pc.

The problem of existing very wide systems was studied by Prof. V. A. Ambartsumian (1954). The first list, consisting of 5 wide Trapezium-type systems, was also compiled by him. Here it should be noted that the spectral-type in these systems is known for primary stars only. Unfortunately, nobody has dealt with this problem. Quite recently Gyolbudaghian (1983) has found 11 wide systems in Puppis.

* Presented by G. N. Salukvadze.

Astrophysics and Space Science **142** (1988) 79–83.

TABLE I

No.	BD	α_{1950}	δ_{1950}	d (kpc)	m_v	Sp.	r_1 (pc)	r_2 (pc)
1	+60°502	$2^h28^m53^s$	+61°14′	1.4	7.ᵐ83	O5	2.6	5.7
	+60 504	2 29 01	+61 09	1.3	8.11	O4		
	+60 507	2 29 31	+61 18	2.1	8.42	O5		
2	−05 1319	5 32 55	−05 27	0.6	5.08	O9.5	0.4	1.7
	−05 1315	5 32 48	−05 25	0.8	5.13	O6		
	−05 1320	5 32 58	−05 27	0.7	6.39	B1		
	−05 1325	5 33 03	−05 18	0.6	6.83	B1		
3	−04 1184	5 32 53	−04 27	0.4	6.22	B2.5	0.7	2.1
	−04 1186	5 33 02	−04 224	0.6	6.31	B2.5		
	−04 1183	5 32 53	−04 31	0.7	6.56	B1.5		
	−04 1190	5 33 37	−04 27	0.7	7.14	B2		
4	+1331124	6 05 40	+13 59	1.0	6.94	B1	0.8	1.1
	+1331120	6 05 27	+13 59	0.8	8.13	B2.5		
	+13 1123	6 05 37	+13 56	1.4	8.78	B2		
5	+05 1283	6 29 16	+04 59	1.4	6.73	O6	1.6	4.7
	+04 1302	6 29 29	+04 52	1.7	7.26	O5		
	+05 1282	6 29 12	+05 04	1.5	7.59	O8.5		
	+05 1279	6 28 58	+05 04	1.1	7.91	B1		
	+04 1291	6 28 41	+04 52	1.9	8.17	O8		
	+05 1286	6 29 30	+05 00	1.9	8.18	O9		
	+04 1295	6 28 54	+04 53	1.2	9.29	B2		
	+05 1281	6 29 12	+04 58	1.3	9.36	B2		
6	+04 1361	6 35 49	+04 39	1.4	7.14	B0III	0.8	1.6
	+04 1360	6 35 43	+04 40	1.4	8.19	B0.9		
	+04 1363	6 35 57	+04 40	1.8	8.70	B1		
7	−57 3508	10 33 54	−57 58	2.1	6.51	B0II	1.3	1.7
	−57 3499	10 33 47	−57 56	2.2	6.71	B0.51		
	−57 3502	10 33 48	−57 59	2.0	7.30	B1II		
8	−57 3533	10 34 13	−57 57	(1.9)	8.2	BIIII	2.8	3.8
	−57 3540	10 34 21	−58 01	2.7	8.3	B0.5II–III		
	−57 3516	10 33 59	−58 00	2.4	8.6	BIIII		
9	−60 3102	11 33 35	−61 19 37″	1.4	8.05	B2Iy	1.8	5.2
	−60 3128	11 33 51	−61 19	1.5	8.46	B2Iy		
	−60 3157	11 34 11	−61 17	1.7	8.54	B2III		
	−60 3136	11 33 56	−61 11	1.6	8.57	B2Iy		
	−60 3145	11 34 00	−61 22	1.9	9.07	B2Iy		
10	−59 4551	12 50 47	−60 03	1.4	8.01	B0.5	1.5	3.0
	−59 4552	12 50 46	−60 07	2.0	8.35	B1III		
	−59 4564	12 50 57	−60 08	2.1	9.07	B2III		
11	−62 3096	13 11 28	−63 19	1.2		O9		
	−62 3079	13 10 30	−63 19	1.8		O9		
	−62 3090	13 10 56	−63 07	0.8		B2		
12	−33 11875	17 12 02	−33 29	0.8	5.53	O8	4.1	4.5
	−33 11887	17 12 33	−33 40	1.2	6.55	O9		
	−33 11867	17 11 42	−33 41	1.4	7.92	B1		

Table I (continued)

No.	BD	α_{1950}	δ_{1950}	d (kpc)	m_v	Sp.	r_1 (pc)	r_2 (pc)
13	−32° 12574	17h15m04s	−32° 16′	0.9	8m.15	B2	1.4	2.7
	−32 12593	17 15 39	−32 17	1.1	8.73	B2		
	−32 12576	17 15 08	−32 21	1.1	8.74	B2		
14	+61 2364	22 51 57	+62 19	1.0	9.07	B1	1.7	3.3
	+61 2365	22 52 21	+62 23	1.2	9.20	B1		
	+61 2366	22 52 39	+62 20	1.1	9.72	B1		
15	+61 2373	22 54 48	+62 27	0.9	7.65	O7	1.6	3.9
	+61 2136	22 54 33	+62 36	1.1	7.74	B0		
	+61 2372	22 54 44	+62 21	0.8	8.78	B1		

We intended to search for wide Trapezium-type multiple systems in which all the components are of O–B2 spectral-type since a large percent of Trapezium-type real systems are encountered among this spectral-type.

The SAO catalogues sample of O–B2 spectral-types with astrophysical data was performed at the Soviet Stellar Data Center on our request. Altogether 2601 such objects are found. Further the stars were arranged on the computer Nairi-2 according to increasing right ascentions and in the square of $\pm 15'$ around each star all stars were picked out.

On the whole this procedure was done for all 2601 stars from our list. Then the listings were carefully examined and all the sample groups were extracted. Finally 60 groupings were revealed. The distances between the components and position angles were computed. Consequently, the ratios of the greatest distances to the smallest ones were estimated. It was found that many groupings do not suffice the definition of Trapezium-type multiple systems and they are called ε Lyrae; some others consisting of several components involve the groupings of young O–B2 stars.

Hence, the data on 15 Trapezia left in the final list, given in Table I, are as follows: stellar numbers according to BD, coordinates of the stars making the Trapezia, their distances, a visual magnitude, their spectra, minimum and maximum distances between the Trapezia components.

The distance moduli to the stars forming the Trapezia were determined by a known formula $m - M = 5 \lg r - 5 + A(r)$, where $A(r)$ is the interstellar extinction. In order to compute $A(r)$ the relation $A(r) = R(N - V)_0$ was used. For R the value of 3.1 was adopted. M_v and $(B - V)_0$ values for known spectral types in the MK system were adopted from Schmidt-Kaler (1985).

For stars forming System No. 8 (Table I) the distances are taken from papers by Rubin (1982) and Neckel (1967). As the errors in distance estimates by the photometric method amount to 30%, it can be thought that the dispersions in distances of the stars in each wide system (Table I) is not high and they are almost at the same distance.

It follows from the above assumption that in general Trapezium-type multiple systems represent groups of stars formed together. Very likely most of wide systems found by us are real Trapezia.

The fact that wide systems exist favours the suggestion of a positive total energy of Trapezium-type multiple systems. It is clear that, with a negative total energy, the configuration of multiple stars conforming to the Trapezium definition would not be satisfied.

References

Ambartsumian, V. A.: 1954, *Bull. Bjurakan Obs.* **15**, 3.

Salukvadze, G. N.: 1978, *Bull. Abastumani Astrophys. Obs.*, No. 49, 39.

Salukvadze, G. N.: 1985, *Astrophysics (Astrofizika)* **22**, 97.

Gyolbudaghian, A. L.: 1983, *Astrophysics* **19**, 747.

Ochsenbein, F.: 1980, *Inf. Bull. Strasbourg Stellar Data Center* **19**, 74.

Schmidt-Kaler, Th.: 1965, 'Zustandgrössen und Zustanddiagramme der Sterne', in *Landolt-Börnstein, Zahlenwerte und Functionen aus Wissenschaft und Technik*, Berlin, p. 251.

Rubin, V. C., Burley, I., Kiasatpoor, A., Klock, B., Pease, G., Ruschheidt, E., and Smith, C.: 1982, Astron. J., 67, 491.

Neckel, Th.: 1967, *Landessternwarte Heidelberg-Königstuhl Veröffentl.* **19**, 1.

POVEDA – I think it is appropriate to recall that some years ago in collaboration with C. Allen and C. Worley we made a study of the separations among pairs of stars in as many Trapezia as we could find reliable data in a long base time. In that study modeling about 40 Trapezia with sufficient data we did not find a single case of expansion.

We found, however, that occasionally one star will move out of the Trapezia as for instance was the case of star E in Orion Trapezium. We concluded, therefore, that there is no evidence of positive energy in Trapezia, not even for the best documented case, the Trapezium in Orion.

SALUKVADZE – We have studied the kinematics of 15 Trapezium-type multiple stars with O–B2 spectral-type primary stars.

Besides our own photographic observations we have used the measurement results of relative positions of the multiple stars published in different catalogues of double stars. In most cases the time interval of the observational material is more than 100 years.

The results of calculations and the plots of dependence of distances from the epoch of observation indicate that out of 15 Trapezia investigated by us, 14 show the expansion.

A similar work is done by Allen, Poveda, and Worley. They have considered 44 Trapezia from Prof. Ambartsumian's list and a full expansion was found in none of them. Only in 16 Trapezia one or a few components showed systematic removal from the primary star.

Out of 15 Trapezia picked out by us, in common with C. Allen *et al.*, relative motions were found in three Trapezia ADS 2843 (component B), ADS 13374 (component C), and ADS 14831 (component C).

In our opinion the reason of divergence in the results consists in the following:

(1) Allen *et al.*, considered the Trapezia of all spectral types and, what really matters, out of 44 Trapezia studied, 13 ones are not included in our Catalogue, i.e., they are not Trapezium type systems, 6 ones are Trapezia of late and unknown spectral types. From the remaining 25 Trapezia 15 ones are the Trapezia under investigation.

(2) The plots of the paper by Allen *et al.*, show that the observational data obtained by 1965 were used to construct them. We used the published observations up to 1980. It is clear that the final observations are of a decisive value in drawing conclusions on the relative motions of the Trapezia components.

The results of osur investigations of the Trapezia kinematics have led us to a main conclusion as follows:

Investigations of Trapezia based on the treatment of the observational data confirm the suggestion on the instability of certain parts of the Trapezium type multiple stars.

At the same time it should be noted that at present for the majority of stars the observational precision is not enough in order that the relative motions of their components be determined.

ABT – What is the difference between a large trapezium system and an early-type cluster with very few members? Does it depend on the geometry of the stellar positions, as assumed by Poveda *et al.* and Ambartsumian?

SALUKVADZE – If there are such galactic clusters with ~ 10 members, then the difference between such clusters and Trapezium-type wide systems depend on their geometric configuration. As I have mentioned the term of Trapezium-type multiple systems was first introduced by V. A. Ambartsumian as a particular class of multiple stars in which the ratio of the greatest distances to the smallest ones between the components is less than 3.

However, when we consider the internal motion of stars in multiple systems with negative total energies, we say that they are similar to galactic clusters and differ from them only by that the number of stars in them is much less. Consequently in determining the relaxation time for Trapezium type systems, we use a formula deduced by V. A. Ambartsumian for stellar clusters.

THE KINEMATICS OF TRAPEZIUM-TYPE SYSTEMS

A. I. YATSENKO

Main Astronomical Observatory, Academy of Sciences of the Ukrainian SSR, Kiev, U.S.S.R.

Summary. In this paper the results of the research of the stars proper motions Trapezium components are reported. They are: the galactic coordinates of the solar aprx and the Sun velocity ($L_{\odot} = 43 \pm 18°$, $B_{\odot} = +28 \pm 13°$, $V_{\odot} = 13 \pm 4$ km s^{-1}), the dispersion of peculiar velocities in the direction of the galactic coordinates for the above mentioned stars ($\sigma_l = \pm 11$ km s^{-1}, $\sigma_b = \pm 7$ km s^{-1}). The attained accuracy of the proper motions ($\pm 0.005''$ yr^{-1}) is shown to be insufficient to the study of internal space motions in these systems. At present the work to increase the relative proper motions accuracy for multiple system components and to improve reductions from the relative to absolute proper motions, is being carried out in the Main Astronomical Observatory (Academy of Sciences of the Ukrainian SSR). The new catalogue of the AGK3 stars is composed now in the vicinity of the galactic equator in order to improve reductions from the relative to absolute proper motions. The r.m.s. errors of the proper motions, obtained in the AGK3 system, are $\pm 0.005''$ yr^{-1}.

Starting from the late forties of this century the kinematics and dynamics researches became very actual (Blaauw, 1952; Markarjan, 1953; Parenago, 1954; Achundova, 1957) in connection with the works by Ambartsumian (1950, 1955). Special attention was paid lately to the researches of the internal motions in the cores of the O–B associations, in particular in the Trapezium-type systems (Duboshin *et al.*, 1971; Allen *et al.*, 1974; Salukvadze, 1981).

Since 1956 regular observations of the Trapezium-type systems according to Ambartsumian's (1954) list have been led on the double long-focus astrograph (5500/400). Eighty-eight plates with mean epochs difference of 18 years were used for obtaining the relative proper motions of 950 stars (Yatsenko, 1978, 1980, 1981), including 45 components of 16 Trapezium-type system. But the proper motions themself do not clear up the question of dynamics or kinematics of the Trapezia. However, if we know a distance to these systems, the tangential velocities of the stars can be computed and such characterization can be obtained as: the components of solar motion relative to the centroid of Trapezia, the peculiar velocity dispersion of the Trapezia stars and the velocities within the multiple system. The distances to the Trapezia were taken from the paper by Echevarria *et al.* (1979). Besides, if we want to get reliable centroid characteristics we should have Trapezia covering either the whole or the greater part of the galactic equator arc. For this purpose we combined the data of the absolute proper motions of 68 stars of 34 Trapezium-type systems, covering 236° in galactic longitude within $\pm 10°$ of galactic latitude.

The solar-motion components were computed from the tangential velocities of these stars, corrected for differential galactic rotation:

$$\begin{aligned} X \sin l - Y \cos l &= V_l \cos b - r(B \cos b + A \cos 2l \cos b)\,, \\ X \cos l \sin b + Y \sin l \sin b - Z \cos b &= V_b + \tfrac{1}{2} rA \sin 2l \sin 2b\,, \end{aligned} \tag{1}$$

Astrophysics and Space Science **142** (1988) 85–88.

where r (kpc) is the distance from the Sun to the Trapezium; the Oort constants $A = 15$ km s^{-1} kpc^{-1}, $B = -10$km s^{-1} kpc^{-1}; $V_l \cos b$, V_b represent the tangential components of the solar motion relative to the centroid of the Trapezia in the galactic coordinates system.

By solving equations of condition (1) the following values were obtained for the solar-motion components: $X = +8.4 \pm 4.5$ km s^{-1}, $Y = +7.8 \pm 3.8$ km s^{-1}, $Z = +6.0 \pm 2.9$ km s^{-1}. These values give the following coordinates for the apex and velocity of solar motion: $L_\odot = 43 \pm 18°$, $B_\odot = +28 \pm 13°$, $V_\odot = 13 \pm 4$ km s^{-1}.

After peculiar velocity components of these Trapezia were corrected for the influence of random observational errors, the following peculiar-velocity dispersion were obtained: in galactic longitude, $\sigma_l = \pm 11$ km s^{-1}; in galactic latitude, $\sigma_b = \pm 7$ km s^{-1}. The accuracy of the relative proper motions of stars ($\pm 0.005''$ yr^{-1}), found from our observations, is too small for the determination of the internal star velocities inside the Trapezia, because the multiple systems have distances to the Sun from 200 to 2000 pc. However, we could find that two of eight Trapezia are optical systems. The first one is ADS 14526. In relative velocities of the components have mean absolute value of 14 km s^{-1}. This result is supported by spectroscopic and photometric studies of ADS 14526 (Echevarria *et al.*, 1979). The same can be said about ADS 11179, where the C-member is an optical component. Its velocity relative to components A and B of the system is 100 km s^{-1}, if C is regarded as at the same distance as A and B from the Sun ($r = 226$ pc).

At present the researches of multiple systems are developed on the following directions. On the one hand, the observations of the Trapezium-type systems are continued with the long-focus astrograph for the purpose of increasing the quantity of plates, epochs difference and accuracy of determination of the relative proper motions. On the other hand, in order to improve the reductions from the relative to absolute proper motions we started composing proper motions catalogue of AGK3 stars. For this aim the plates from Soviet program of North Sky Photographical Survey will be used (Kolchinskij and Onegina, 1979). They were obtained in Kiev with the double wide-angle astrograph (2000/400). The proper motions of all stars of the AGK3 list will be obtained for the galactic equator region ($|b| \leq 10°$) in the North part of the sky. To analyse possible accuracy of the new catalogue a pilot-program was made, which contained the reduction on five plates, obtained in 1983 and 1984. The plate centres are on the same parallel ($\delta = 60°$), and their useful fields half overlap each other. On the average every of 5 plates has the images of ≈ 160 AGK3 stars, obtained with two exposures (30^s, 16^m). In general, in the region with celestial coordinates $\alpha = 23^h12^m - 0^h48^m$, $\delta = +58° - +62°$, 505 AGK3 stars were studied. The common reduction of the measured coordinates of the images of two exposures was made separately for every plate. The reduction equations contained the linear, square and cubic terms dependent on the coordinates and also terms dependent on the magnitude (the magnitude dependent correction and the coma effect) (Ivanov *et al.*, 1987). The star positions obtained by different exposures were averaged with appropriate weights, dependent on the measurement accuracy. The star positions by epoch 1985.0 and their proper motions were obtained for solving the systems of equations, taken with appropriate weights:

$$\alpha_{1985.0} + \mu_\alpha(T_i - 1985.0) = \alpha_i\,, \qquad (2)$$

where α_i, the AGK2, AGK3 positions or the one computed by one-two wide-angle astrograph plates; T_i, the appropriate epochs; $\alpha_{1985.0}$ and μ_α, the unknowns. The analogous set of equations was computed for coordinates. Naturally, the system of the catalogue obtained in such a manner is identical to the AGK3 system. Yet, the random errors of the proper motions and positions on epoch 1985.0 are lesser to that of the data taken from AGK3. The estimates of random errors of new catalogue data are contained in Table I obtained both by single and by all overlapping plates.

TABLE I
Mean errors of new catalogue

Plate	$\varepsilon_\alpha \cos\delta$	ε_δ	$\varepsilon_{\mu_\alpha} \cos\delta$	ε_{μ_δ}
219	0".16	0".21	0".0046	0".0058
220	0.22	0.23	0.0062	0.0063
221	0.23	0.20	0.0065	0.0057
222	0.20	0.03	0.0057	0.0065
523	0.23	0.25	0.0064	0.0071
Mean errors for one plate	0.21	0.23	0.0059	0.0063
Mean errors of catalogue	0.19	0.19	0.0053	0.0052

The estimates were obtained by solving the sets of Equations (2) and were confirmed by comparing the new catalogue with AGK3RN data. Systematic differences were not detected between positions and proper motions in the new and AGK3 catalogues. After the consummation of the proper motions catalogue for the total Milky Way region of the northern part of the sky it would be used for the research of the multiple Trapezium-type systems. In addition it may be used for the statistical studies and search of the double and multiple wide systems in this region.

References

Achundova, G. W.: 1957, *Izv. Glav. Astron. Obs. Akad. Nauk SSSR* **21**, 83.

Allen, Chr., Poveda, A., and Worley, Ch. E.: 1974, *Rev. Mexicana Astron. Astrof.* **1**, 101.

Ambartsumian, V. A.: 1950, *Izv. Akad. Nauk SSSR (Ser. Fiz.)* **14**, 15.

Ambartsumian, V. A.: 1954, *Soobshch. Byurakansk. Obs. Akad. Nauk Arm. S.S.R.* **15**, 3.

Ambartsumian, V. A.: 1955, *Observatory* **75**, 72.

Blaauw, A.: 1952, *Bull. Astron. Inst. Neth.* **11**, 405.

Duboshin, G. N., Rybakov, A. I., Kalinina, E. P., and Kholopov, P. N.: 1971, *Soobshch. Gos. Astron. Inst. Sternberga* **175**, 3.

Echevarria, J., Roth, M., and Warman, J.: 1979, *Rev. Mexicana Astron. Astrof.* **4**, 287.

Ivanov, G. A., Kur'yanova, A. N., and Yatsenko, A. I.: 1987, *Kinematica i fizika neb. tel.* **3**, No. 2, 82.

Kolchinskij, I. G. and Onegina, A. B.: 1979, *Astrometr. Astrofiz., Vyp* **39**, 57.
Markaryan, B. E.: 1953, *Soobshch. Byurakansk. Obs. Akad. Nauk Arm. S.S.R.* **11**, 2.
Parenago, P. P.: 1954, *Trudy Gos. Astron. Inst. Sternberga* **25**, 3.
Salukvadze, G. N.: 1981, *Soobshch. AN Grus. S.S.R.* **104**, 41.
Yatsenko, A. I.: 1978, *Astrometr. Astrofiz. Vyp* **35**, 9.
Yatsenko, A. I.: 1980, *Astrometr. Astrofiz. Vyp* **40**, 51.
Yatsenko, A. I.: 1981, *Astrometr. Astrofiz. Vyp* **45**, 63.

A SURVEY OF THIRD COMPONENTS IN SPECTROSCOPIC AND ECLIPSING SYSTEMS

T. J. HERCZEG

University of Oklahoma, Norman, Oklahoma, U.S.A.

and

Remeis Sternwarte, Bamberg, W. Germany

Summary. Multiplicities recorded in catalogues as well as in the current literature are considered. The basis of the statistical treatment is about one half of the material collected in the *Seventh Catalogue* of spectroscopic binaries by A. Batten and his collaborators, subjected also to a magnitude limit at $V = 6.8$, altogether 310 systems. It is hoped that this sample of brighter stars represents close and also wider systems with a reasonable degree of completeness.

Incidence of higher multiple systems, in particular of ab–c-type triple stars, turns out rather high: the ratio of double to multiple systems is about 2 : 1, in terms of the directly observed data. Allowing for some more obvious observational bias, this ratio approaches 1 : 1. The number of individual components in multiple systems is almost certainly superior to that of double star components.

An investigation in progress, concerning multiplicities in the specific classes of W UMa-systems and cataclysmic variables, is briefly described.

1. Introduction

In this short note or interim report we wish to deal with triple systems of the type ab–c, that is, internally stable systems; Trapezium-type multiples are not involved. In particular, we consider triple (and also multiple) systems where the close pair is a spectroscopic or eclipsing binary with orbital dimensions typically of the order of 1–2 AU or less. For a study of the fragmentation processes in protostellar nebulae, frequency of such pairs and the distribution of their orbital separations are of some importance; the available material referring to close pairs and their distant companions is both larger and more homogeneous than the material furnished by visual triple systems.

2. The Material

For counting of these triple (or: at least triple) systems, our basic source was the *Seventh Catalogue of the Orbital Elements of Spectroscopic Binary Systems* (Batten *et al.*, 1978); it is particularly well suited to study close binaries with distant companions because it brings all the relevant information about visual multiplicity. It is true, the *Catalogue* only lists spectroscopic binaries *with orbits*, but for the brighter systems we may assume without reservation that once a spectroscopic binary was discovered, the orbit will be soon calculated. Furthermore, in searching for close pairs, it is entirely sufficient – among the brighter stars – to look at the spectroscopic binaries in the first place; there is hardly a bright eclipsing variable without spectroscopic observations. For this reason

Astrophysics and Space Science **142** (1988) 89–95.

we have chosen $V = 6.8$ as magnitude limit. This means that our basic material is close to that contained in the *Bright Star Catalogue*. Going beyond about the 7th mag, even approximate completeness of the spectroscopic material rapidly deteriorates; also, binary discovery becomes strongly biassed toward easily recognizable types of objects such as W UMa-stars, cataclysmic variables or evolved semi-detached systems with deep eclipses.

3. Observational Bias

It is important to point out that in this group of stars recognition of spectroscopic or visual binary nature were essentially independent discoveries. Spectroscopically, most if not all bright stars have been investigated for radial velocity above the range of 5–10 km s^{-1} with a good chance of discovering variable Doppler-shifts. On the other hand, recognition of visual double stars mainly resulted from large systematic surveys by W. Struve, S. W. Burnham, R. G. Aitken, or the observers in South Africa. Occasional discoveries of new bright spectroscopic or visual binaries may remind us of the lacking completeness of the material; yet it remains a limited but remarkably sound statistical material.

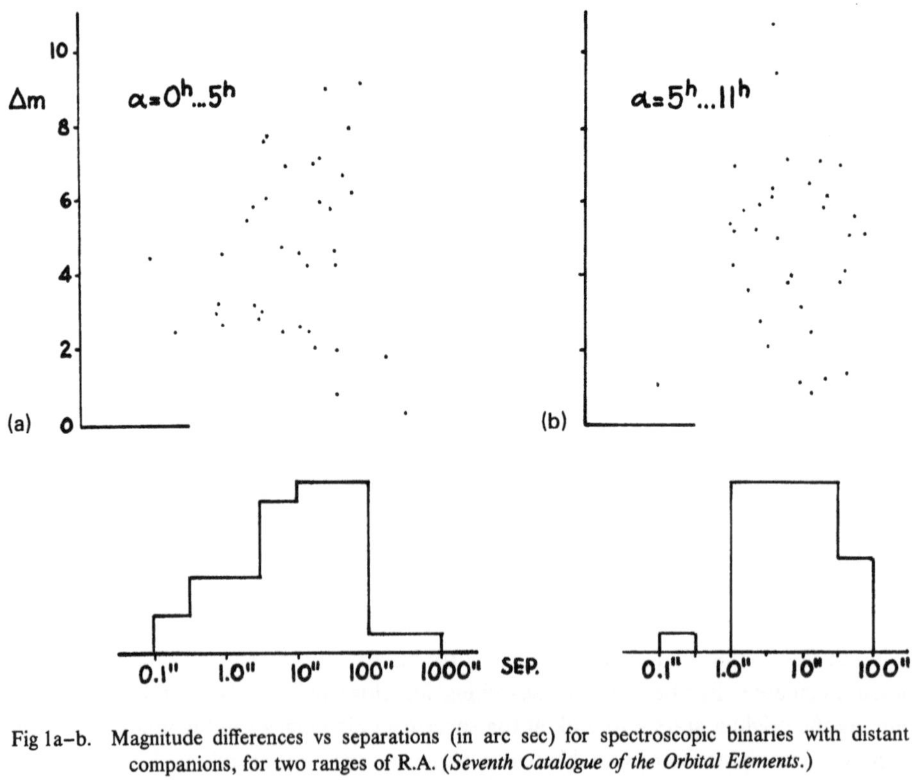

Fig 1a–b. Magnitude differences vs separations (in arc sec) for spectroscopic binaries with distant companions, for two ranges of R.A. (*Seventh Catalogue of the Orbital Elements.*)

There is one obvious bias still to be expected but relatively easy to account for: among these ab–c-type triple systems, it is prefereably the *brighter* visual component whose close binary nature will be recognized.

Various bias of observation have been carefully treated, for instance, in a recent work by Halbwachs (1983). We are concerned primarily with the dependence of visual discovery upon closeness of the pair and the magnitude difference between the components. Plots like Figures 1(a) and 1(b) are useful to judge the *relative* completeness of the sample, normed to the best recognized triple systems, such as objects with separations between 20″–40″ and magnitude differences less than 4. Data about absolute completeness is difficult to come up with. In this sense, our 'final' numbers about the frequency of multiples will represent lower bounds.

After eliminating – as well as it could be done (see below) – the merely optical companions of a close binary, I have also tried to estimate the projected linear separations between *ab* and *c*. To do this, we need lacking better information, the photometric parallaxes; we obtained them from the usually quite well determined spectral types and the Landolt–Börnstein *Tables* for absolute visual magnitudes (Schmidt-Kaler, 1982); for the class of Am-stars we used, as a crude mean value, $M_V = +1.5$. In spite of the considerably uncertainty of the individual data, the linear dimensions of the triple systems obtained in this way (see Figure 2) may contribute to the important problem: how distant are the most distant companions?

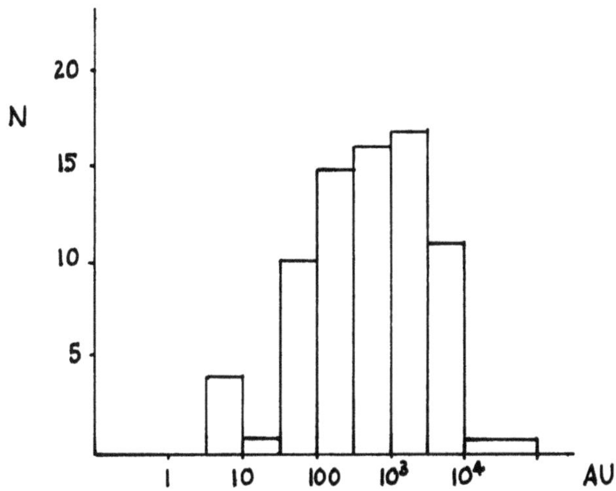

Fig. 2. Histogram of projected linear separations for 76 spectroscopic binaries with distant companions. The distribution appears uniform between about 100 and 3000 AU (separations on a logarithmic scale.) The drop beyond 10^4 AU may be influenced by rejecting very distant companions as merely optical ones. Several probably physical pairs with c.p.m. (e.g., 56 Tau and the binary 53 Tau) are not plotted.

4. Optical Companions

A particular difficulty in connection with double star counts is the question of merely optical companions. Even at the risk of stating the obvious, we dwell upon the procedure applied here. While in a catalogue it is perfectly legitimate to remark only: 'probably optical' or 'optical?', we have to come up with *decisions* for these cases.

As a rule, companions under $10''$ separation have been counted as physical (referring to the Poisson statistics of the distribution of nearest neighbours in the plane of the sky) and, somewhat arbitrarily, above $60''$ separation as optical, except, of course, if information about the relative motion of the components suggested otherwise.

For binaries with projected angular separation, say, between $10''$ and $30''$, the order of magnitude of the linear separation may be between 500 and 6000 AU, supposed the distances range from 50 to 200 pc, rather at the low end of their distribution. The corresponding orbital velocities (for masses around $3\,M_{\odot}$) turn out to be about 0.5 km s^{-1} to 2 km s^{-1}; projection on the plane of the sky further reduces the motion. It is easy to see that at typical distances these values correspond to very slow angular motion, perhaps $1''$ per century, or no observable relative motion at all. These are the 'fixed' pairs Abt and Levy (1976) call cpm pairs, as compared with the 'visual' ones where orbital motion is discernible. Without discussing questions of terminology, we note only that as a rule such fixed pairs have been counted as physical pairs, except if a closer look at the system did supply some further, decisive information. We briefly mention a couple of cases as *illustration*.

The spectroscopic binary HD 112486 is the brightest member of the triple system ADS 8710, with *AB* ($3.7''$) and *AC* ($124''$) forming 'probably optical' pairs. *AB* is a binary known since 1830, and the micrometric measures suggest very slow motion, about $\Delta\theta \sim 6°$ cent.$^{-1}$, $\Delta\rho \sim 0.3'$ cent.$^{-1}$; this corresponds to a tangential velocity of about 1.5 km s^{-1}, compatible with orbital motion. Since the proper motion is, according to the *Boss Catalogue*, $7.5''$ cent.$^{-1}$, it is clear that the couple moves together in the plane of sky and *B* must be counted as physical companion. *C* is beyond the separation limit set for our counts and the observations are not numerous enough to reverse the conclusion of an optical companion.

The textbook case of an optical companion is ADS 6251C (Procyon C): this component is moving, according to micrometric measurements since 1836, at about $1''$ yr^{-1} in a direction around 35° with respect to *AB*, while the p.m. of *AB* is $1.24''$ yr^{-1} in 214°.

After a case-to-case deliberation, more than $\frac{1}{4}$ of the possible distant companions have been rejected as optical ones. In several cases the question was settled only by the 60 arc sec 'guillotine' and I think it possible that hereby a few distant, physical companions have been disregarded.

5. Light Time Orbits

Before presenting the results of our statistics based on the *Seventh Catalogue*, we may mention other, less direct ways of detecting a third component around a close binary system. This is either periodic variations of the system velocity (γ-velocity) or periodic light-time effects in the timing of minima*. These effects can be very useful, particularly in the period range of 15–30 years corresponding, roughly, to a range of orbital semi-major axes of 8–15 AU. There are very few known visual system in this range owing to the difficulty of detecting and reliably measuring separations below 0.1″. Speckle interferometry, reaching down to 0.02″–0.03″, may help to bridge this gap. Unfortunately, the number of systems with well-documented light-time effects is less than 10, probably even less than 6 and has not reached statistical significance yet. Moreover, in all but two cases I know of (IU Aur, AH Cep), the multiplicity of the system had been already known from spectroscopic or astrometric studies, before the light-time orbit was calculated as, for instance, in the case of the triple systems Algol, EE Peg, or VW Cep.

6. Star Counts

The counts resulted in the following numbers – considered was the R.A. interval 0^h–13^h in the *Seventh Catalogue* (*op. cit.*), comprising very nearly $\frac{1}{2}$ of the catalogue material. The magnitude limit was, as already mentioned, $V = 6.8$. Excluded were 170 entries, including the Trapezium in Orion. The rest contains:

172 spectroscopic binaries without distant companion detected,
121 binaries with distant companion, among these
31 classified as optical (that is, 26%), and
17 multiple systems ($n \geq 4$), with 43 individual components.

If we add the systems with optical companions to the spectroscopic doubles, the final count becomes:

203 binaries,
90 triple systems of type ab–c,
17 multiple systems.

The fraction of triples and multiples is 0.346, about twice as large as the generally accepted value near $\frac{1}{6}$. Moreover, this ratio may turn out much more to the favour of the triple systems, if we take two obvious observational bias in consideration:

(1) The deficiency of triple systems with short period orbits of *C* around *AB* (separations near 10 AU, see Figure 2) is in all likelihood a consequence of observational selection; we are approaching the domain of the almost inaccessible visual binaries, at ~0.01″ to ~0.05″. *Assuming* that the (logarithmic) distribution of separations can be described as approximately constant, we may raise the frequency of triple systems by 20%.

(2) Among the ab–c systems we found in the *Seventh Catalogue*, in 83 cases the close pair was the brighter component, and in only 7 cases the fainter one. It is conceivable that processes of star formation, the mode of fragmentation in the proto-stellar nebula lead to this asymmetric distribution. Yet if we tentatively ascribe this unevenness to

* About the nodal precession effect, see Mayor and Mazeh (1987).

rather obvious *observational selection*, we may nearly double the number of observed triple systems and, in rough figures, we reach about the same frequency for close spectroscopic binaries alone as for close binaries in triple and multiple systems; the multiple systems may even preponderate.

Against this conclusion one could perhaps argue in the following way:

There may be a large number of brighter stars which, although apparently single, are actually spectroscopic binaries with low (a few km s^{-1}) velocity amplitudes. Results of comprehensive surveys of solar type stars (Abt and Levy, 1976) or A-type stars (Abt, 1961, 1965), moreover, the successful treatment of several difficult cases in Griffin's series of photoelectric radial velocity determinations (see his 1983 synopsis of 50 papers), do suggest that numerous unrecognized binaries may exist even among brighter stars. Whether their number could reach the necessary 200 or so, to circumvent the tentative conclusion just stated, remains questionable.

7. A Progress Report

Finally, closing the report I would like to mention a ramification of these counts of triple systems: the extension of the statistics to fainter magnitudes, dealing with two well-defined subclasses of close binaries, to *W UMa-type systems* and to *cataclysmic variables*. As to the first group, a surprising number of better studied systems have distant, third companions: W UMa itself, *i* Boo, VW Cep, AK Her, ER Ori, AH Vir, to mention a few, not to speak of the visual system ADS 9537 where both components are W UMa-type binaries (BV and BW Dra). A closer look seems, indeed, to indicate a somewhat elevated percentage of distant companions but the find remains marginal, not very convincing.

The case of cataclysmic variables is more intriguing. Work in progress strongly suggests here a lack of triple systems, with third companions probably under 10%. This may be a consequence of the past and continued mass-loss from the systems, weakening the gravitational binding between the cataclysmic pair and the distant companion. It can be, however, merely a result of observational selection: most cataclysmic variables are relative faint objects, around 12th, 13th mag, and an M-type or white dwarf companion may easily be of 17th or 18th mag only. Thus completing such a study also includes looking up these objects on copies of the Sky Surveys.

I hope to present a report on this extension of the present study soon.

References

Abt, H. A.: 1965, *Astrophys. J. Suppl.* **11**, 429.
Abt, H. A. and Levy, S.: 1976, *Astrophys. J. Suppl.* **30**, 273; see also **36**, 241 (1978).
Abt, H. A.: 1961, *Astrophys. J. Suppl.* **6**, 37.
Batten, A. H., Fletcher, J. M., and Mann, P. J.: 1978, *Publ. Dominion Astrophys. Obs.* **15**, No. 5.
Griffin, R. F.: 1983, *Observatory* **103**, 273.

Halbwachs, J. L.: 1983, *Astron. Astrophys.* **128**, 399.
Hoffleit, D. and Jaschek, C.: 1982, *The Bright Star Catalogue* (4th ed.), Yale Univ. Press, New Haven.
Mayor, M. and Mazeh, T.: 1987, *Astron. Astrophys.* **171**, 157.
Schmidt-Kaler, Th.: 1982, in K. Schaifers and H. H. Voigt (eds.), *Landolt-Börnstein, Numerical Data and Functional Relationships in Science and Technology*, New Series, Group VI, Vol. 2b, Springer-Verlag, Berlin.

MAZEH – Dr Mayor of the Geneva Observatory and myself published a paper about a study of 25 close spectroscopic binaries early this year. We found that between $\frac{1}{3}$ and $\frac{2}{3}$ of these binary stars are in multiple systems.

HERCZEG – This investigation refers to the observationally difficult systems with separations between the components in the range of 10 to 100 AU. If the details can be corroborated by further data and the sample made somewhat larger (it is still a small-number-statistics), it seems to support clearly the extrapolated distribution of separations that follows from the data presented in the lecture.

GLIESE – Among the 32 wide binaries/multiple systems nearer than 25 parsecs there are two thirds (22) systems with more than 2 components (3 to 6). Spectroscopic binary components are included.

HERCZEG – I am glad to learn this additional information confirming the data proposed in the lecture.

ABT – May I ask Dr Mazeh regarding his sample of 25 binaries in which he found that $\frac{1}{3}$–$\frac{2}{3}$ have a third component, did that include visual companions?

MAZEH – Yes, the study included visual binaries.

IMAGING APPROACH TO MULTIPLE STAR RECOGNITION: THE ALGORITHM, THE RESULTS

L. BORRIELLO

CSATA – Valenzano (Bari), Italia

Summary*. The data collected by HIPPARCOS satellite can be used in an imaging approach, like in VLBI field, to reconstruct the stars system image and to extract the astrometric parameters.

The method is able to treat multiple star systems to determine the stars position and their intensities. It is not able to process variable stars, very faint stars and systems with high orbital motion.

This paper is devoted to give a short description of the approach and to present a complete outline of the implemented algorithm.

Results related to cpu time consuming and to reduction results are given with particular emphasis on detectable separation limits and parameters precision.

* The full paper may be found in: *Third Fast Thinkshop – Processing of Scientific Data from the ESA Astrometry Satellite HIPPARCOS*, P. L. Bernacca and J. Kovalevsky (eds.), Padova, 1987, pp. 189–197.

Astrophysics and Space Science **142** (1988) 97.

PHYSICAL PARAMETERS OF THE MASSIVE MULTIPLE SYSTEM ETA ORIONIS

C. WAELKENS and P. LAMPENS
Astronomisch Instituut, Katholieke Universiteit Leuven, Belgium

Summary. The bright massive object η Orionis is at the same time a totally eclipsing system, a spectroscopic triple system, a speckle triple system and a visual binary. These circumstances offer an exceptional opportunity for determining the physical parameters of massive stars. Our observations show that the published elements are wrong. We present a new determination of several parameters of the multiple system and propose new visual observations.

1. Introduction

The advent of high-spatial-resolution techniques is closing the gap between so-called 'close' and 'wide' binaries, and so giving a new impetus to one of the oldest disciplines in stellar astronomy. The study of multiple stars has gained much from recent technical progress, and important results may be expected from detailed studies of such systems. Our knowledge of stellar parameters and of star formation owes much to the observation of binary and multiple stars. In this paper we illustrate that in this context it is still useful to carefully observe even the brightest stars, as our understanding of their multiplicity is sometimes highly inaccurate.

Eta Orionis (m_v = 3.33) is a remarkable multiple star. It is a visual triple system, the brightest components (AB) of which are separated by 1.6″. Component A is a spectroscopic triple (AaAbAc). The pair AaAb is eclipsing. The system AabAcB has recently been resolved by speckle interferometry (McAlister, 1976). Finally, one of the four components of the system AB is a short-period photometric variable (Chambliss, 1978; Koch, 1978).

In the present paper, we summarize our new observational results on η Orionis. A more detailed account, with more emphasis on the short-period variability, will appear elsewhere (Waelkens and Lampens, 1988).

2. Photometric Observations and Analysis

We have obtained extensive photometric observations of η Orionis with the Geneva photometer attached at the Swiss Telescope at La Silla Observatory in Chile, from late 1980 to late 1985. We were motivated by the possibility that the system may offer a unique opportunity of monitoring a pulsating early-type star which undergoes total eclipses (Koch *et al.*, 1980), but now this possibility seems unlikely. However, it turns out that our data imply significant corrections on the parameters of the system as they were published in the recent literature.

* Based on observations collected with the Swiss Telescope at the European Southern Observatory, La Silla, Chile.

Astrophysics and Space Science **142** (1988) 99–101.

Observations outside eclipse show that the short period suggested earlier is an alias of the true period, which amounts to 0.43208 days. This period, and the absence of colour variations, render the pulsation hypothesis unlikely. More probably, the variable itself is a close binary or a severely distorted single object. Since the short-period modulation increases its amplitude during eclipse, it is most probably associated with either Ac or B.

Both eclipses are approximately 20% deep. It appears from simultaneous spectroscopic data that secondary eclipse is (nearly) total, while star Aa is still prominent during primary eclipse. Both components thus have similar temperatures, but different sizes, component Aa being the largest. It also is clear that, unlike earlier suggestions, the pair Aab contributes more to the integrated light than the two other components. It is then desirable to reconsider the determination of the parameters of the four components.

3. Determination of Stellar Parameters

The pair Aab being a double-lined binary which undergoes total eclipses, the masses of both components can be determined. The radii follow from the duration of eclipses and an estimate of the brightness ratio. From the colours, the effective temperatures and reddening can be estimated rather accurately, and with the radii the absolute magnitudes. The visual absorption is found to be $A_v = 0.31$ mag. We then find a distance modulus $(m - M)_0 = 7.58$. The results are summarized in Table I.

TABLE I
Parameters for the four components of η Orionis, assuming that $(m - M)_0 = 7.58$

Component	m_v	$M/M_\odot$	$R/R_\odot$	M_v
Aa	4.44	13.9	7.0	−3.45
Ab	5.08	11.9	5.2	−2.8
Ac	4.98	13.4[a]		−2.9
B	4.98			−2.9

[a] Tentative value, assuming coplanarity of orbits and provisional spectroscopic elements.

The distance can in principle be determined in a more direct way from a comparison of the speckle data and the spectroscopic results for the system AabAc. Determined this way, the distance modulus turns out to be 8.41 mag. There is thus a clear discrepancy here, which we attribute mainly to the inaccuracy of the spectroscopic orbit. Indeed, the most significant second period in Žižka and Beardsley's (1981) velocity data turns out not to be the cited value of 9.219 years, but rather 9.51 years. With such a period, the discrepancy essentially disappears and more recent speckle data (McAlister and De Gioia, 1979) are better accounted for.

4. Concluding Remarks

According to McAllister (1985), η Orionis is 'the most massive system for which masses and luminosities have been determined directly, without reference to a temperature calibration'. Unfortunately, the current data are not precise enough for achieving the precise parameter determination that this remarkable system deserves. We urge that new interest be devoted to η Orionis by both spectroscopists and speckle observers. With the high spatial precision currently attainable, even the study of the visual period (estimated period about 1000 years) may be within reach.

Acknowledgement

This project was supported by the Belgian Fund for Collective Fundamental Research, under project 2.0119.83.

References

Chambliss, C. R.: 1978, *Inf. Bull. Var. Stars*, No. 1398.
Koch, R. H.: 1978, *Inf. Bull. Var. Stars*, No. 1418.
Koch, R. H., Hrivnak, B. J., and Bradstreet, D. H.: 1980, *Bull. Am. Astron. Soc.* **12**, 452.
McAlister, H. A.: 1976, *Publ. Astron. Soc. Pacific* **88**, 957.
McAlister, H. A.: 1985, *Ann. Rev. Astron. Astrophys.* **23**, 59.
McAlister, H. A. and DeGioia, K. A.: 1979, *Astrophys. J.* **228**, 483.
Waelkens, C., Lampens, P.: 1988, *Astron. Astrophys.* (in press).
Žižka, E. R., Beardsley, W. R.: 1981, *Astron. J.* **86**, 1944.

ABT – I hope that your final system distance will place the system within the Orion Association because it is a member of the Association.

WAELKENS – The distance modulus we derived is 7.58 may, i.e., at the lower limit of the most recently determined value for the distance to the association, which is 8.03 ± 0.46.

WORLEY – I am very sorry that H. A. McAlister was unable to attend this meeting. As I understand he and his colleagues have devloped an algorithm and hardware which will allow resolution of the quadrant ambiguity in speckle observations.

In the last decade, there have been a surprising number of naked-eye stars resolved as binary or multiple systems. These include Betelgeuse (which may be triple), and others such as η and θ Vir, ε Lyr, and β Cyg. I hope that, when you complete your study of η Ori, you might consider studying the nearby system δ Ori, which has recently been resolved as a $0\farcs25$ visual pair. It was discovered to be an eclipsing binary by Stebbins in 1912, and has more distant companions. It is a multiple system deserving attention!

MAZEH – Which of the stars is the pulsating one?

WAELKENS – From the photometry, it seems that the short-period variable is not one of the eclipsing components, but one of the redder stars Ac or B. The arguments are: (a) the amplitude of the short variation increases during eclipse, (b) the integrated color is redder at maximum of the short-period variation.

However, we are not sure that the short-period photometric variation is due to pulsation, and there is *spectral* evidence that the eclipsed component Ab is a pulsating star!

VALLS-GABAUD – Which are the individual masses you have obtained?

WAELKENS – We find masses of 14 and 12 solar masses for Aa and Ab. The mass of Ac may be about 13 solar masses.

A SEARCH AND CLASSIFICATION OF MULTIPLE SYSTEMS IN THE GENERAL FIELD

J. P. ANOSOVA*

Astronomical Observatory, Leningrad State University, Leningrad, U.S.S.R.

Summary. The general problems are formulated, and the classification of the multiple star and galaxy systems is proposed. The dynamical states of eight nearby triple stars are ascertained.

1. Introduction. The General Problems

The general problems in the study of multiple systems of objects (stars, galaxies, etc.) are the following ones: (1) the discovery of multiple systems in the general galactic or metagalactic field or inside star or galaxy clusters; (2) the subdivision of the discovered multiples into chance (optical) groups of objects and non-chance ones; (3) the education of the systems with components connected physically; (4) the division of such multiple systems into ones with components connected dynamically and ones without such connection; (5) the division of the systems with connected components into stable and unstable ones; (6) the study of the dynamical evolution of the unstable systems: the estimations of their life-time; the analysis of the processes of formation, evolution, and disruption of a smaller multiplicity subsystems; the tracing of the trajectories of the components' relative motions, etc. At present, such a study can be carried out by means of numerical methods.

One must solve the above formulated problems according to the indicated sequence. The necessary condition for solving these problems is the availability of the three-dimensional coordinates and velocities, and the relative masses for all components in a multiple system, i.e., the availability of the data for these objects obtained by the complex of astrometrical and astrophysical observations. The problems enumerated require diverse accuracy levels in obtaining the data observed for their solution, besides that, the requirements to the accuracy level grow from problem (1) to problem (6). The complete solution of all problems enumerated above for multiple stars is possible only at the highest accuracy level attainable at present for the observed data obtained by the ground observations as well as by the space observatories.

2. Classification of Multiple Systems

The multiple systems of objects (stars, galaxies, their groups or clusters, etc.) with any number of components might be divided into three basic types from dynamics' point of view:

(1) the multiple systems in which all or a few components are by chance in the system

* Presented by S. Vereshchagin.

Astrophysics and Space Science **142** (1988) 103–108.

region of the phase space; for these systems, the criterion of isolation from the background objects in the phase space is not fulfilled; one may qualify such systems as chance groups of objects or optical systems;

(II) the multiple systems which are not such chance groups; they are isolated from the background objects in the phase space; their components have some common features; however, the gravitating forces between components are small or approximately equal to the regular forces of the environment; for these systems, the criterion of isolation from the background objects in the coordinate space is not fulfilled; one may qualify such systems as non-chance groups of objects. One might divide these systems into two classes

(IIa) the systems which are completely isolated in the velocity space (e.g., moving star clusters);

(IIb) the systems which are partially isolated in the velocity space (e.g., in one or two of velocity components, like some of the well-known O. Eggen's kinematical groups);

(III) the multiple systems for which the criteria of isolation from the background objects are fulfilled in the coordinate space; in these systems the irregular gravitating forces between their components which are n on their total number essentially exceed the regular forces of the environment; one may qualify such systems as multiples with a physical connection between components.

One might divide the systems of this type into two classes:

(III-1) the multiple systems in which there is no dynamical connection among all their components or a part of them; their total energy is non-negative $E \geq 0$. One might also divide such systems into two subclasses:

(III-1a) the multiple systems in which there is no genetic connection amongst all their components or a part of them: all those n objects in system which are single rapidly pass by each other or a number $s(s < n)$ of single components and subsystems inside this system rapidly pass by each other;

(III-1b) the multiple systems in which there is a genetic community of components but also there is a 'recession' of components ($E > 0$) – e.g., according to the well-known conception of V. A. Ambartsumian;

(III-2) the multiples in which there is a dynamical connection between components, their total energy is negative $E < 0$; one might also divide such systems into two subclasses:

(III-2a) the multiple systems in which there is no genetic community inside any part of components: $s(s < n)$ of single components and subsystems with a smaller multiplicity inside a system slowly ($E < 0$) pass by each other;

(III-2b) the multiples in which there is a genetic community amongst all components (a co-formation), such systems may belong to one of the two dynamical types:

(III–I)* dynamically stable multiples;

(III–II)* dynamically unstable ones.

The proposed classification of multiple systems might be clearly shown by a block-scheme (Table I).

TABLE I

A classification of multiple systems – a block-scheme

A classification of the state of the multiple systems

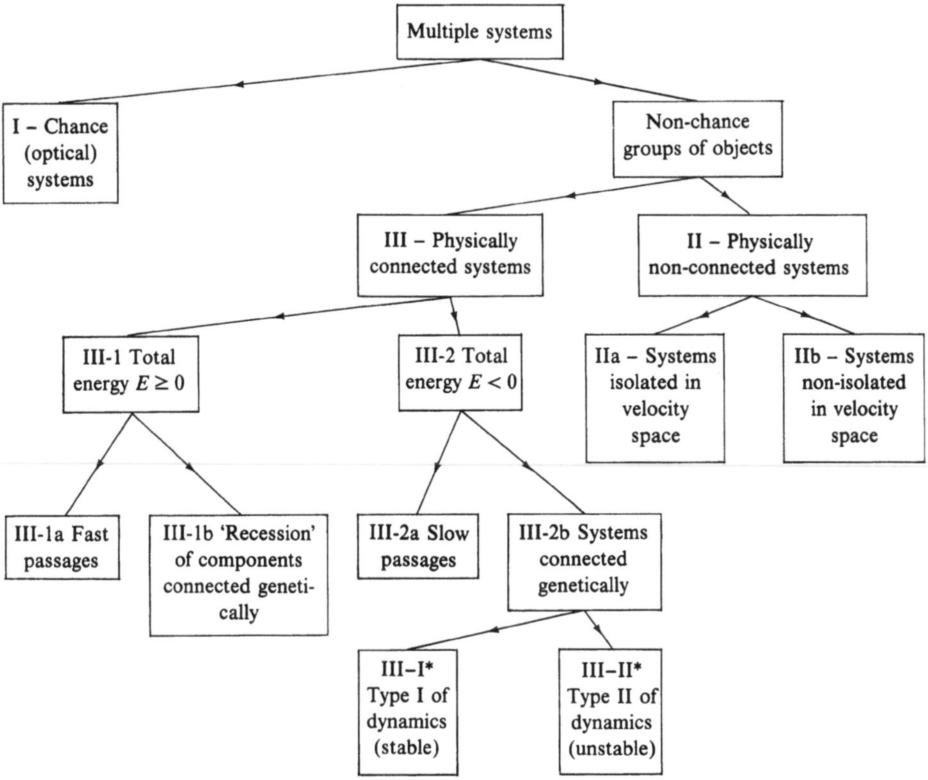

3. Criteria of Classification

In order to recognize the systems of type I – the optical systems or chance groups of objects) among the observed multiple stars or galaxies one might use a statistical criterion (see Anosova, 1987) which consist in obtaining a probability P of this system being such one, and in estimating an expectation EX of the number of such chance groups in some phase volume Σ. One assumes for a calculation of P and EX that the objects under study are randomly distributed inside the Σ. Generally speaking, this criterion may be generalized for every distribution of the phase density of objects. If the values P and EX obtained are too big ($P \sim 1$, $EX \sim N/n$, where N/n is the maximum number of the systems with multiplicity n in the volume Σ), then this system is a probably chance system. Otherwise ($P \sim 0$, $EX \leq 1$), such a multiple system is a probably non-chance group of objects and it belongs to the multiples of types II or III.

In order to recognize the systems of types II and III impartially one might use an analogous statistical criterion of isolation of the multiple systems in the coordinate space (see Anosova, 1987). If such a criterion is not fulfilled, a system belongs to type II; otherwise, this multiple system is one with components connected physically.

4. Application of Statistical Criterion to the Multiple Stars

In the Index catalogue of double and multiple stars (Jeffers *et al.*, 1963), there are 20% wide multiple systems which have large angular separations $2' \lesssim \rho \lesssim 100'$ between components. Amongst these triple stars, one selects seven systems located at shorter distances from the Sun than $R = 20$ pc and having the total complex of observed data for all their components. Using the proposed statistical criterion, one estimates for the multiple stars under consideration the probability P that component C, that is the most distant one from the primary component A, will be by chance in its environment, and the expectation EX of the number of such optical systems inside the sphere of radius $R = \max(r_A, r_C)$ with the following parameters of the stellar field (see, e.g., Wielen, 1974): the radius $R = 20$ pc, the mean local star number density $\nu = 0.12$ stars in 1 pc^3, the mean velocity 20 km s^{-1}, and the peculiar velocity standard 20 km s^{-1}. The results of calculations are the following ones:

ADS	α Cen	20390 S3142	10058	11853	48	6175 (Castor)
ρ'	1.3×10^2	77	9.4	6.9	5.5	1.2
R (pc)	1	20	700	50	15	14
P	7.0×10^{-8}	7.6×10^{-8}	0.30	2.6×10^{-5}	4.0×10^{-10}	8.0×10^{-10}
EX	1.1×10^{-4}	1.3×10^{-4}	5.6×10^5	0.68	2.6×10^{-7}	1.3×10^{-6}

It appears that the first two multiples, and the last two ones are certainly physical systems; ADS 10058 is a probably optical one; the star ADS 11853 is on optical quadruple one, it is a projection of two binaries.

5. Classification of Dynamical States of the Certainly Physical Multiple Systems

The main characteristics of dynamical states in the certainly physical multiples are the values of its total energy E and the relative energies $E_{ij} (i, j = 1, 2, \ldots, n; i \neq j)$ amongst every possible combinations of components. The sign of energies E and E_{ij} determine the existence or absence of dynamical connection between the components and, correspondingly, the classes of the states of multiples.

In order to recognize the two types of dynamics in multiple systems (stable or unstable) with the components connected dynamically one must use the criteria for stability of the systems consisting of N gravitating bodies developed in celestial mechanics (e.g., see Timoshkova and Kholshevnikov, 1982; Anosova, 1985).

In order to estimate the dynamical state of a star or galaxy system observed one must take into account some uncertainties in the data observed, i.e., take into account the errors in an initial state of this system.

In this paper, in order to take into account this effect one propose to use the method of statistical tests – a method of variation of the data observed in the limits of their uncertainties for the components of multiple stars. For every system, one might make N tests varying all parameters (or some of them in turns) in order to define every time the new quantities for these parameters and to calculate from them the energetic characteristics E, E_{ij} which determine every time the class of the dynamical state of this multiple system. For a large number of tests ($N \sim 10^3$–10^4), the frequency of realization of the classes of dynamical states determines the probabilities of these states for a multiple system under consideration.

This paper investigates eight triple stars with the necessary information observed for their components: ADS 1630, 2926, 6175, 6650, 6811, 7114, 9626, and 9909. The results of using the above method of simultaneous variations of all parameters observed within the limits of their errors are the following ones:

Name	ADS	r (pc)	Probabilities			$E \pm \sigma_E$	$E_{bin} \pm \sigma_{E_{bin}}$
			$(E>0)$	$(E<0)$			
				stable	unstable		
γ And	1630	80	0.00	0.53	0.47	-117 ± 25	-104 ± 19
α Gem	6175	14	0.02	0.12	0.86	-18 ± 10	-17 ± 10
ζ Sco	9909	22	0.00	0.20	0.80	-53.1 ± 8.6	-44.3 ± 7.5
ζ Cnc	6650	19	0.29	0.71		-12 ± 26	-31 ± 17
μ Boo	9026	29	0.11	0.89		-10.0 ± 8.4	-19.6 ± 5.4
24 Cnc	6811	50	0.30	0.70		-50 ± 140	-90 ± 110
ι UMa	7114	16	0.20	0.80		-4.6 ± 4.5	-2.6 ± 2.6
–	2926	100	0.97	0.03		$+23 \pm 23$	$+15 \pm 23$

where the unit system: the distance unit 0.01 pc, the mass unit 2×10^{33} g, the time unit 10^8 s is adopted; there are the probabilities of dynamical states of these systems, the mean quantities of their energetic characteristics – the total energy E and energy E_{bin} of a close pair and the mean quantities σ_E and $\sigma_{E_{bin}}$ for the boundaries of the confidence intervals at level 1σ for these values are given.

One may formulate the results obtained as follows:

(1) The triple stars ADS 1630, 6175, and 9909 located, respectively, at distances of 80, 14, and 22 pc from the Sun are certainly dynamically connected systems with the probability 0.98, the variations of their energies are 0.2–0.6.

(2) The other triples have large variations of the total energy 1–3. Therefore, one cannot obtain for them values of the total energies which are reliable enough values to

qualify their dynamical states. That is the reason why the problem of physical and dynamical connection between components in these star systems may be solved only after a closer definition of the observed data on the components.

References

Anosova, J. P.: 1985, *Totals Sci. Techn. Ser. Astrophysics* **26**, 57.

Anosova, J. P.: 1987, *Astrofizika* **27**, 3, 555.

Jeffers, H., van den Bos, W., and Greeby, F.: 1963, *Index Catalogue of Visual Double Stars*, Univ. Calif., Publ. Lick Observ., No. 21.

Timoshkoya, E. I. and Kholshevnikov, K. V.: 1982, *Totals Sci. Techn. Astrophys.* **20**, 3.

Wielen, R.: 1974, *Veröffentl. Astron. Rechen-Inst. Heidelberg* **A86**, 395.

FOURTH SESSION

OBSERVATIONS OF WIDE PAIRS AND THEIR RESULTS

Chairman: Z. KOPAL

VISUAL BINARY SEPARATIONS AS FUNCTIONS OF PRIMARY TYPES, AGES, AND LOCATIONS

Introductory Lecture

H. A. ABT

Kitt Peak National Observatory, Tucson, Arizona, U.S.A.

Summary. We studied samples from the *Bright Star Catalogue* of B1–B7, F3–G2, and KM dwarfs for frequencies of visual companions and their separations. The frequency of detected companions does not vary significantly along the Main Sequence. But the most frequent separations and the limiting separations vary markedly along the main sequence: the limiting separations are 42000 AU for the B stars to 800 AU for the M dwarfs. This confirms results found by Öpik in 1924. Stars off the Main Sequence (giants, white dwarfs) show consistent results, as do members of Trapezium systems, hierarchical multiple-star systems, the widest known cataloged binaries, and binaries in a moving cluster (Ursa Major Stream).

We propose a scenario in which binaries in open clusters can have large separations, but after they leave the cluster environments, chance encounters with field stars and molecular clouds will disrupt all but the more compact systems. The comoving pairs discussed by Luyten, Eggen, and others are seen as members of disrupting clusters and still retaining the common cluster motions. However, the velocity dispersions within clusters ensure that the comoving pairs have similar, but not identical, motions, whereas members of truly bound binaries would have motions that are indistinguishable. Appropriate measurements should distinguish between wide binaries and the comoving pairs, which can have separations up to the outer diameters or more of open clusters.

1. Introduction

This study is to determine how the maximum separations of double- and multiple-star systems depend upon the masses or ages of the primary stars and upon their locations. We will look at published data for stars on the Main Sequence (B stars, FG dwarfs, and KM dwarfs) and stars off the Main Sequence (giants, white dwarfs). These data will be shown to give a consistent picture of the decrease of maximum separations with age (or mass). We show confirmatory results from Trapezium systems, from a new study of wide field double stars, and from binaries in moving groups. Finally we summarize these results in terms of a suggested evolutionary model for binaries and comoving pairs.

In this discussion we will consider only visual binaries and common-proper-motion companions. The former are defined as pairs with known orbital elements or evidence of orbital motion; the latter, which are usually wider pairs, show only common motions because the probable binary periods exceed the duration of observations by at least several orders of magnitude. We will not include spectroscopic or interferometric binaries because we are interested mainly in maximum separations. That is, we ask the question: are there upper limits to binary separations and are they functions of the ages of the binaries?

A byproduct of this study is whether visual binaries show significant differences in frequencies, depending on the nature of the primaries. Among spectroscopic binaries some stars nearly or always have companions (e.g., common novae, dwarf novae,

Astrophysics and Space Science **142** (1988) 111–121.

metallic-line stars) and some rarely have companions (e.g., Be stars, high-velocity dwarfs, giants). Does such a diversity occur for visual binaries?

2. Solar-Type Dwarfs

We start with the brighter solar-type stars (F3–G2 IV or V) because they have been well observed. We considered those with $V < 5.5$ mag and $\delta > -20°$ in the new *Bright Star Catalogue* (Hoffleit and Jaschek, 1982). These include 90 primaries. The frequencies of separations of their visual companions are shown in Figure 1, grouped in bins of factors

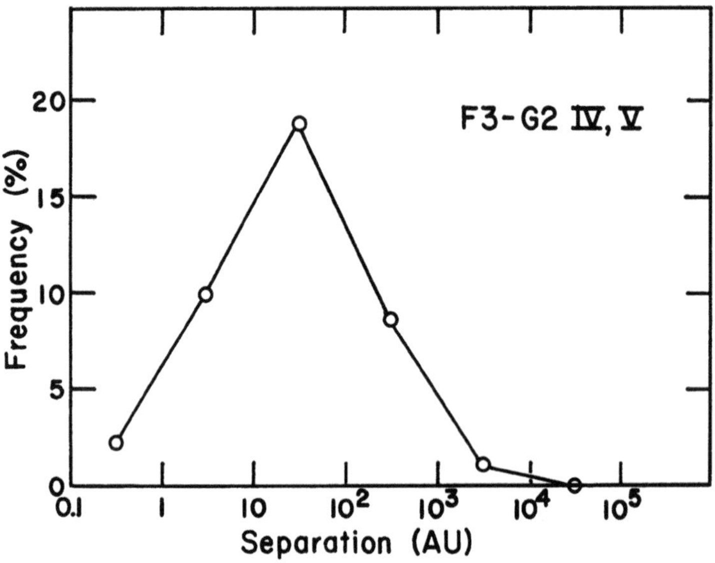

Fig. 1. For a sample of 90 northern solar-type stars brighter than $V = 5.5$ mag, we used the data in the *Bright Star Catalogue* (Hoffleit and Jaschek, 1982) to determine the frequencies of various separations for the visual companions. The errors are about 25% of the ordinates. The most frequent separation is 40 AU and the largest separation is 1090 AU.

of 10 in separation. Note the peak frequency at 40 AU; Kuiper (1935) obtained a similar result. But note that there is an upper limit near 10^3 AU; there is only one binary (τ^1 Hya = HR 3759) in this sample with a larger separation, but that separation is only 1090 AU.

3. B-Type Stars

Now let us consider a younger sample (or more massive stars): namely, the B1–B7 III, IV, or V stars in the new *Bright Star Catalogue*. We selected all such stars to the catalog limit but with $\delta > -30°$. More significantly, we limited the stars to those not labelled as being in open clusters or associations. That is an important criterion because in clusters it is difficult to know for what separations the binaries stop and we are simply including other cluster members. The sample includes 191 primaries.

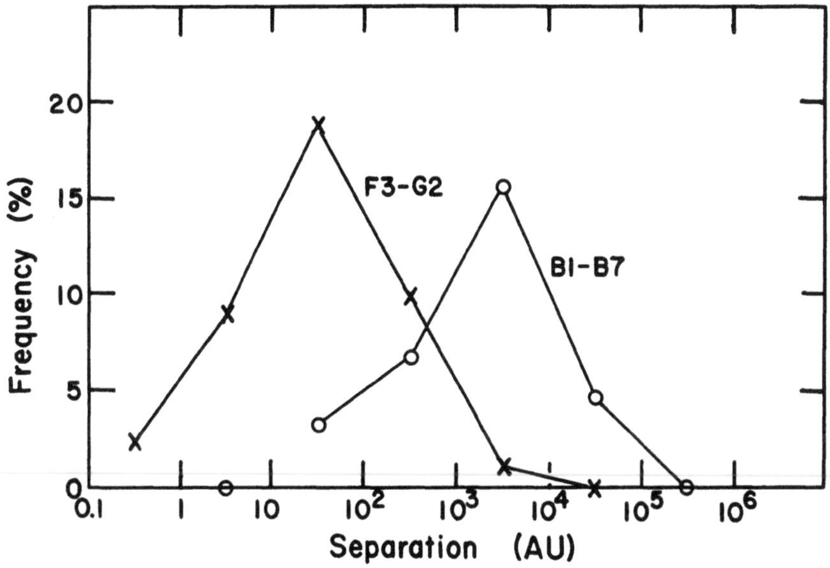

Fig. 2. A sample of 191 northern B1–B7 III, IV, or V stars in the *Bright Star Catalogue* is studied for the frequencies of separations (circles) of their visual companions. These stars are not members of known clusters or associations. The peak is at 5100 AU and the maximum separation is 42000 AU. The results for the F3–G2 dwarfs (Xs) are taken from Figure 1.

The frequencies of various separations are shown in Figure 2, compared with the data for the F3–G2 stars taken from Figure 1. There are large differences. First, the peak occurs at 5100 AU, rather than the 40 AU for the F3–G2 stars. At first this might seem to be an observational selection effect in that the B stars are more distant than the FG stars, so that a given angular separation will correspond to a larger linear separation (in AU) for the B stars. In fact, the average distances of the B and FG stars are 282 and 22 pc, respectively. But for the peak FG stars the average separations are 2″.8 and for the B stars 23″.4. The B stars in the next smaller bin (10^2–10^3 AU) have a mean separation of 1″.4, which is easy to distinguish, so the peak is not an observational selection effect. Thus whereas the B stars are 10 times more distant and one would expect them to be in visually more compact systems, the opposite is true: they are less compact by a factor of 10 in mean angular separations or 100 in linear separations.

It is very difficult to avoid the conclusion that B stars occur in systems with larger separations than the FG stars.

Secondly, Figure 2 shows that the upper limit to B-star binaries is about 10^5 AU, or actually 42000 AU, which is a factor of 40 greater than for FG dwarfs.

Third, note that the areas under the two curves are nearly the same, so the total frequencies of visual binaries, namely 30 ± 4% for the B stars and 41 ± 7% for the FG dwarfs, does not vary significantly between the two groups.

4. Late-Type Dwarfs

Let us discuss stars at the other end of the Main Sequence: namely, the K and M dwarfs.

For these we can use the data by Kuiper (1935). He showed that a sample of 132 K dwarfs has a maximum separations of 8000 AU while a sample of 59 M dwarfs has a maximum separation of 800 AU. His sample of A stars has a maximum separation of 60 000 AU. Thus we see a general progression of maximum separations from 42 000 AU for B stars to 800 AU for M dwarfs.

Somewhat surprisingly, similar results had been derived by Öpik (1924)! He found that (without allowance for interstellar absorption) the maximum separations of binaries varied from 112 000 AU for O stars to 1700 AU for GK dwarfs.

Kuiper's frequency of companions to the K dwarfs is 26%, which is only slightly less than the 41% for the F3–G2 dwarfs and 30% for the B1–B7 stars.

5. Giants

Let us consider stars off the Main Sequence. We choose a sample of 161 normal G, K, and M stars with classifications by Keenan and Pitts (1980) and luminosity classes

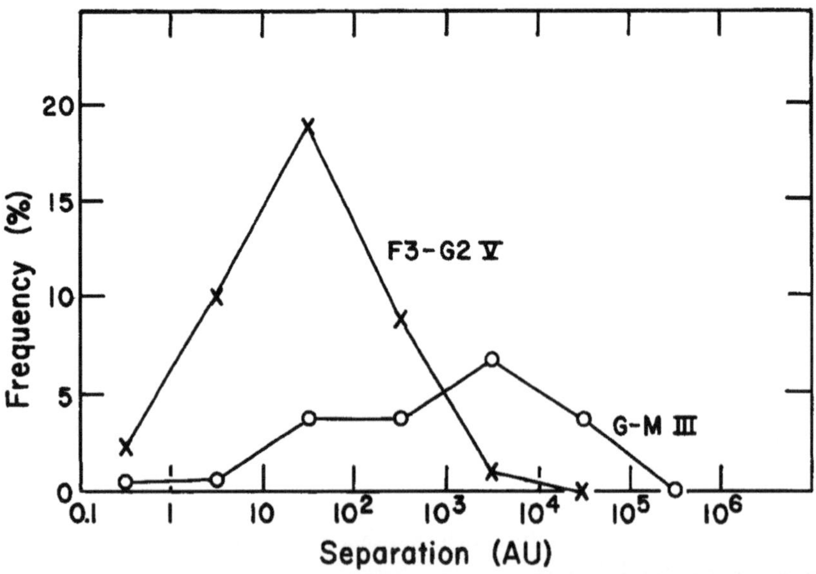

Fig. 3. For a sample of 161 normal G, K, and M giants classified by Keenan and Pitts, the frequencies of separations of the visual companions are shown, compared with the results of Figure 1 for solar-type dwarfs. The curve for the giants is shifted as one would expect for their predecessors of type about A0V.

of II–III, III, or III–IV. All the stars are listed in the *Bright Star Catalogue* and our data on companions that are likely to be physical ones are taken from that catalog and other standard sources.

The frequencies of various binary separations are shown in Figure 3. These stars evolved from Main-Sequence stars of about A0, so it is not surprising that the peak frequency at 3000 AU is nearly the same as the 5100 AU for the B1–B7 stars and the maximum separation of 25000 AU is nearly that (42000 AU) of the B1–B7 stars. Thus the progenators of these giants fit into the progressiosn along the Main Sequence given in the previous sections.

Finally, note that the area under the curve for the giants is less than for the FG dwarfs, indicating total frequencies of 19 and 41%, respectively. That difference is easily understood because the giant primaries brightened visually by 1.5 mag since they left the Main Sequence, so the fainter Main-Sequence companions are more difficult to detect.

6. White Dwarfs

Greenstein (1986) has recently surveyed the visual binaries that include white dwarfs. He found a deficiency of visual binaries, even at separations of 1–30″, but a maximum separation of 10^4 AU. Such a separation is not inconsistent with that of the intermediate Main-Sequence stars from which they originated.

7. Intermediate Conclusions

From Sections 2–6 we derive the following two conclusions:

(1) The total frequency of detected visual companions does not vary significantly along the Main Sequence around an average of about 32%.

(2) There is a steep dependence of maximum double-star separations along the main sequence from about 42000 AU for B primaries to about 800 AU for M dwarfs.

In the next three sections we add confirmatory data for the second conclusion from other studies. In the last two sections we will discuss the apparent conflict between the second conclusion and the existance of much wider comoving pairs.

8. Trapezium Systems

A published study (Abt, 1986) of Trapezium systems showed them to be invariably relatively young ($< 10^{8.9}$ yr). Also, both the Trapezium and hierarchical systems have maximum dimensions that decrease rapidly with increasing primary types. Figure 4 shows those data with a curve that conservatively represents the upper limits.

The weakness of that study is that the upper separation limits might represent catalog limits, rather than physical ones. Conservative catalogers, such as Aitken (1932), left out wide doubles on the grounds that too many of them are merely optical pairs. For instance, at $V = 7.5$ mag he adopted a limit of 20″; for B0V stars that corresponds to

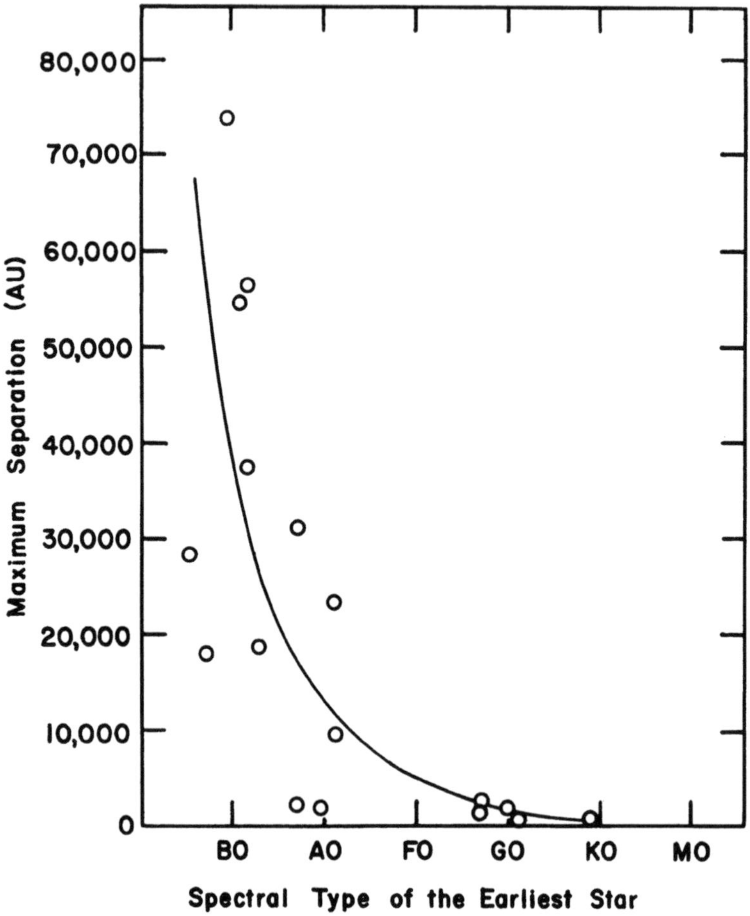

Fig. 4. A figure taken from Abt (1986) that shows how the separations from the primaries of the outermost physical companions in Trapezium and hierarchical systems differ for Main-Sequence primaries of various types. The curve is a conservative representation of the upper envelope. (Reprinted from the *Astrophysical Journal.*)

a separation of 35 000 AU while for a G0V star the limit corresponds to 800 AU. Thus we need a new study that is free of this weakness. Such a study is described in the next section.

9. Wide Field Visual Binaries

In a new study, to be published in full elsewhere, I studied a sample of 141 primaries with 147 companions listed in the Aitken catalog and with separations in excess of 10″. Furthermore, the companions show common proper motions or orbital motions during observing intervals of at least 30 years. Most of these companions are third stars that violate Aitken's limits; that is, if the primary has a close companion that falls within his separation limits, he often included the published data on third stars that violated his limits. Thus for the 77 companions discussed below, the average separations were 48″ and seven exceeded 100″.

In this study it was found to be important to have photoelectric photometry of the components, not so much to determine the reddening as to have good magnitudes. The magnitudes of the fainter ($V > 9$) components in visual systems are based upon BD and CD magnitudes that are systematically listed to be too bright by 0.5–1.7 pag at $V = 10$–13 mag and with a typical scatter of 0.7 mag (Abt, 1986).

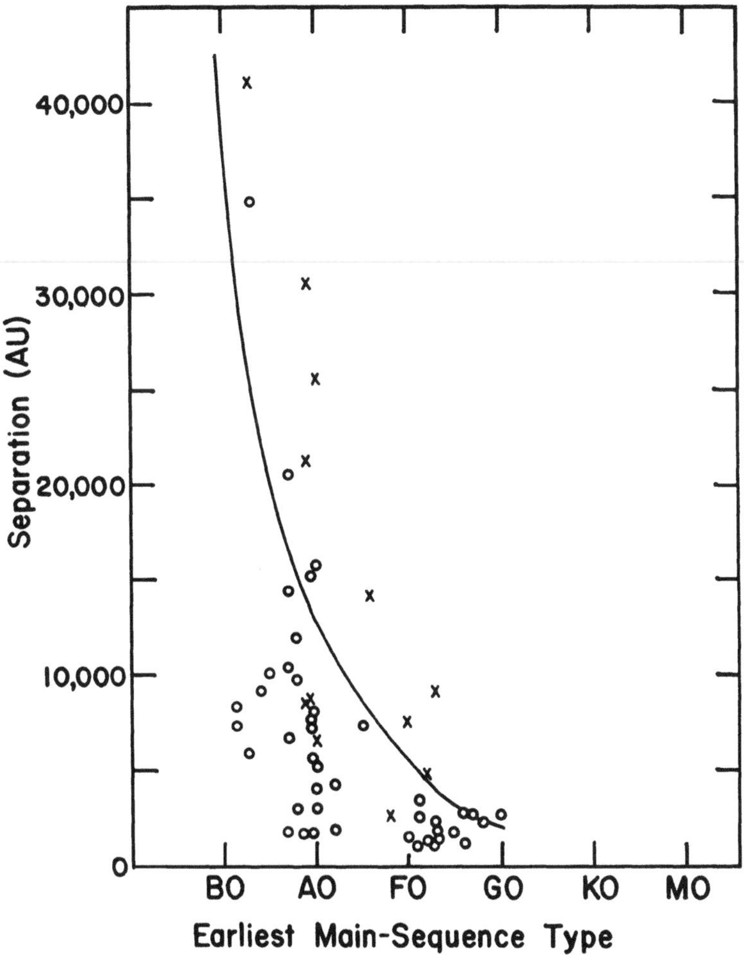

Fig. 5. The linear separations of distant (> 10″) companions to primaries of various spectral types. The physical companions are represented by circles and the optical ones by Xs, plotted as though they had the same distances as the primaries. The curve is taken from Figure 4. Only those companions more distant than 1000 AU are plotted.

I classified the 288 components and from spectroscopic distance moduli for the 77 companions with photoelectric photometry, found 63 to be physical companions and 14 to be optical ones. Figure 5 shows the companions with separations in excess of 1000 AU. The open circles represent the physical companions and the Xs the optical ones, plotted as if they were at the distances of their primaries. The curve is taken from Figure 4.

We see in Figure 5 that all the companions far above the conservatively-drawn line are optical ones. That means that the limits represented roughly by the curve are physical ones, rather than catalog limits.

Thus the real upper limits for double- and multiple-star separations are approximately 30000 AU at B5, 10000 AU at A5, 3500 AU at F5, and less than 3000 AU at G0. These numbers are consistent with the limits quoted in Sections 2–6, and with Öpik's (1924) results if some allowance is made in his data for interstellar absorption.

10. Binaries in Moving Groups

We now start to face the apparent discrepancy between the upper limits (as a function of age or primary mass) quoted above for cataloged double- and multiple-star systems and the evidence of comoving pairs given by Vandervoort (1968), Lü and Upgren (1973), Eggen (1986), Luyten (1987), and others. The comoving pairs often have separations of 1–20 pc (2×10^5–4×10^6 AU), which far exceeds the separation limits stated above. We suspect that the comoving pairs are related to the moving groups discussed in many

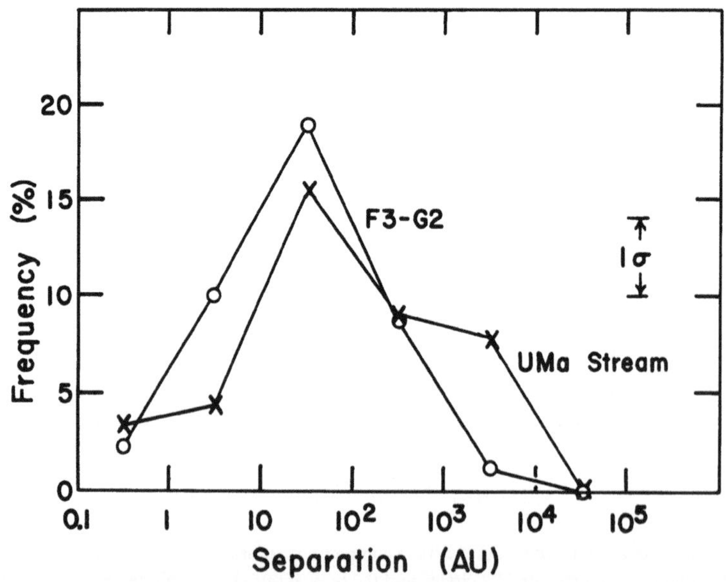

Fig. 6. The frequencies of various separations (Xs) of the visual companions to 89 primaries in the Ursa Major Stream. For comparison the data (circles) on field solar-type stars are taken from Figure 1.

papers by Eggen (e.g., 1965). Therefore, let us discuss first the visual binaries found in one moving group: the Ursa Major Stream.

Although the Ursa Major Stream was recognized more than a century ago, the modern major studies include Roman (1949) and Eggen (1950). We assumed the member identifications given in the *Bright Star Catalogue* and its evidence for visual duplicity for 89 primaries of types A0V to K2V. The frequencies of various binary separations are shown in Figure 6, compared with the data for solar-type field stars taken from Figure 1. The similar positions of the peaks is due to the heavy overlap in types. The extension to larger separations (maximum of 4400 AU) is due to the inclusion of the A-type stars in the Ursa Major Stream. The areas under the curves are similar, indicating similar visual binary frequencies, namely 41% and 40% for the F3–G2 field stars and A0–K2 Ursa Major Stream stars, respectively.

We, therefore, see no significant differences between visual binaries in the field and in a moving group.

11. Discussion

In the inner parts of open clusters the average separations of stars is about 0.9 pc (180000 AU) and there are occasional exchanges of distant companions among primaries. When a cluster is disrupted because it no longer has a negative energy, the binaries in it no longer have as frequent chances to acquire additional companions from other cluster stars. In fact, the more distant companions will be pulled away in chance encounters with field stars and, especially, with molecular clouds, so the gradual decrease of maximum separations with time starts. Even if a multiple-star system were left alone by neighbouring stars and molecular clouds, the gradual evolution of such a system involves close pairs becoming closer and the distant companions leaving. Theoretical calculations by Ambartsumian (1954), Heggie (1975, 1977), Retterer and King (1982), and others of the effects of random encounters with field stars predict maximum separations after several billion years of about 10000 AU. This general picture seems to explain the observed decrease (Figures 4 and 5) is maximum separations with average age of the primaries.

Moving clusters seem to be open clusters in the process of being disrupted. For a while they may retain a nucleus cluster, as does the Ursa Major Stream with its Ursa Major Cluster, but with time the members stars lose their proximity and easy identification with each other. However, during this time some of the cluster members retain common cluster motions. Although the cores of well-known clusters (Pleiades, Hyades, Coma, Praesepe) tend to have diameters of about 5 pc, they all have outlying members, so cluster members (even before disruption of the clusters) can be 10–20 pc apart. The stars in a cluster have similar – but not identical – space motions; the velocity dispersions in open clusters are of the order of 1 km s^{-1} (Mathieu *et al.*, 1986). Thus we note in Eggen's careful study, presented at this meeting, of five sets of comoving stars, their proper motions are similar but differ by about 0$''$.02 yr^{-1} or more than the measuring errors in proper motions. We thus identify the comoving pairs discussed by Vandervoort and others as individual members of clusters undergoing disruption.

How can we distinguish between bound members of binary systems and comoving members of partially-disrupted clusters? Members of wide bound binaries should have identical proper motions because if their separations are, say, 10 000 AU or more, the orbital motion is less than 1 km s^{-1}; at a distance of, say, 100 pc the orbital motions corresponds to a difference in proper motion of less than $0\overset{''}{.}001$ yr^{-1}. Thus wide bound binaries should have differences in radial velocities or proper motions that are less than our current measuring errors, whereas the members of comoving pairs may have differences of several km s^{-1} or several $0\overset{''}{.}01$ yr^{-1}. It would be interesting to make radial-velocity measurements on suitable (sharp-lined) binaries and comoving pairs to check this distinction.

But both bound binaries and members of comoving pairs are useful because they generally share common origins and, therefore, can be used, as we do with cluster members, to check evolutionary models.

References

Abt, H. A.: 1986, *Astrophys. J.* **304**, 688.

Aitken, R. G.: 1932, *New General Catalogue of Double Stars Within 120° of the North Pole*, Carnegie Institution of Washington, Publ. No. 417.

Ambartsumian, V. A.: 1954, *Byurakan Ap. Obs. Contr.* **15**, 1.

Eggen, O. J.: 1950, *Astrophys. J.* **111**, 414.

Eggen, O. J.: 1965, in A. Blaauw and M. Schmidt (eds.), *Stars and Stellar System*. V. *Galactic Structure*, Chicago Univ. Press, Chicago, p. 111.

Eggen, O. J.: 1986, *Astron. J.* **92**, 125.

Greenstein, J. L.: 1986, *Astron. J.* **92**, 859, 867.

Heggie, D. L.: 1975, *Monthly Notices Roy. Astron. Soc.* **173**, 729.

Heggie, D. L.: 1977, *Rev. Mexicana Astron. Astrophys.* **3**, 169.

Hoffleit, D. and Jaschek, C.: 1982, *The Bright Star Catalogue*, 4th rev. ed., Yale Univ. Press, New Haven.

Keenan, P. C. and Pitts, R. E.: 1980, *Astrophys. J. Suppl.* **42**, 541.

Kuiper, G. P.: 1935, *Publ. Astron. Soc. Pacific* **47**, 15, 121.

Lü, P. K. and Upgren, A. R.: 1973, *Astrophys. J.* **180**, 91.

Luyten, W. J.: 1987, *Astrophys. Space Sci.* **142**, 17 (this issue).

Mathieu, R. D., Latham, D. W., Griffin, R. F., and Gunn, J. E.: 1986, *Astron. J.* **92**, 1100.

Öpik, E.: 1924, *Publ. Tartu Univ. Obs.* **25**, No. 6.

Retterer, J. M. and King, I. R.: 1982, *Astrophys. J.* **254**, 214.

Roman, N. G.: 1949, *Astrophys. J.* **110**, 205.

Vandervoort, P. O.: 1968, *Astrophys. J.* **152**, 895.

WEINBERG – First, thank you, Dr Abt, for an excellent talk. I have several comments and a question. I have a warning about the remarks section of the *Yale Bright Star Catalog*: by comparing the identifications in YBS with the cluster membership lists by Mermilliod, I have found that the identifications in the remarks section to be incomplete. Also, in my work with Shapiro and Wasserman, we found that initially powerlaw distributions in separation evolve to new powerlaw distributions. Since it seems unlikely that a given spectral type forms binaries with a particular separation, we should be careful about assigning the given spectral type with a unique disruption lifetime. Finally, since the ages of B stars are short compared to dissolution time for the clusters in which they are presumably born, could your distribution for the B stars reflect the location of star formation and not the distribution of true binaries?

ABT – I agree with your first comment: it is a lazy technique to take data from the *Bright Star Catalogue*, rather than to go back to the original publications. Actually I checked the comments in the BSC with the IDS and some other sources, rather than accept them blindly.

I also agree with your second point. I was only suggesting upper limits for bound systems for each Main-Sequence type.

If your third point is correct, why are there so many B stars outside of clusters and associations?

OBLAK – From a study by Dworak on eclipsing binaries published in *Acta Astronomica*, in 1985, it results that the luminosity class IV appears only for separations between around 40 and 80 $R_{\odot}$. The author concludes that there should be a greater number of binaries in field stars or components of visual double stars for the luminosity class IV than for the others. His work concerns all spectral types.

Can you give a confirmation of this result from your statistical study?

ABT – I was concerned primarily with the largest separations in visual pairs, so relatively small separations would provide no conflict.

WEIS – Your conclusions concerning M-dwarfs were based on the data of Kuiper which are half a century old and are dominated by what you define as visual binaries. A more modern compilation, such as Luyten's NLTT catalogue tends to be dominated by what you define as C.P.M. pairs, and these give a very different picture regarding the maximum separations of M-dwarf pairs. A significant fraction ($\sim 5\%$) of M-dwarf C.P.M. pairs appear to have separations exceeding 5000 AU.

ABT – One of the central problems for this conference is to understand the apparent discrepancy between real upper limits to the separations of bound double stars and the existence of much wider co-moving pairs. We need to determine the limits of real binaries without confusing them with stars having similar (but not exactly the same) motions because they originated from the same cluster.

POVEDA – I find quite interesting and reassuring that such a different set of stars leads to a limiting separation of solar type binaries similar to the one we found for the sample of nearby binaries in Gliese's catalogue, i.e., $\cong 1000$–1500 AU.

WASSERMAN – (1) What do you assume for the initial distributions of semi-major axes for the different spectral types?

(2) Does tidal shearing (e.g., differential rotation) play any role in your view in separating unbound, comoving pairs?

(By the way, Dr Abt, if they are B-stars they may not live long enough (~ 10 yr)).

ABT – (1) They were not assumed, but rather derived.

(2) There are various mechanisms (close encounters, molecular clouds, etc.) to disrupt comoving pairs. I hope that you will study that problem and give us the answers.

KOPAL – I wonder whether you would care to classify the characteristics of the curves you exhibited on the screen by stellar populations?

ABT – Although I do not find that the frequencies of visual binaries are sensitive to the nature of primaries among Population I stars, that does not seem to hold for Population II stars. High-velocity dwarfs seem to be deficient in both spectroscopic and visual binaries. That can be understood in that in globular clusters the binaries should sink toward their central regions and the single stars are preferentially ejected.

PHOTOELECTRIC OBSERVATIONS OF CPM STARS IN THE *BVRI* SYSTEM*

K. D. RAKOS

Institute for Astronomy, Vienna, Austria

and

O. G. FRANZ

Lowell Observatory, Flagstaff, Arizona, U.S.A.

Summary. A photometric program to estimate the frequency of close unresolved companions in visual binary systems has been developed. The sensitivity of the detection depends on the accuracy of the differential photometric measurements of binary components. Unresolved companions up to seven magnitudes fainter than the primary are detectable in the *BVRI* photometric system.

1. Introduction

Modern theories of star formation predict that multiplicity should be the natural configuration in which stars are found. The observed binary frequency along the main sequence is one of the interesting parameters useful in obtaining the luminosity distribution function for multiple star systems. This function is very poorly known, and the results of several investigations – due to extreme selection effects – are still controversial.

Recently, visual binaries have been used to estimate the luminosity distribution of close unresolved companions in a binary system (Rakos, 1985). The most probable mass ratio $q = Mb/Ma$ found was 0.5. Comparable results have been published for spectroscopic binaries by Abt and Levy (1976, 1978). According to their calculations, for biaries with periods less than 100 years the number of companions with 1.2 solar masses corresponds to the van Rhijn function. The number of companions of only 0.6 solar masses drops significantly in comparison to this function. Trimble (1978) claims a bimodal distribution with peaks near $q = 0.2$–0.3 and 0.9–1.0.

Visual binaries of interest have been usually selected according to apparent magnitude and separation. Such selection introduces systematic errors (Lutz and Kelker, 1973).

Using common-proper-motion stars selected according to the ratio between angular separation and proper motion will provide a more complete statistical sample according to the volume of space (Halbwachs, 1986).

Three-hundred and twenty-six CPM binaries have been selected in such a way from the AGK3 catalogue. The proportion of optical pairs in the sample is estimated to be 1%.

* Presented by K. D. Rakos.

Astrophysics and Space Science **142** (1988) 123–129.

2. Observations

We have started a program of accurate photometric observations in the *BVRI* Cousins System for these stars to estimate the luminosity distribution function for possible third companions.

An extension of the photometric observations into the infrared would, of course, be even more suitable for the detection of companions of very low temperature, but the availability of large telescopes for such a program is very unlikely.

The *BVRI* system can be used at small telescopes with an accuracy sufficient to allow detection of close companions with magnitude differences as large as 7 mag. The *I*-filter has its maximum transmission at 8300 Å, corresponding to the maximum in the radiation curve of a blackbody at 3500 K (M2V star). Therefore, the colour index $(V - I)$ will be a very good indicator of the excess radiation caused by an invisible companion. In this way we will be able to reach the 'bottom' of the Main Sequence and to get the luminosity distribution of stars fainter than $M_v = 14$. In the Figure 1 we can

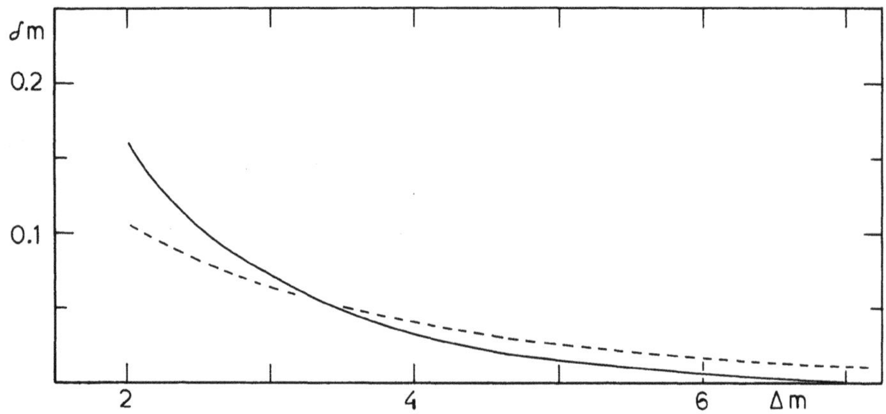

Fig. 1. The solid line is the difference (δm) of the combined magnitude and the magnitude without an unresolved companion as a function of the brightness difference (Δm) between companion and the component A. The dashed line is the colour excess (δm) in the $(V - I)$ index as a function of Δm.

see the colour excess in the $(V - I)$ index as a function of the Δm, the brightness difference between one component of the visual binary and its invisible companion (dashed line). The solid line is the difference of the combined magnitude and the magnitude of the component alone.

Table I shows an example for three different companions, a, b, and c, of a component A of the visual pair A, B.

Suppose the component B is a single Main-Sequence star and its colours $(B - V)$ and $(V - I)$ correspond to the absolute magnitude of this component. The magnitude difference ΔV between the components gives the difference of the absolute magnitude

TABLE I

	a	b	c
		In solar masses	
Mass	0.6	0.4	0.21
Spectral type	K7V	M2V	M4.5V
M_v	8.25	10.90	13.50
$(B - V)$	1.32	1.53	1.74
$(V - I)$	1.53	2.19	3.03

of A and the colour indices according to the (M_v, colour index) diagram. The deviation of the measured colour from that predicted from the magnitude difference ΔV is the colour excess. It can be used to estimate M_v and other parameters of the unresolved companion.

TABLE II

	A	A + a	A + b	A + c
			Composite	
Spectral type	K2	K2 + K7	K2 + M2	K2 + M4.5
M_v	6.30	6.13	6.28	6.30
$(B - V)$	0.88	0.94	0.89	0.88
$(V - I)$	0.93	1.04	0.96	0.94

According to Table II we can expect to be able to detect the companion c only if the accuracy in the measured brightness difference and colours is very high, on the order of 0.001 mag. For differential photometry this is a high but achievable goal.

For visual pairs closer than 12 arc sec, the solid-state area scanner will be used (Rakos, 1981) on the 1.5 m telescope at the Institute for Astronomy in Vienna. The observations consist usually of 50 scans recorded individually in the memory of the computer for each filter. The integration time can be selected and is on the order of 1 second. In Figure 2 we show two scans for ADS 8119 made under different seeing conditions. Better seeing means higher resolution and higher peaks in the image profiles. The scans are first sorted according to seeing and then reduced in the usual way (Rakos *et al.*, 1982). Table III shows the results for the first ten scans sorted according to seeing conditions. Three-decimal accuracy can be achieved even during poor photometric nights.

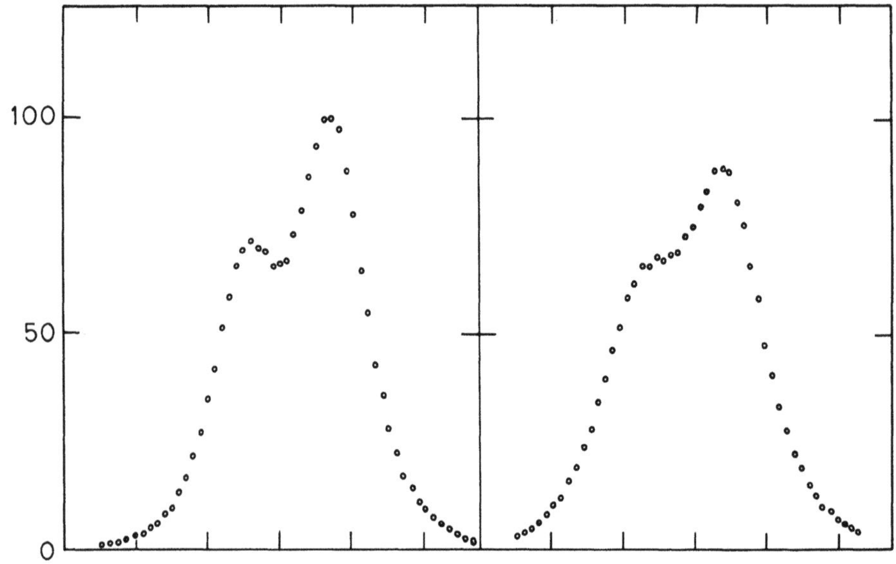

Fig. 2. Two scans for ADS 8119 under different seeing conditions. The first scan has a height of 100, the second only of 89.

TABLE III

Scan height	Separation[a] (arc sec)	Δm (mag)
100	2.649	0.387
98	2.646	0.388
97	2.646	0.392
94	2.646	0.388
94	2.646	0.387
93	2.642	0.387
91	2.649	0.393
91	2.651	0.385
89	2.651	0.385
89	2.642	0.394

[a] Relative values based upon approximate scale.

3. Results

In September and October 1986 we observed at Lowell Observatory, Flagstaff Arizona, 73 of the selected sample of CPM stars with separations greater than 12 arc sec. Their magnitudes and colours, transformed to the *BVRI* system, are presented in Table IV.

TABLE IV

AGK3	HD/BD	ADS/IDS	$\alpha_{(1950)}$	δ	V	$(B-V)$	$(V-R)$	$(R-I)$
+66 0015	+65 0014		$00^h09^m14^s$	66°04.′5	10.06	0.38	0.30	0.29
+66 0014	+65 0012		00 09 13	66 05.0	10.85	0.56	0.30	0.32
+15 0058	+15 0094		00 34 12	15 48.8	10.10	0.58	0.35	0.33
+15 0057	+15 0094		00 34 11	15 49.3	11.30	0.73	0.50	0.35
+04 0137	6480	903B	01 03 16	4 38.6	7.23	0.41	0.26	0.29
+11 0131	8127		01 18 10	11 11.8	8.14	0.57	0.35	0.33
+11 0132	+10 0167		01 18 14	11 11.7	10.05	0.80	0.43	0.41
+31 0132	8624	1134C	01 22 50	31 17.2	8.23	0.66	0.37	0.37
+31 0130	8610	1134A	01 22 45	31 17.5	8.27	0.65	0.34	0.35
+07 0162	8949		01 25 45	7 42.2	6.25	1.12	0.61	0.54
+07 0163	8956		01 25 50	7 42.0	8.00	0.52	0.28	0.38
+09 0155	10352	1328A	01 38 40	9 51.7	8.79	0.49	0.28	0.30
+09 0154	10351	1328B	01 38 39	9 51.9	8.92	0.53	0.28	0.31
+28 0257	14082	1752A	02 14 31	28 30.9	7.03	0.48	0.27	0.29
+28 0256	+28 0382	1752B	02 14 30	28 30.7	7.75	0.59	0.34	0.34
+15 0205	14786		02 20 35	15 11.4	8.84	0.58	0.32	0.34
+15 0204	14772		02 20 32	15 11.3	9.35	0.67	0.36	0.37
+24 0227	16246	1982A	02 34 07	24 25.8	6.50	0.37	0.22	0.23
+24 0226	16232	1982B	02 34 04	24 25.9	7.10	0.46	0.26	0.29
+07 0353	20430	2451AB	03 14 46	7 28.4	7.37	0.52	0.29	0.31
+07 0354	20439		03 14 52	7 30.5	7.74	0.58	0.30	0.32
+53 0486	37394		05 37 17	53 27.7	6.23	0.85	0.46	0.44
+53 0487	233153		05 37 27	53 28.3	9.87	1.46	0.97	1.05
+47 1214	153525	10288C	16 56 19	47 26.1	7.91	0.91	0.53	0.54
+47 1216	153557	10288A	16 56 30	47 26.3	8.06	0.95	0.52	0.50
+06 2083	159481	10638A	17 32 22	6 03.4	7.42	0.51	0.28	0.25
+06 2084	+06 3456	01638B	17 32 22	6 03.8	10.55	0.86	0.45	0.42
+27 1700	163970	17538+2751A	17 55 47	27 51.2	8.90	0.57	0.30	0.33
+27 1699	+27 2922	17538+2751B	17 55 47	27 51.0	9.90	0.97	0.50	0.45
+28 1759	167215	18091+2813A	18 11 01	28 14.5	8.15	0.51	0.29	0.31
+28 1761	167216	18091+2813B	18 11 04	28 13.8	8.28	0.50	0.29	0.32
+21 1797	341868	18101+2126A	18 12 16	21 26.2	9.19	0.88	0.49	0.42
+21 1798	341869	18101+2126B	18 12 17	21 26.1	9.77	0.64	0.33	0.31
+45 1487	176071	18535+4522A	18 54 54	45 26.4	8.40	0.49	0.26	0.31
+45 1488	+45 2798	18535+4522B	18 55 14	45 28.4	9.34	0.57	0.31	0.27
+34 1826	180286	12215A	19 12 39	34 28.5	7.25	0.43	0.20	0.23
+34 1825	+34 3475	12215BC	19 12 37	34 28.2	10.10	1.10	0.63	0.60
+50 1407	186408	12815A	19 40 29	50 24.5	6.00	0.65	0.36	0.33
+50 1408	186427	12815B	19 40 32	50 24.0	6.25	0.66	0.36	0.34
+33 1831	187013	12913A	19 44 32	33 36.6	4.96	0.47	0.27	0.26
+33 1832	225732	12913B	19 44 34	33 36.7	8.58	1.02	0.61	0.52
+37 1855	226250	19484+3746A	19 50 12	37 54.0	9.60	0.67	0.37	0.33
+37 1856	37 3651	19484+3746B	19 50 13	37 59.9	9.78	0.73	0.37	0.35
+06 2688	192344	20093+0618A	20 11 43	6 26.2	7.75	0.69	0.36	0.38
+06 2687	192343	20093+0618B	20 11 42	6 25.5	8.04	0.67	0.33	0.33
+06 2691	192623		20 13 10	6 05.5	9.26	0.85	0.44	0.43
+06 2690	+05 4460		20 13 02	6 04.9	10.45	0.63	0.38	0.37
−02 1184[a]	194766	13868B	20 24 51	−2 17.0	7.48	0.49	0.31	0.28
+14 2238	195631		20 29 41	14 06.3	8.39	0.64	0.30	0.27
+14 2237	+13 4427		20 29 41	14 05.1	10.13	0.74	0.34	0.38
+30 2374	+30 4470	21290+3031A	21 31 11	30 44.3	10.04	0.66	0.40	0.32

Table IV (continued)

AGK3	HD/BD	ADS/IDS	$\alpha_{(1950)}$	δ	V	$(B-V)$	$(V-R)$	$(R-I)$
+30 2373	+30 4470	21290+3031B	21h31m07s	30°44′.2	10.61	0.63	0.34	0.36
+00 2712	205525	21308+0014A	21 33 22	0 27.3	8.76	0.60	0.31	0.32
+00 2713	+00 4751	21308+0014B	21 33 23	0 27.2	9.98	0.83	0.45	0.42
+06 2957	205811	15147A	21 35 14	6 23.6	6.17	0.05	0.01	0.02
+06 2956	+05 4829	15147B	21 35 14	6 24.2	7.68	0.35	0.20	0.25
+18 2249	210275		22 06 40	18 14.5	8.53	0.36	0.22	0.24
+18 2248	+17 4697		22 06 35	18 14.6	10.46	0.80	0.45	0.44
+37 2207	211797	15828A	22 16 44	37 31.1	6.14	0.26	0.15	0.17
+37 2206	+37 4537	15828B	22 16 44	37 30.8	8.78	0.54	0.31	0.35
+56 1594	214238	16097A	22 33 43	56 36.1	7.63	0.67	0.38	0.36
+56 1595	+56 2814	16097B	22 33 45	56 36.3	9.74	0.78	0.44	0.42
+67 1041	216606	22490+6727A	22 50 56	67 43.4	6.93	0.47	0.20	0.22
+38 2387	+37 4734	16402A	22 55 27	38 24.4	9.71	0.46	0.39	0.23
+38 2388	+37 4734	16402A	22 55 27	38 24.5	10.34	0.56	0.32	0.32
+47 2022	218739	16557A	23 07 41	47 41.3	7.20	0.63	0.36	0.32
+47 2021	218738	16557B	23 07 39	47 41.2	7.93	0.90	0.49	0.48
+45 2134	220445	16720A	23 21 06	45 31.1	8.61	0.92	0.51	0.46
+45 2135	+44 4400	16720B	23 21 07	45 30.9	9.38	1.03	0.62	0.54
+16 2620	+15 4865		23 38 42	16 30.4	9.76	0.58	0.29	0.33
+16 2619	+15 4865		23 38 41	16 30.8	10.19	0.62	0.31	0.33
+10 3290	224994	17180AB	23 59 52	10 30.0	8.50	0.53	0.30	0.31
+10 3289	224984		23 59 48	10 30.5	8.70	0.57	0.30	0.32

[a] AGK2.

The observations were carried out with the 24-inch Morgan reflector. Twelve standard stars were observed for reduction and calibration purposes. The mean extinction coefficients were found to be:

$$k(V) = 0.24\,, \qquad k(B-V) = 0.08\,, \qquad k(V-R) = 0.06\,, \qquad k(R-I) = 0.05\,.$$

The coefficients of the colour transformation equations,

$$(B-V)_0 = b0 + b1(B-V)_{\text{observed}}\,,$$

$$(V-R)_0 = c0 + c1(V-R)_{\text{observed}}\,,$$

$$(R-I)_0 = d0 + d1(R-I)_{\text{observed}}\,,$$

remained constant for all nights. Their values are:

$$b0 = -0.407 \pm 0.005\,, \qquad b1 = 0.995 \pm 0.003\,,$$

$$c0 = -0.065 \pm 0.003\,, \qquad c1 = 0.992 \pm 0.005\,,$$

$$d0 = 0.901 \pm 0.006\,, \qquad d1 = 0.820 \pm 0.006\,.$$

The coefficients $a0$ and $a1$ in the equation $V - V_0 = a0 + a1(B-V)_0$ are slightly variable from night to night, mainly because of occasional cirrus clouds.

The accuracy of the observations was unfortunately influenced by bad weather and is on the order of 0.01 or 0.02 mag throughout Table IV. We will repeat and continue these observations in September and October (usually the months with the best observing weather in Flagstaff) of 1987. We expect to complete this program in two or three years.

References

Abt, H. A. and Levy, S. G.: 1976, *Astrophys. J. Suppl.* **30**, 273.
Abt, H. A. and Levy, S. G.: 1978, *Astrophys. J. Suppl.* **36**, 241.
Halbwachs, J. L.: 1986, *Astron. Astrophys. Suppl. Ser.* **66**, 131.
Lutz, T. E. and Kelker, D. H.: 1973, *Publ. Astron. Soc. Pacific* **85**, 573.
Rakos, K. D.: 1981, *Lowell Obs. Bulletin* **IX** (No. 167), 220.
Rakos, K. D.: 1985, *Astrophys. Space Sci.* **110**, 143.
Rakos, K. D. *et al.*: 1982, *Astron. Astrophys. Suppl. Ser.* **47**, 221.
Trimble, V. L.: 1978, *Observatory* **98**, 163.

ABT – Do you generally find that members of these pairs indeed have common origins? That is, do they have common distances, ages, and other characteristics?

RAKOS – It seems that much more than 1% of these pairs are optical.

HALBWACHS – The proportion of optical systems among these CPM binaries was verified from radial velocities, for the pairs with ρ/μ less than 1000 years, I had considered all the measurements that were available, and I had not found any pair with a significant discrepancy between the RV of the components. For another subset with ρ/μ in the range 1000 to 3500 years, I had estimated that the proportion of optical systems should be 40%. I am now measuring the RV of these stars and the proportion of physical systems seems to be compatible with this estimation.

ON THE WIDE BINARIES IN SELECTED AREA 57[†] *

T. MAZEH

School of Physics and Astronomy, Tel Aviv University, Tel Aviv, Israel

and

D. W. LATHAM

Harvard-Smithsonian Center for Astrophysics, Cambridge, Mass., U.S.A.

Summary. We present some results from a preliminary analysis of a new radial-velocity survey of 244 stars in Selected Area 57, complete for stars with V brighter than 12.0 mag. We have considered all pairs with angular separations between 2 and 600 arc sec, but find only one probable physical binary with a separation larger than 100 arc sec. These data provide a stronger constraint on the distribution of wide binaries in the Galaxy than was previously available. Our results suggest that binaries with separations larger than 0.1 pc are rare.

1. Introduction

The seminal work of Bahcall and Soneira (1981) has stimulated considerable interest in the use of the bright stars in Selected Area 57 (SA57) to study the distribution of wide binaries in the Galaxy. In that work they investigated close pairs of stars in the sample of 244 stars with V brighter than 12.0 mag in the Weistrop (1972) catalog, centered on the North Galactic Pole. They found 19 pairs with angular separations less than 120 arc sec, a statistically significant excess compared to what would be expected from random accidental alignments. Therefore, they predicted that several of the pairs would prove to be physically associated. Using new radial-velocity measurements of the 36 member stars in the 19 pairs, Latham *et al.* (1984) concluded that six pairs were indeed physical binaries, with two additional pairs possibly being binaries, thus confirming the Bahcall and Soneira (1981) prediction. They concluded that all the other candidate pairs were not binaries, on the basis that the radial velocities of the two components differed by more than 1 km s^{-1} for each pair.

In this paper we report on the first results from a complete radial-velocity survey of 244 stars in the Bahcall and Soneira (1981) sample. The new data allow us to extend the analysis to both smaller and larger angular separations, from 2 to 600 arc sec. The extension to wider angular separations can improve our understanding of the distribution with separation of wide binaries in the Galaxy. This is of particular importance, as discussed recently by Bahcall *et al.* (1985) and Weinberg *et al.* (1987).

In this paper we describe briefly the observational survey, discuss some ambiguities which arise when radial-velocity data are used to determine which pairs are binaries, and present some very preliminary results on the wide-binary distribution, taking into account the selection effects pointed out recently by Wasserman and Weinberg (1987).

[†] The research reported here is based in part on observations made with the Multiple Mirror Telescope, a joint facility of the Smithsonian Institution and the University of Arizona.

* Presented by T. Mazeh.

Astrophysics and Space Science **142** (1988) 131–138.

2. Observations

A total of more than 3500 radial-velocity exposures were obtained for the 244 stars in the Weistrop catalog with V brighter than 12.0 mag. The instrument and techniques used are described by Latham (1985). Identical instruments were used on the 1.5-m Wyeth reflector at the Oak Ridge Observatory in Harvard, Massachusetts, and the 1.5-m Tillinghast reflector at the Fred L. Whipple Observatory on Mt. Hopkins in Arizona. For some of the observations a nearly identical system on the Multiple-Mirror telescope was used.

All stars in the sample were observed several times over a span of five years. In this way we could discover spectroscopic binaries with relatively long periods. The estimated error for an individual measurement is about 0.7 km s^{-1} for the late-type stars. As multiple observations were accumulated, we observed more frequently any star with an r.m.s. variation larger than 1 km s^{-1}, to facilitate searches for periodicities and to allow solutions for orbital elements. We estimate that our threshold for detection of spectroscopic binaries is an amplitude larger than about 1.5 km s^{-1}. For periods shorter than about two years our detection of spectroscopic binaries should be nearly complete for this amplitude threshold.

At our telescopes on nights of good seeing we detected close companions to three stars in the sample, using the intensified television viewing systems. We estimate that we have not missed any visual companions with angular separations larger than 2 arc sec, and within two or three magnitudes of the primary.

The detailed results of the survey will be published elsewhere (Latham *et al.*, 1989). We report here only on some implications for wide binaries based on the data taken through the 1986 season.

3. Can Radial Velocities be Used to Prove a Wide Pair is not Bound?

Radial velocities can provide a method for estimating whether or not a wide pair is a physical binary. The method is possible for our sample because the velocity dispersion here, roughly 23 km s^{-1} (Latham *et al.*, 1989), is much larger than the orbital velocity in a wide binary, which is normally expected to be considerably less than 1 km s^{-1}. Thus the probability that *any* two neighbouring stars in our sample will have the same radial velocity within 1 km s^{-1} by chance is small, but not zero, and the probability grows as larger separations and, therefore, larger numbers of neighbours are considered. Indeed, in their pioneering attempt to determine which pairs in SA57 are binaries, Latham *et al.* (1984) adopted 1 km s^{-1} as the threshold for the velocity difference above which a pair was presumed to be unbound. To see why this threshold was deemed reasonable, consider a binary with a circular orbit. The expected velocity difference is

$$\Delta V_{\text{cir}} = 0.3 \left\{\frac{M}{2\,M_{\odot}}\right\}^{1/2} \left\{\frac{\Re}{0.1\ \text{pc}}\right\}^{-1/2} \text{km s}^{-1},$$

where M is the total mass of the system, $M_{\odot}$ is the solar mass, and $\mathfrak{R}$ is the binary separation. For eccentric orbits the velocity difference can be somewhat larger, but can not exceed $\sqrt{2}\,\Delta V_{\text{cir}}$. In general the actual radial-velocity difference will be smaller than ΔV_{cir}, because of projection effects.

However, even in a bound wide pair the velocity difference can exceed 1 km s^{-1} if there are additional unseen companions to either of the stars. For example, consider a bound wide pair consisting of two stars A and B. Suppose that star A has a close unseen companion, *Ab*, with orbital period *P*. We denote, therefore, the primary star of A by *Aa*. For a circular orbit, the relative velocity of *Aa* and *Ab* is

$$\Delta V_{\text{cir}} = 6.4 \left\{\frac{M}{1\,M_{\odot}}\right\}^{1/3} \left\{\frac{P}{100\text{ yr}}\right\}^{-1/3} \text{km s}^{-1}\,.$$

The difference between the velocity of Aa, Va, and the velocity of the center of mass of the system Aa and Ab, Vab, is

$$Va - Vab = \Delta V_{\text{cir}}(Mb/M)\,,$$

where Mb is the mass of *Ab*. The *radial* velocity of *Aa* with respect to that of the center of mass of *Aa* and *Ab* is $Va - Vab$ multiplied by a projection factor. Therefore, For *P* of the order of 100 years, and Mb/M of the order of 0.25, *Va might* differ from the radial velocity of B by about 2 km s^{-1}, despite the fact that A (which includes *Aa* and *Ab*) and B are physically bound.

We note that in the case discussed here it would be difficult to discover the real reason for the radial-velocity difference between A and B. For a system at distance *d*, the angular separation between *Aa* and *Ab* can be as large as

$$\Delta\theta = 0.2 \left\{\frac{M}{1\,M_{\odot}}\right\}^{1/3} \left\{\frac{P}{100\text{ yr}}\right\}^{2/3} \left\{\frac{d}{100\text{ pc}}\right\}^{-1} \text{arc sec}\,,$$

depending on the projection factor. Such a small separation is almost impossible to detect visually unless the system is nearby (though it might be detectable with modern speckle imaging or other interferometric techniques). If *Aa* and *Ab* are Main-Sequence stars, a mass ratio Ma/Mb greater than 3 would imply that *Ab* is more than 5 mag fainter than *Aa*. Thus, it would be very difficult to detect the signature of Ab in the combined spectrum.

We suggest, therefore, that a difference as large as 2 km s^{-1} between two stars of a wide pair is not sufficient to prove that the system is not bound. Although bound systems with this large a velocity difference are rare and should not affect drastically any statistical conclusion about a sample of stars, each specific system must be interpreted with caution. A special case, which was discovered in the course of our extended study of SA57, is discussed in the next section.

4. The System W11449/W11450

W11449 and W11450 appear as two stars separated by 8 arc sec, and are included as one of the 19 pairs in the Bahcall and Soneira (1981) list. Latham *et al.* (1984) reported a velocity difference of 0.5 km s^{-1} between these two stars, and suggested that they are physically bound. Subsequently we observed W11449 and W11450 intensively. Based on a much more extensive set of data, we now find the radial velocity of W11449 to be -2.22 ± 0.07 km s^{-1}, and that of W11450 to be -0.20 ± 0.09 km s^{-1}. Thus, with a difference of 2.02 ± 0.12 km s^{-1}, it is very unlikely that W11449 and W11450 can be bound, unless there are additional stars in the system.

A close examination of the correlation plots for the various spectra of W11450 revealed a weak secondary peak that was occasionally separated enough from the primary peak to be detected, thus indicating the presence of another star in the system. After dozens of spectra had been obtained, we were able to find an orbital solution for the periodic modulation of the secondary, with a period of 3.345 ± 0.25 days and amplitude of 31.3 ± 0.3 km s^{-1}. Yet the primary peak showed no sign of any periodic modulation in velocity, with a two-sigma upper limit of 0.8 km s^{-1}! We concluded that W11450 could not simply be two stars in orbit around each other. Instead the fainter star responsible for the secondary peak is itself a single-lined spectroscopic binary, and this close pair is in a wider orbit around the bright star responsible for the primary peak. In turn this system of three stars, collectively appearing as W11450, is in a still wider orbit around the fourth star, W11449. Further support to this model is given by the relative velocities. Within the subsystem W11450, the bright star responsible for the primary peak, A in Figure 1, has a relative radial velocity of $+2.0$ km s^{-1}, while the

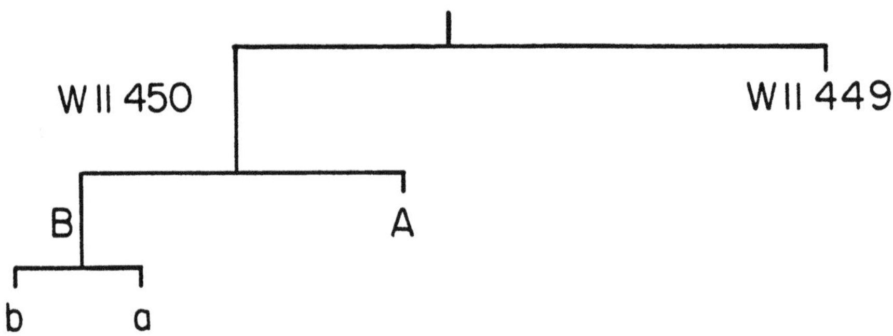

Fig. 1. The mobile diagram for the system W11449/W11450.

gamma velocity for the close pair, Ba-Bb, has a relative velocity of -2.4 km s^{-1}. The average of these two velocities, which is our best estimate for the velocity of the subsystem W11450, is close to the velocity of W11449. The reason that we were able to decipher this special case is because W11450Ba disclosed itself as a spectroscopic binary with a relatively large amplitude. Thus W11449/W11450 is an example of a complex system where a cursory examination of the radial velocities and a naive interpretation could lead to the incorrect conclusion that the stars are not physically associated.

Using the distance of 50 pc estimated for W11449/W11450 (Latham *et al.*, 1984) and adopting some reasonable assumptions about the orbits, we concluded privately that W11450Ba should be separated by about 0.1 arc sec from W11450A. Subsequently we discovered that this prediction had been confirmed many years ago, and that W11449/W11450 is also known as the triple system ADS 8811, with an observed visual separation 0.3 arc sec for the close pair (Aitken, 1932).

5. Preliminary Discussion of the Distribution of Wide Binaries in SA57

With our radial-velocity survey of a complete sample in SA57 in hand, we can now search for wide binaries over a large range of angular separations. We choose to consider all pairs of stars with separations smaller than 600 arc sec, deriving for each pair the observed radial-velocity difference and estimated error. For each of the candidates we calculated the probability that it is really bound, as follows. The real velocity difference of the pair was assumed to be within a range of radial velocities centered on the observationally derived velocity. The probability distribution of the velocity difference was assumed to be normal, with the observed standard deviation. For each value of the velocity difference we calculated the probability that the pair is bound using the Wasserman and Weinberg (1987) prescription, assuming the total mass of the system is two solar masses. The distance to each pair was estimated using the observed colours and assuming the stars to be on the Main Sequence. Finally, the total probability that the pair is bound was derived by integrating over the velocity-difference range, with the normal distribution as a weighting function. From all the possible candidates we choose as *probable* physical binaries only those pairs with probability to be bound larger than 25%. In Table I we list all the probable binaries, including the new relatively close pairs which were discovered visually. The table includes updated results for the probable binaries observed by Latham *et al.* (1984). The angular separations were calculated from

TABLE I

The probable binaries in SA57

		V_{rel} (km s^{-1})	Sep. (arc sec)	d (pc)	Sep. (pc)	P_{bound}
W12096N	W12096S	1.9 ± 1.2	2	225	0.002	0.35
W44496N	W44496S	0.9 ± 0.7	2	175	0.002	0.65
W41083N	W41083S	0.3 ± 0.4	5	275	0.007	0.75
W11449	W11450AB	0.2 ± 0.2	8	50	0.002	0.95
W50670	W50671	0.5 ± 0.4	19	225	0.02	0.40
W25138	W25140	0.6 ± 0.2	96	50	0.02	0.30
W23082	W23083	0.1 ± 0.4	25	300	0.04	0.50
W43227	W43228	0.4 ± 0.3	39	225	0.04	0.40
W22942	W23067	0.3 ± 0.5	490	50	0.1	0.25

the accurate positions (Latham *et al.*, 1989). We note that only one probable binary with angular separation larger than 120 arc sec was found, and in fact this pair has a low probability of being bound.

To estimate the shape of the distribution with separation of the wide binaries in the Galaxy, we compare the cumulative distribution of the probable binaries with various theoretical cumulative distribution functions, convolved with the selection effects of the sample. These effects are different from those of Latham *et al.* (1984) because our separations range between 2 and 600 arc sec, while the previous range was between 10 and 120 arc sec. Therefore, we recalculated the selection effects of the present sample, following Wasserman and Weinberg (1987). The preliminary results of our calculations are presented in Figure 2. We plot the observed cumulative distribution for *all* the

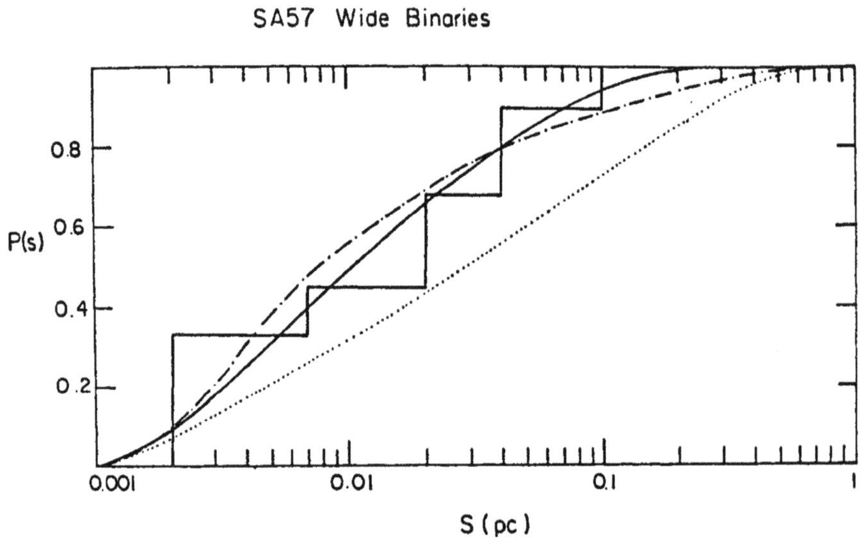

Fig. 2. Observed and theoretical cumulative distributions of wide binaries versus separation. The SA57 observations are plotted as a solid stepped line. The dotted curve is the theoretical distribution for a mild power-law dropoff ($\lambda = 1.0$), the dashed curve for a steeper power-law dropoff ($\lambda = 1.4$), and the solid curve for a power-law dropoff ($\lambda = 1.0$) with an exponential cutoff at 0.1 pc.

probable binaries of Table I, together with theoretical distributions using two different power-law dropoffs, $n(s) = s^{-\lambda}$, where s is the separation. We also plot a theoretical distribution for a power-law dropoff with an exponential cutoff at 0.1 pc, of the form

$$n(s)\,ds = s^{-\lambda} \exp^{-(s/0.1\ \mathrm{pc})}\,ds\,.$$

Of the three theoretical distributions, the one with the cutoff fits the observed distribution the best. The distribution with a mild power-law dropoff ($\lambda = 1.0$) gives the poorest fit.

The theoretical distributions plotted in Figure 2 are based on a very simple-minded *preliminary* treatment of the selection effects in our sample. In particular, the density of stars of different types as a function of height above the galactic plane and the percentage of wide binaries within different types of stars were only crudely estimated. Also, no corrections have been made for accidental pairs. We, therefore, regard the distributions plotted in Figure 2 as indicative only of what may be revealed by future analysis.

The two theoretical distributions which fit the data reasonably well in Figure 2 both imply that the number of very wide physical binaries, with orbital semi-major axes larger than 0.1 pc, is rather small. These distributions suggest that less than 5 to 10% of the binaries with semi-major axes larger than 0.001 pc are very wide. In contrast, the distribution derived using the mild power-law dropoff predicts a much larger fraction of very wide binaries, with about one third of all wide binaries being very wide. To confirm this suggestion, further analysis of our data will be carried out. In addition, proper motion observations could add valuable information in the identification of physical pairs.

As already noted by Weinberg *et al.* (1987), studies of wide binaries have an indirect bearing on the question of the existence of Nemesis, the purported distant companion of the Sun that has been proposed as being responsible for periodic mass extinctions. The apparent deficiency of binaries with orbits as wide as is needed for Nemesis, with semi-major axis of about 0.4 pc, implies that Nemesis would be a very rare and special case in the Galaxy.

Acknowledgements

We thank John Bahcall for bringing this research problem to our attention and for stimulating discussions and continuing encouragement. We have also benefited from discussions with Ira Wasserman and Martin Weinberg. Although this is not a data paper, the results reported here would not have been possible without the work of the team who helped with the observations and data reductions, especially Bob Davis, Bob Stefanik, Dick McCrosky, Susan Tokarz, Ed Horine, Jim Peters, and Skip Schwartz.

References

Aitken, R. G.: 1932, *New General Catalog of Double Stars within 120° of the North Pole*, Carnegie Inst. Washington, p. 749.

Bahcall, J. N. and Soneira, R. M.: 1981, *Astrophys. J.* **246**, 122.

Bahcall, J. N., Hut, P., and Tremaine, S.: 1985, *Astrophys. J.* **290**, 15.

Latham, D. W., 1985, in A. G. D. Phillip and D. W. Latham (eds.), 'Stellar Radial Velocities', *IAU Colloq.* **88**, 21.

Latham, D. W., Tonry, J., Bahcall, J. N., Soneira, R. M., and Schechter, P.: 1984, *Astrophys. J. Lett.* **281**, L41.

Latham, D. W., Mazeh, T., Davis, R. J., Stefanik, R. P., and McCrosky, R. E., 1989, in preparation.

Wasserman, I. and Weinberg, M. D.: 1987, *Astrophys. J.* **312**, 390.

Weinberg, M. D., Shapiro, S. L., and Wasserman, I.: 1987, *Astrophys. J.* **312**, 367.

Weistrop, D.: 1972, *Astron. J.* **77**, 366.

UPGREN – When you look toward a galactic pole, you are looking backward in time, in the sense that the average age of stars increases quickly with distance. Have you looked at the frequency of these pairs as a function of distance (from the galactic plane) and thus as a function of age? As your sample increases if you continue this work, you may find a variation with age.

MAZEH – The number of confirmed wide binaries in the sample is still small, and a detailed study is not possible.

STUDIES OF A SAMPLE OF CPM STARS

J.-L. HALBWACHS
Observatoire de Strasbourg, France

Summary. A sample of CPM stars selected from the AGK2/3 is used to investigate the following points: (a) What is the distribution of mass ratios of binaries with wide separations?, and (b) what is the proportion of wide binaries among stars. The distribution previously derived from the visual binaries of the *Bright Star Catalogue* seem to be an acceptable solution to question (a). On the basis of this, the proportion of stars that have a companion with a separation between 300 and 10000 AU is found to be about 13%. Another possibility is that the distribution of mass ratios is similar to the initial mass function derived by Miller and Scalo (1979) for stars lighter than the Sun. The proportion of wide binaries with mass ratios greater than 0.16 should then be around 17%.

1. Introduction

In a previous paper (Halbwachs, 1986b), a sample of 326 CPM pairs was selected from the AGK2/3 (Lacroute and Valbousquet, 1974), using the following criteria: (a) the proper motions of the stars were larger than 0.050″/y, (b) the components of each pair had proper motions compatible at the 5% level, and (c) the ratio between the apparent separations and the proper motions was less than 1000 years (in fact, this ratio refers to the time during which the displacement of a binary was equal to the separation between the components). The number of optical pairs was evaluated at around 4. These few unidentified systems have been neglected in this paper and the sample has been used in order to investigate the intrinsic properties of wide binaries.

2. The Selection Effects

Besides the aforementioned criteria used to discriminate between physical and optical pairs, other selection effects may come from the failure to detect stars under certain conditions.

2.1. Failure to detect faint stars

If we consider all the stars in the AGK2/3 that are brighter than a given photographic magnitude, m_{pg}, their number, $N(m_{pg})$, may be stated by the law:

$$\ln N(m_{pg}) = 1.135 m_{pg} + 0.4472\,. \tag{1}$$

However, this relation does not apply to stars fainter than 9.15 mag, a phenomenon that is certainly due to incomplete detection.

The probability of detecting a star with a given magnitude is the ratio between the true number of stars with the same magnitude and the theoretical number derived from Equation (1). The logarithm of this probability fits with six linear functions, the limits of which are given in Table I.

Astrophysics and Space Science **142** (1988) 139–143.

TABLE I

Probability of detecting a star with magnitude m_{pg}

m_{pg}	<9.15	9.6	10.4	11.0	11.5	12.0	12.5
$\log P(m_{pg})$	0	−0.16	−0.57	−1.06	−1.67	−2.49	−3.55

In order to correct this effect, we also need to know the probability of detecting the two components of a binary. Since the probability of detecting a single star probably depends on its position, only systems with primaries brighter than 9.15 mag should be retained in the sample. This way, the probability of detecting a system is the same as the probability of detecting the secondary components, as given in Table I.

2.2. Failure to Detect Close Binaries

The detectability of a close binary is related to the size of the images of the components: if the stars are bright, the images are large and overlapping. It appears from the distribution of the 326 systems in the plane apparent separation (ρ) versus magnitude, that the detection is complete when

$$\rho > R_1 + R_2, \quad \text{with} \quad R = (13 - \min(10, m_{pg}))/7 \tag{2}$$

(the indices 1 and 2 refer to the primary and the secondary components).

3. Calculation of the Visual Magnitudes

The photographic magnitude of the AGK2/3 must be translated into visual magnitude m_v in order to derive the distribution of the differences of magnitude between the components Δm_v. The difference between visual and photographic magnitude depends on the spectral type, which may be represented by the absolute magnitude of the star M_v assuming that it is on the Main-Sequence (MS). We find the following relations:

– for B9 to K0 type stars:

$$m_v = m_{pg} - 0.18125 M_v + 0.27125\,, \tag{3}$$

– for stars later than K0:

$$m_v = m_{pg} - 0.295 M_v + 0.931\,. \tag{4}$$

Stars earlier than B9-type are not numerous, and they follow a relation with a steep slope, making systems with primary components earlier than B9 unsuitable for the sample. Also, for systems with dwarf primaries later than K0, the secondaries are still later and it could be hazardous to extrapolate Equation (4). For that reason, only systems with B9 to K0-type primary components have been retained.

4. Selection of Systems with Dwarf Primary Components

The systems with components on or near the Main-Sequence are of particular interest because their mass ratios are related to Δm_v. Moreover, more luminous systems are probably not very numerous in the sample, due to the range in the proper motions that were considered. In order to reject systems with primaries far from the Main Sequence, we have computed their tangential velocities, assuming the primaries are dwarfs. We have kept only systems with tangential velocities in the range of 10 to 100km s^{-1}. This precaution should be sufficient for stars earlier than G-type. For G-type stars, the luminosity of giants is much greater than it is for dwarfs, and all giants must be rejected. But this difference is not great enough to allow for a selection based upon kinetic criteria. For that reason, all the systems with G-type primaries were rejected. In the case of K0 stars, it is possible to select dwarfs by using the H parameter defined by Luyten (1922). According to Ochsenbein and Egret (1981), all giants will be eliminated by keeping only stars with H greater than 10. After all of these selections, the sample contains 96 CPM binaries.

5. Generation of a Synthetic Sample

The AGK2/3 contains 5467 stars that are not secondary components of a binary systems and that fulfill the same conditions as the primary components of the 96 systems that remain in the sample. These stars were then considered to be true primary components, and synthetic binaries were generated by assuming that they obey a given distribution of Δm_v or of mass ratios q and a given distribution of separations s. The mass ratios were translated into Δm_v, using the mass-luminosity relation

$$\Delta m_v = -10 \log q . \tag{5}$$

The apparent separations and the secondary photographic magnitudes were derived, and the weight of each systems has been computed from Table I and from the secondary magnitude. Only systems that satisfy the same selection conditions as the systems of the true sample have been kept in the synthetic sample. The distributions of primary absolute magnitudes, primary photographic magnitudes, mass ratios, Δm_{pg} and apparent separations were derived for the synthetic sample and they are compared to the distributions of the true observed sample. In the case of incompatibility, another distribution of Δm_v or another distribution of s has to be used. If they are compatible, these two distributions probably refer to binaries in space, and the numbers of systems that contain the synthetic sample and the true sample may be used to derive the proportion of binaries among stars.

6. The Results

The binaries in the sample all have separations between 300 and 10000 AU. The distribution of $\log s$ may be constant on the range 2.5 to 4, but the slightly decreasing distribution found by Halbwachs (1986a) is also valid.

The distribution of Δm_v derived from MS visual binaries by Halbwachs (1986a) is acceptable at the 5% level. Assuming the distribution of $\log s$ found in the same study, the proportion of MS binaries with $\log s$ between 2.5 and 4 and with $\Delta m_v < 8$ mag is 13%.

The distribution of mass ratios derived from spectroscopic binaries by Halbwachs (1987) is rejected at the 1% level. This distribution contains too few systems with small mass ratios.

The distribution of mass ratios derived from Salpeter's initial mass function (1955), that is $q^{-2.35}$, provides a distribution of Δm_{pg} that is acceptable at the 10% level; but the distribution of primary photographic magnitudes is rejected at the 1% level since it contains too few systems with faint primaries. In a previous paper, Trimble (1988) investigated the same sample of 326 binaries. She found that the intrinsic distribution of mass ratios resembles $q^{-3.2}$. This is incompatible with our statistics, since it should provide a distribution of primary magnitudes even flater than that corresponding to Salpeter's function.

The distribution of mass ratios derived from the initial mass function found by Miller and Scalo (1979) for masses lighter than 1 solar mass, that is $q^{-1.4}$, is acceptable at the 10% level. If the distribution of $\log s$ is constant, then the proportion of stars with a companion with $q > 0.16$ and $2.5 < \log s < 4$ should be 17%. Another consequence of this result is that wide binaries may be obtained by a random combination of stars of the same age, as was initially suggested by Öpik (1924).

Acknowledgements

I would like to thank Dr H. Zinnecker for valuable discussions and Mme C. Holtzer for correcting the English. The travel expenses were supported by the 'Comité National Français d'Astronomie'. The subsistence support provided by the Local Organizing Committee on Belgian funds, is gratefully acknowledged.

References

Halbwachs, J. L.: 1986a, *Astron. Astrophys.* **168**, 161.
Halbwachs, J. L.: 1986b, *Astron. Astrophys. Suppl. Ser.* **66**, 131.
Halbwachs, J. L.: 1987, *Astron. Astrophys.* **183**, 234.
Lacroute, P. and Valbousquet, A.: 1974, *Astron. Astrophys. Suppl. Ser.* **16**, 345.
Luyten, W. J.: 1922, *Publ. Astron. Soc. Pacific* **34**, 156.
Miller, G. E. and Scalo, J. M.: 1979, *Astrophys. J. Suppl. Ser.* **41**, 513.
Ochsenbein, F. and Egret, D.: 1981, *Bull. Inf. C.D.S.* **20**, 60.
Öpik, E. J.: 1924, *Publ. Obs. Astron. Univ. Tartu* **XXV**, No. 6.
Salpeter, E. E.: 1955, *Astrophys. J.* **121**, 161.
Trimble, V.: 1988, *Astron. Nachr.*, preprint.

VALLS-GABAUD – I think you couldn't use one power law to represent the Miller-Scalo (1979) initial mass function, as it is divided into 3 subsets in mass.

It is just a first order approximation for the upper mass range.

HALBWACHS – The part of the IMF with exponent – 1.4 was derived for masses less than 1 solar mass. The primary components of my sample are on average A- or F-type stars, and they have effectively masses heavier than this limit, but the secondaries are generally below. I agree, however, that it should be more accurate to use all the IMF instead of a single part.

WIDE BINARIES AND OLD DISK POPULATION GROUPS*

O. J. EGGEN

*Cerro Tololo Inter-American Observatory, National Optical Astronomy Observatories, Cerro Tololo, Chile***

Summary. Distinguishing wide binaries from contiguous members of young disk population superclusters is not possible with present definitions. Members of old disk population groups have non-parallel space motions and the question, 'when is a wide binary' is more easily approached.

A previous discussion of young disk binaries indicated that with present definitions, wide binaries cannot be distinguished from contiguous members of young disk superclusters. In the old disk and halo populations, survivors of these superclusters only rendezvous in the solar neighbourhood if the V (direction of galactic rotation) vectors of the space motions have a dispersion of less than a km s^{-1}, although the U (away from the galactic centre) and W (North Galactic Pole) components may reflect the dispersion inflicted by whatever force disrupted the original parallel space motions of the supercluster members. Because parallel space motion is a reasonable definition of a wide binary, such systems can be isolated in the old disk population by a criterion more stringent than that for group members. The purpose of this discussion is to point out several possible double and multiple systems on the basis of (1) common proper motion and (2) the consistency of photometric luminosities obtained from 4 colour intermediate band and Hβ or $R - I$ photometry. The consistency of the binary label can also be tested by the metallicity (including CN) parameters but this assumes that such consistency is a real requirement. In most cases the missing element is the radial velocity of the individual components and it is hoped that these comments will prompt supply of this element.

1. System I

HD 62850, 63581A and B, 64185. The photometric parameters give a mean modulus of $3.00 \pm 0.19(\delta)$ mag and the stars are represented in Figure 1 by clear circles, compared with the Hyades main sequence (continuous curve). The proper motions for HD 62850 are improved GC values and those for HD 64185 are improved FK4 values. Both components of HD 63581 are in the *Sydney and Cape Zone Catalogues*. All proper motions discussed here are on the FK4 system with precessional corrections. The photometric values of the metallicity, P[Fe/H], give a mean of $+0.20 \pm 0.08(\delta)$ dex and the dispersion is the same as that found from discussions of members of such clusters as the Hyades. The values of C_m lead to ΔC_m (Hyades-System I) = $+0.010$ mag. The

* Presented by H. A. Abt.

** Operated by the Association of Universities for Research in Astronomy, Inc., under contract with the National Science Foundation.

Astrophysics and Space Science **142** (1988) 145–151.

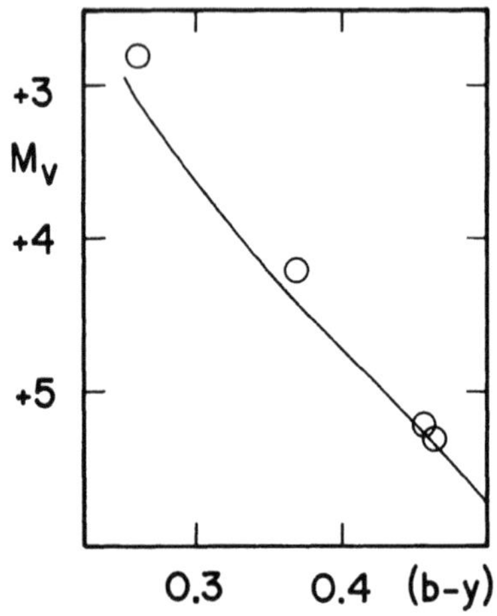

Fig. 1.

System I

HD	V	$B-Y$	M_1	C_1	R	$R-I$	Sp. T	C_M
62850	$7^{m}.21$	$0^{m}.368$	$0^{m}.191$	0.460	$\beta = 2.611$		G2/3 V	0.124
63581A	8.18	0.453	0.272	0.442	$7^{m}.84$	$0^{m}.256$	G8/K1 IV/V	0.152
63581B	8.30	0.457	0.289	0.437	7.95	0.261	G8/K1 IV/V	0.170
64185	5.79	0.258	0.168	0.583	$\beta = 2.684$		F3 IV/V	0.177

	$\alpha_{(1950)}$	$\beta_{(1950)}$	μ_α	μ_δ	S
62850	$07^{h}41^{m}44^{s}.0$	$-59°10'46''.5$	$-0''.048$	$+0''.148$	GC (Rev)
63581A	07 45 24.1	−59 41 32.2	−0.066	+0.145	Cape + Syd.
63581B	07 45 26.2	−59 41 15.6	−0.064	+0.144	Cape + Syd.
64185	07 48 23.4	−60 09 30.6	−0.056	+0.153	FK4 (Rev)
	62850–64185	0.9 pc			
	62850–63581	0.5 pc			
	63581A–B	0.003 pc			
	63581–64185	0.4 pc			

	$[M_1]$	$[C_1]$		P[Fe/H]	M_V	Mod.
62850	$0^{m}.301$	$0^{m}.386$		$+0^{m}.30$	$+4^{m}.2$	$3^{m}.0$
63581A	0.408	0.351		+0.10	+5.05	3.1
63581B	0.426	0.346		+0.20	+5.1	3.2
64185	0.425	0.531		+0.20	+3.05	2.75
			Mean	+0.20		3.00
			σ	± 0.08		± 0.19

radial velocity determination is available only for HD 64185 (HR 3070). The spatial separations of the components range from 0.003 pc for the components of HD 62581 to nearly 1 par sec for HD 62850 and 64185.

2. System II

Herschel 2621 = HD 111908 = +8.265ABC. I suspect Herschel catalogued this wide system mainly because of its appearance; he called it 'An elegant triple star, equatorial triangle'. Because of the separations involved it was dropped from most double star catalogues, except that of Burnham. One component was discovered to be an R star with values of $V = 9.99$ mag and $B - V = 1.01$ mag. The V magnitude adopted here was obtained in the intermediate band system and the difference undoubtedly arises from the peculiar energy distribution in the spectrum of this carbon star; the m_{48} magnitudes in the DDO system are 9.91 mag for A and 9.83 mag for the carbon star. The A component is of type K2/3 III. There is some confusion in the literature and star A has occasionally been observed under the assumption that it is the carbon star. The B component was later classified as a CH-like star with a radial velocity of $+7$ km s^{-1} and a very poor determination of ± 19 km s^{-1} from a single spectrogram is also available for A (although the description does not preclude the possibility that this applies to the C component). The A component occurs in many meridean circle catalogues and the proper motion is well determined; the motions of B and C are based on the AGK3. Figure 2 shows members of the cluster M67 (dots) in the V_0,

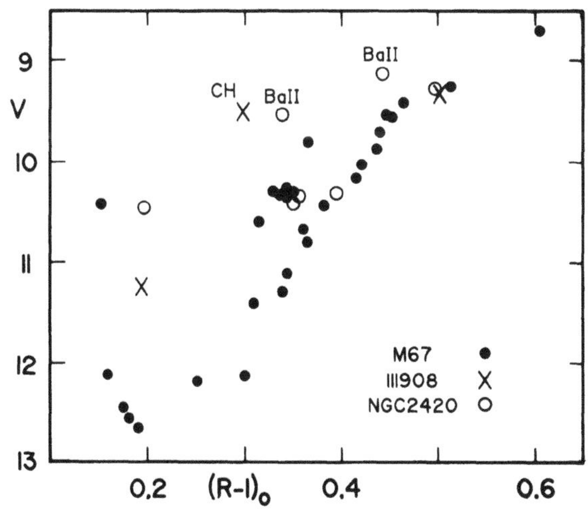

Fig. 2.

$(R - I)_0$ plane. The cluster is at nearly the same declination but 4 hours west of Herschel 2621 and the components of the latter system are represented by crosses in the figure. At $b \sim 70°$, the system is assumed to be unreddened so, judging from the reddest component (A) in Figure 2, its distance may be the same as that of the cluster. Also shown in Figure 2 are the few members of NGC 2420 for which I have photometry, with the distance adjusted to that of M67. This cluster contains two Ba II stars, labeled in the figure; the CH-like, B component of Herschel 2621 is also labeled. The A component of Herschel 2621 is almost identical to star A of NGC 2420 (clear circle) in Figure 2 and is intermediate between the two red giants in M67 (+12.1924 and

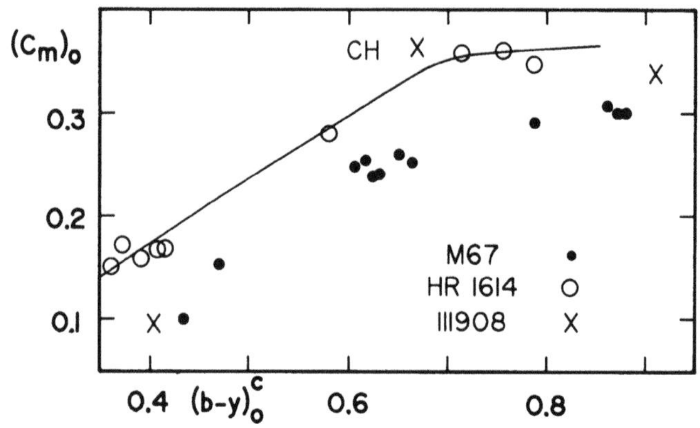

Fig. 3.

System II

Herschel 2621 = +8.2654 = HD 111908

	ΔX	ΔY		μ_α	μ_δ	
				(0.″001)		
AB	35″	3″	A	−14	+36	Meridian
AC	30″	−24″	B	−20	+24	AGK3
			C	−18	+40	AGK3

	V	$B-Y$	M_1	C_1	R	$R-I$	45–48	42–45	41–42
A	9.35	0.864	0.702	0.428	9.79	0.500	1.224	0.934	0.349
B	9.50	0.674	0.483	−0.161	9.20	0.300	1.300	1.248	0.278
C	11.19	0.363	0.118	0.385	11.12	0.195	0.992	0.544	0.021

M67	M_V							
+12.1924	−0.4	0.867	0.763	0.540	0.510	1.342	1.261	0.270
103	−0.05	0.797	0.707	0.407	0.465	1.291	1.151	0.270

NGC 2420				
A	0.500	1.307	1.012	0.267

	111908	M67	NGC 2420
P[Fe/H]	−026	−0.05	−0.47

No. 103). If the three components of system II are physically related, component C would be a blue straggler, like some members of M67 and NGC 2420 in Figure 2. The photometric parameters of components A and C give $P[\mathrm{Fe/H}] = -0.26$ dex, which is intermediate between the values of -0.05 dex for M67 and -0.47 dex for NGC 2420. The M67 members are represented by dots in the $(C_m)_0$, $(b-y)_0^c$ plane of Figure 3; $(b-y)_0^c = (b-y)_0 - 0.18[\mathrm{Fe/H}]$. Also, some members of the strong CN, HR 1614 group are represented by clear circles and the three components of Herschel 2621 by crosses. Although the CH-like star, component B, indicates a CN strength similar to that of the HR 1614 group members, the other two components are very similar to M67 members in this respect. The available data gives a space motion of $(U, V, W) = (+116, +80, +58)$ km s^{-1} for a modulus of 9.6 mag and clearly this will be an important system if accurate radial velocities confirm its physical nature. The mean separation of AB and AC is 0.15 pc.

3. System III

HD 15987 and 16094. The proper motions of both stars are from the *Cape and Yale Zone Catalogues*. The photometric parameters of HD 16094 (F/3 IV/V) give $P[\mathrm{Fe/H}] = -0.50$ dex, $M_V = +4.1$ mag and the star is compared with members of the Wolf 630 group and NGC 752 in the following tabulation. The modulus obtained from this companion is used to place both components in Figure 4 (clear circles), which also contains a schematic Main Sequence and giant branch for the cluster M67. The components are separated by 2.9 pc. No radial velocity determinations are available.

System III

	HD							
A	15987	8.58	0.504	0.187	0.542	8.20	0.303	G5/6III/IV
B	16094	9.86	0.321	0.130	0.441	$\beta = 2.635$		F3/5IV/V

	$\alpha_{(1950)}$		$\delta_{(1950)}$	μ_α	μ_δ
A	$2^h31^m8^s.1$		$-31°56'22''.9$	$+0''.010$	-0.104
B	2 32 10.9		-30 47 8.9	-0.023	-0.108
		2.9 pc			

	M_V	$B-Y$	$[M_1]$	$[C_1]$	β	$P[\mathrm{Fe/H}]$	
B	+4.1	0.321	0.226	0.377	2.636	−0.50	
HR 314	+4.05	0.316	0.235	0.367	2.630	−0.25	Wolf 630 group
752–88	+3.8	0.330	0.232	0.378	–	−0.53	NGC 752

4. System IV

HD 22625 and 22700. The proper motions are from the *Yale and Cape Zone Catalogues*. The photometric parameters for HD 22625 (F0V) give $P[\mathrm{Fe/H}] = 0.00$ dex, $M_V = +2.85$ mag and the resulting modulus is used to place both stars in Figure 4

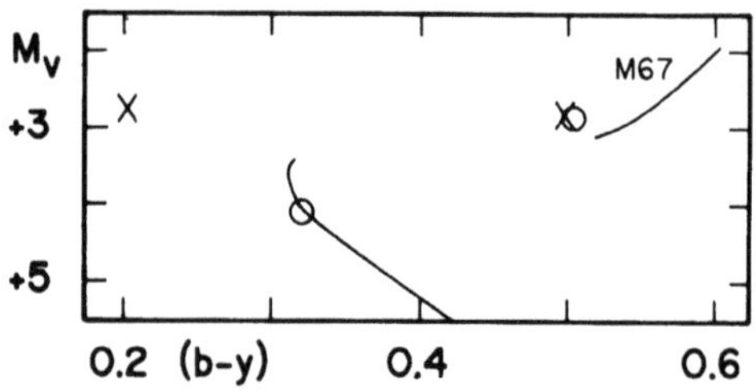

Fig. 4.

System IV

	HD	V	$B-Y$	M_1	C_1	R	$R-I$	Sp.
A	22625	9.23	0.205	0.169	0.718	$\beta = 2.738$		FO V
B	22700	9.13	0.498	0.247	0.507	8.80	0.285	G8 IV

	$\alpha_{(1950)}$	$\delta_{(1950)}$	μ_α	μ_δ
A	$03^h33^m21^s.9$	$-69°11'01''.9$	$+0''.089$	$+0''.100$
B	03 34 00.2	-68 51 43.5	$+0''.072$	$+0''.096$

Sep. = 1.1 pc

Star	M_V	$B-Y$	$[M_1]$	$[C_1]$	β	P[Fe/H]
22625	+2.85	0.205	0.231	0.677	2.738	0.0
M67-3012	+2.6	0.162	0.239	0.799	2.814	−0.05

					$R-I$	
22700	+2.7	0.498	0.396	0.407	0.285	0.0
15987	+2.8	0.504	0.338	0.442	0.300	−0.5

$V_T = 115$ km s^{-1}

(crosses). HD 22625 is compared with the blue straggler M67–3012, and HD 22700 with HD 15987 (System III), in the following tabulation. No radial velocity observations are available but the tangential velocity is 11 km s^{-1}. The two components are separated by 1.1 pc.

5. System V

HD 115136 and 115061. The (*UBVRI*) observations were obtained at the Lick and Mount Wilson Observatories, 30 years ago, and the DDO indices are synthesized from available Copenhagen photometry. The latter photometry indicates solar abundances of both Fe and CN. The proper motions of both stars are very well determined and the radial velocities are based on 3 Mount Wilson spectra of each object. The inequality of the visual magnitude indicates that the brightest star may be an AGB object. The space motion listed in the following tabulation, based on a modulus of 5.8 mag, suggests that the stars may be members of the 61 Cyg group, although fitting the fainter component to the M67, giant sequence would give the larger modulus of 6.0 mag. More accurate and extensive photometry is needed. A modulus of 6.0 mag leads to a spatial separation of 0.14 pc where as membership in the 61 Cyg group gives 0.11 pc.

System V

	V	$B-V$	$U-B$	$R-I$	45–48	42–45	41–42	C_M	Sp.
115136	6.54	1.1325	1.06	0.405	1.213	0.971	0.228	0.254	K2 III
115061	6.98	1.145	1.08	0.410	1.219	0.990	2.243	0.274	K2 III

	$\alpha_{(1950)}$	$\delta_{(1950)}$	μ_α	μ_δ	Rad. vel. (km s^{-1})
115061	$13^h11^m18^s$	$+67°34'23''$	$-0''.160$	$+0''.007$	+4.0
115136	13 11 45	+67 33 08	−0.160	+0.006	+4.2

	0.11 pc	
Mod 5.6	(M67)	$(U, V, W) = (+83, -53, +10)$ km s^{-1} 61 Cyg $= (+89, -53, -8)$ km s^{-1}

Another method of isolating possible wide binaries in the old disk population is to examine common proper motion companions to members of groups in that population. Perhaps the most obvious choice for this procedure is the HR 1614 group (4×10^9 yr old) in which both giants and dwarfs show the strongest known CN strengths. HD 159807 is a K-type subgiant member of this group with $M_V = +3$ mag. This object is also in one of the possible common proper motion groups (No. 33a) isolated in *Yale Zone Catalogue* (−40° to 50°). The well determined proper motion is shared by another member of that proper motion group, HD 166796, which is also a K-type subgiant ($M_V = +3.4$ mag) if a member of the HR 1614 group and, like HD 159807, shows a large excess of CN. A radial velocity has been published for HD 166976, by Phillip Lu (*IAU Colloquium*, No. 76) but the tabulated data for other stars in that source is so obviously garbled that the velocity cannot be trusted until it appears in a more reliable publication. This method of selecting wide binaries will be discussed in more detail elsewhere.

ORBIT DETERMINATION FROM A SHORT ARC THROUGH THE COEFFICIENTS OF POWER SERIES IN TIME

N. WIETH-KNUDSEN

Frederiksberg, Denmark

Summary. Attention is called to the method(s) of orbit determination as mentioned in the heading, deriving the values of the coefficients to be applied, by adjustment by least squares, of the power series in either rectangular coordinate, to the values observed, of this, – as initiated in the past by several authors, among others by Silva and De Caro in the 30's and 40's, and critically studied by the present author in the 50ies, with merely a short note published (*Lund Ann.* **12**, C1–C7 (1953)).

The avail of modern computers for overcoming the numerical trouble by taking into account the terms of higher order, than those until the fourth necessary for the orbit computation – an account unevitable for a successful accomplishment, as shown by the author – may justify a resumed interest in those methods, as the only ones allowing to extract just the information contained in the observations of such a short arc – without any hypothesis in advance.

1. Introduction

At the middle of this century, and in the decades in advance, several authors were engaged in the subject here dealt with. Those, the efforts of which shall be presented, are presented upon Table I, with the underlinings applied further on to distinguish between the contributions of each of them. (In *Ann. of Lund*, **12**, an entire list of references is given, which the author thinks to be complete, at least until 1950.)

As dealing with wide pairs, the relative positions of which have been determined either photographically or by equal means, we may assume those to be described by rectangular coordinates (x, y) in the plane of projection, the celestial tangential plane, either of a precision as comparable to that of the other one. However, for our purpose it is convenient to have the coordinates transformed to a system with the x-axis through some 'mean' point of the arc observed, and counting the time t (the epochs of observation) in a scale with zero when the companion is passing that point. The Italian authors recommend to use a unit of time greater than a year, say a century or half a century. The Thiele–Innes constants (A, B, F, G), ordinarily referred to the 'regular' system of coordinates first mentioned, may of course be described in the latter system, too, and will be assumed to be so in the following, with applying no other, particular, notation, as that is merely a matter of transformation.

As soon as a relative motion is noticed, the coordinates can be expressed as linear functions of time; when curvature is appearing, a quadratic term (in the x-coordinate) can be added; hence, with increasing analytical complexity of the motion, that leads to, by least squares to approximate this with polynomes with an increasing number of terms, polynomes that are actually the first terms of an interrupted Taylor-series (1) of either function.

Presenting itself is then the idea: might it not be possible to derive from the coefficients of those polynomes, and without any assumption in advance, the best

Astrophysics and Space Science **142** (1988) 153–160.

approximation obtainable under the circumstances given, of orbital elements of the pair in question?

For that purpose at least the coefficients of the fourth order must be determined, as it has been shown, and as we shall see, whence (most of) the former authors broke off the series there. But even in the cases where the fourth-order terms might be determinable with some, not too poor, precision, obviously the following ones may not be just neglectible, and the omission of them may systematically falsify, particularly the fourth order coefficients. This difficulty must be overcome to render applicable the entire procedure.

(A) The motion defined by (1) must obey the law of constant areal velocity. After differentiating either of (1), we can develop the double areal velocity as a power series in t, any coefficient of which beyond the zero term must be zero, as a condition necessary and sufficient. The condition necessary and sufficient for that, is that when the quantities λ, μ, ν, and so on, are defined by those of the Equations (2) (in x) written at left, they must fulfill the analogous conditions in y, written at right. (Obviously λ is the absolute value of the relative (to the radius vector) acceleration in the orbital motion.) Inserting from (2) the derivatives beyond the first in (1), we shall get, instead of two series of coefficients from the second and onwards, merely one series of unknowns λ, μ, ν, and so on, common to the two series.

(B) *The set of quantities* (λ, μ, ν) *is an objective mapping of* (n, e, E_0), which in turn, and as well known, is a bijective mapping of the dynamical elements (P, e, T). For showing the former theorem we need Equations (3). For deriving these, let's intermediately drop the index 0 in (2), and consider λ, μ, ν, and so on, not as – until now, and hereafter, too – bound to time zero, but as functions of time bound to any point (t, x, y) of the curve; when differentiating the first ones (for x'' and y'') of Equations (2), and comparing with the second ones (for x''' and y'''), we get the first of the (3); the three others are shown analogously by further differentiation.

By differentiating twice the ordinary Thiele–Innes ephemeris equations (4), and utilizing (5a), and after some reduction, during which (4) is utilized anew, we get the formulae shown for x'' and y'', which, when compared with the first ones of (2), give (6) Table II, which, be it said, was found by Finsen as early as in 1931. By (3), and reduction work, increasing with increasing order, we get (7), (8), (9), and (10). The system of Equations (6), (7), and (8) proves one way of our Theorem B.

Regarding the other way of that theorem it may be convenient to introduce the quantities α and γ, defined by Silva (and the physical meaning of which, as the relative 'stretching velocity', and 'stretching acceleration', respectively, of the radius vector, is indicated by the two equations in parentheses). Equations (11) and (12) are found by inserting (6), (7), and (8), and reducing. Equation (11) shows $\sin E$ to have always the same sign as α.

To prove (14), insert (11), (12), and (6) in the right-hand member and reduce. For (13), eliminate n by dividing (11) squared, by (6), insert $e \cos E$ from (14), reduce, and solve with respect to $e^2 \sin^2 E$, take the root, considering what is just said about the sign of $\sin E$, and you obtain (13).

The formula for n is found from (6) inserting (14).

The latter formula, together with (13) and (14), giving univocally e and E, prove the other way of Theorem B.

(C.a.) Assuming reasonable approximations to an initial place $(0, x_0, y_0) \approx \approx (0, x_0, 0)$, and to an initial velocity vector projected $(0, x_0', y_0')$, but yet requiring some improvement $(\Delta x_0, \Delta y_0, \Delta x_0', \Delta y_0')$, to be at hand, and a number of normal places or observations to be given, each normal place or observation, by (1) with (2) inserted, will give a pair of equations of condition of the type (16), where it must be noticed that the terms of 5th- and 6th-order in time have been transferred to the members of the equations with the observed values.

In the first approximation the latter terms are neglected, but the values of (λ, μ, ν) from the resulting normal equations, by (13), (14), (n), (9), and (10) give some values of (ξ, o) with which the right members of the equations of condition can be improved, giving corresponding improvements of the right members of the normal equations, and so on by iteration, until (λ, μ, ν) remain reasonably constant in consideration of the standard deviations derived. The procedure described, in fact laborious, will be facilitated when, once for all, forming the reciprocal matrix of the left members of the normal equations for solving them, and the procedure seems adequate for being programmed for electronic computing, what is even the reason for presenting it just at this time.

(C.b.) The Thiele-Innes Equations (4), together with their derivatives (5), form a pair of linear equations giving from (x_0, x_0') (A, F), and from (y_0, y_0') (B, G), as their determinant $-n\sqrt{1-e^2}\,(1 - e\cos E_0)$ is never zero for elliptical motion.

(D) The conditions presented for circular motion evince from the physical meaning of α and γ; the formula for e^2 obtained by squaring and adding (13) and (14), gives the condition for elliptical motion presented.

In *Ann. Lund* **12**, C5, C6, are given formulae for parabolic motion: (6), (7), (8), (9), and (10), and those corresponding to (13)–(14), and (n). For hyperbolic motion merely some formulae have been 'guessed'.

Applications

The obvious questions: has this ever been practiced? and how does it work? can be answered: In full, merely once, as applied to the normal places from which the author derived the hyperbolic orbit published in *Lund Ann.* **12**, A1–A41. It must be emphasized that that is by no means an attempt at reviving a preliminary orbit, now out of date, but merely a check if the method is able to extract the information, true or false, that is actually implied by a given set of normal places, be that properly or improperly formed.

The main course of the iterative approximations is presented upon Table III.

It is seen that the introduction of the terms of the 5th- and 6th-order really means a change in the values derived, particularly for ν, but that even from the fourth approximation the iteration might stop; nothing more happens; the reason for going on yet thrice was merely to learn what would happen.

The actual amounts of those 5th- and 6th-order terms are presented for the 'utmost' normal places, as well as these where they are negligible.

For comparison with the elements finally derived, the hyperbolic ones obtained by the Thiele–Innes–Van den Bos method are presented, too. The accordance seems to the author very satisfactory, in consideration of the entire uncertainty.

Be it said: Perhaps it is not all necessitated to work in the particular coordinate system here applied, but might work in the usual one when the equations of condition are applied in the symmetrical form (16) by De Caro; but the particular time-scale is unavoidable.

This seems to imply that the method actually works, even if unanswered is the question: Will those power series at all converge in mathematical sense for all parts (or of any part) of the orbit? Once in a private communication to the author the late Prof. Van Albada presented his doubt concerning that.

TABLE I

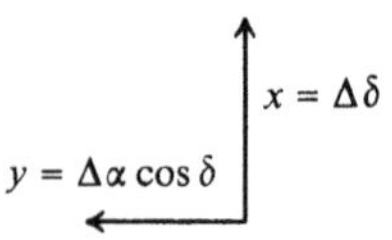

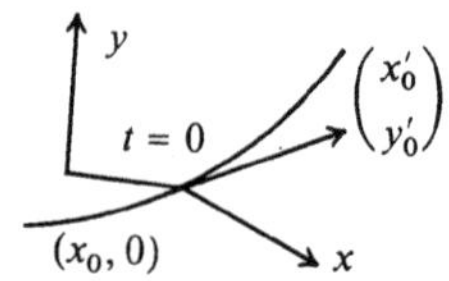

(1) $x = x_0 + x_0't + \frac{1}{2}x_0''t^2 + \frac{1}{6}x_0'''t^3 + \frac{1}{24}x_0''''t^4 \;\vdots\; + \frac{1}{120}x_0'''''t^5 + \frac{1}{720}x_0''''''t^6 + \cdots$

$y = y_0 + y_0't + \frac{1}{2}y_0''t^2 + \frac{1}{6}y_0'''t^3 + \frac{1}{24}y_0''''t^4 \;\vdots\; + \frac{1}{120}y_0'''''t^5 + \frac{1}{720}y_0''''''t^6 + \cdots$

(II) (IV)

$$xy' - yx' = x_0y_0' - y_0x_0' + (x_0y_0'' - y_0x_0'')t + \tfrac{1}{2}(x_0y_0''' - y_0x_0''' + x_0'y_0'' - y_0'x_0'')t^2 + \cdots$$

(2) $x_0'' = -\lambda x_0$ (II) $\quad y_0'' = -\lambda y_0$ (II)

$x''' = -(\mu x_0 + \lambda x_0')$ (II) $\quad y_0''' = -(\mu y_0 + \lambda y_0')$ (II)

$x_0'''' = -(\nu x_0 + 2\mu x_0')$ (II) $\quad y_0'''' = -(\nu y_0 + 2\mu y_0')$ (II)

$x_0''''' = -(\xi x_0 + (3\nu + 2\lambda^2)x_0')$ (IV) $\quad y_0''''' = -(\xi y_0 + (3\nu + 2\lambda^2)y_0')$ (IV)

$x_0'''''' = -(o x_0 + (4\xi + 10\lambda\mu)x_0')$ (IV) $\quad y_0'''''' = -(o x_0 + (4\xi + 10\lambda\mu)y_0')$ (IV)

(3) $\mu = \lambda', \quad \nu = \lambda'' - \lambda^2, \quad \xi = \lambda''' - 4\lambda\lambda', \quad o = \lambda'''' - 7\lambda\lambda'' - 4(\lambda')^2 + \lambda^3$ (IV)

(4) $x = AX + FY = A(\cos E - e) + F\sqrt{1 - e^2}\sin E$

$y = BX + CY = B(\cos E - e) + F\sqrt{1 - e^2}\sin E$

(5a) $E - e\sin E = n(t - T), \quad dE - e\cos E\, dE = n\, dt$

$$x'' = -n^2\frac{1}{(1 - e\cos E)^3}x, \quad y'' = -n^2\frac{1}{(1 - e\cos E)^3}y \qquad \text{(IV)}$$

(I) = Finsen, W. S.: 1931, *Monthly Notices Roy. Astron Soc.* **92**, 47.
(II) = Gilva, G.: 1936, *R. Osserv. Astron. di Padova Publ. e ristampe* **48**, 551.
(III) = De Caro E. and Veca, G.: 1948, *Osserv. Astrofisico – Catania Contributi Astrofisici* **62**.
(I) = Wieth-Knudsen, N.: 1953, *Ann. Obs. Lund* **12**, C1–C7.

TABLE II

$$(6)\quad \lambda = n^2 \frac{1}{(1 - e\cos E)^3} \qquad \text{(I)}$$

$$(7)\quad \mu = -n^3 \frac{3e\sin E}{(1 - e\cos E)^5} \qquad \text{(I)}$$

$$(8)\quad \nu = -n^4 \frac{1 + 2e\cos E + 3e^2(2\cos 2E - 3)}{(1 - e\cos E)^7} \qquad \text{(IV)}$$

$$(9)\quad \xi = n^5 \frac{15e\sin E\,(1 + 2e\cos E + e^2(2\cos 2E - 5))}{(1 - e\cos E)^9} \qquad \text{(IV)}$$

$$(10)\quad o = n^6 \frac{1 + 22e\cos E + 2e^2(52\cos 2E - 53) + \tfrac{3}{2}e^3(59\cos 3E - 103\cos E) + 45e^4(\cos 4E - 10\cos 2E + 10)}{(1 - e\cos E)^{11}} \qquad \text{(IV)}$$

$$(11)\quad \alpha = -\frac{1}{3}\frac{\mu}{\lambda} \;\Big|\; = n\frac{e\sin E}{(1 - e\cos E)^2} \quad (r' = \alpha r) \qquad \text{(II)}$$

$$(12)\quad \gamma = 4\alpha^2 - \frac{1}{3}\left(\lambda + \frac{\nu}{\lambda}\right) \;\Big|\; = n^2\frac{e(\cos E - e)}{(1 - e\cos E)^4} \quad (t'' = \gamma r) \qquad \text{(II)}$$

(II) | (IV)

Table II (continued)

$$(14)\quad e\cos E = \frac{\alpha^2+\gamma}{\lambda} \qquad \text{(IV)}$$

$$(13)\quad e\sin E = \frac{\alpha}{\lambda}\sqrt{\lambda-(\alpha^2+\gamma)} \qquad \text{(IV)}$$

$$n = \sqrt{\lambda\left(1-\frac{\alpha^2+\gamma}{\lambda}\right)^3} \qquad \text{(IV)}$$

$$(16)\quad \Delta x_0 + t\Delta x_0' \;\Big|\; -(\tfrac{1}{2}x_0t^2+\tfrac{1}{6}x_0't^3)\lambda - (\tfrac{1}{6}x_0t^3+\tfrac{1}{12}x_0't^4)\mu - (\tfrac{1}{24}x_0t^4 \;\Big|\; +\tfrac{1}{40}x_0't^5) \;\Big|\; \cdot\, \nu = x - x_0 - x_0't \;\Big|\; +(\tfrac{1}{120}\xi x_0 + \tfrac{1}{60}\lambda^2 x_0')t^5 + (\tfrac{1}{720}o x_0 + (\tfrac{1}{180}\xi + \tfrac{1}{72}\lambda\mu)x_0')t^6 + \cdots$$

$$\Delta y_0 + t\Delta y_0' \;\Big|\; -(\tfrac{1}{2}y_0t^2+\tfrac{1}{6}y_0't^3)\lambda - (\tfrac{1}{6}y_0t^3+\tfrac{1}{12}y_0't^4)\mu - (\tfrac{1}{24}y_0t^4 \;\Big|\; +\tfrac{1}{40}y_0't^5) \;\Big|\; \cdot\, \nu = y - y_0 - y_0't \;\Big|\; +(\tfrac{1}{120}\xi y_0 + \tfrac{1}{60}\lambda^2 y_0')t^5 + (\tfrac{1}{720}o y_0 + (\tfrac{1}{180}\xi + \tfrac{1}{72}\lambda\mu)y_0')t^6 + \cdots$$

(IV) | (III) | (IV) | (III) | (IV)

$$\text{solution:}\quad \begin{matrix} x_0 & x_0' \\ y_0 & y_0' \end{matrix} \quad \lambda \quad \mu \quad \nu$$

$$(5)\quad -n_1\frac{\sin E_0}{1-e\cos_0}A + n\frac{\sqrt{1-e^2}\cos E_0}{1-e\cos E_0} = x_0' \quad \text{(IV)} \qquad -n_1\frac{\sin E_0}{1-e\cos E_0}B + n\frac{\sqrt{1-e^2}\cos E_0}{1-e\cos E_0}G = y_0' \quad \text{(IV)}$$

$$(4)\quad \underbrace{(\cos E_0 - e)A}_{X_0} + \underbrace{\sqrt{1-e^2}\sin E_0}_{Y_0}F = x_0 \qquad \underbrace{(\cos E_0 - e)B}_{V_0} + \underbrace{\sqrt{1-e^2}\sin E_0}_{Y_0}G = y_0$$

$\mu = 0$ and $\nu = -\lambda^2$: circular motion.
$12\lambda^3 - 5\mu^2 + 3\lambda\nu > 0$: elliptic,
$= 0$: parabolic motion,
< 0: hyperbolic.

TABLE III

ADS 11632

6 normal places, each of weight 1, in the interval 1835–1895
10 normal places, each of weight 10, in the interval 1905–1950

Successive values of of starting values	x_0 y_0	x_0' y_0'	λ	μ	ν			
	+17.132	+0″.0180	with x-axis towards P.A. 149°.60 (eq. 2000) zero of time-scale 1905^a.0					
	0.000	+3.0289	unit of time-scale $50^a = 1^q$ (q for quinquagennium)					
Approximation No.			q^{-2}	q^{-3}	q^{-4}	Equations of condition: Terms of the 5th order	6th order	
0	+17.13568 ±689 +0.00694 ±480	+0.14196 ±2088 +3.02068 ±864	+0.26785 ±462	−0.11157 ±876	−0.1811 ±452			
1	+17.13785 +0.00505	+0.113484 +3.024754	+0.265939	−0.144978	−0.13302	−0″.3412 −0.0191	−0″.0895 in x +0.0441 in y	1835^a
2	+17.13798 +0.00397	+0.104944 +3.027145	+0.263933	−0.153120	−0.08147	practically zero	from 1885 to 1925	
3	+17.14068 +0.00326	+0.120473 +3.027280	+0.264789	−0.147369	−0.08104	+0″.0375 +0.0021	−0″.0063 in x +0.0031 in y	1950

Table III (continued)

4	+ 17.14063 + 0.00343	+ 0.122498 + 3.026772	+ 0.265322	− 0.144847	− 0.09299	Hyperbolic elements from the same normal Pls for comparison (merely) Ω = 145°.915 i = + 76.739 ω = − 14.385 e = 1.0439 q = 16″.5477 T = 1871ᵃ.535 ± 4.256
5	+ 17.14057 + 0.00334	+ 0.122674 + 3.026901	+ 0.265145	− 0.145363	− 0.09164	
6	+ 17.14058 + 0.00331	+ 0.122540 + 3.026978	+ 0.265145	− 0.145298	− 0.09066	
7	+ 17.14067 + 0.00335	+ 0.122426 + 3.026887	+ 0.265296	− 0.145325	− 0.09263	
Adopted	+ 17.14058 ± 561 + 0.00332 ± 390	+ 0.122638 ± 17015 + 3.026953 ± 7034	+ 0.265177 ± 3764	− 0.145267 ± 7138	− 0.09111 ± 3684	Ω = 145°.454 i = + 75.438 ω = − 20.612 a = 65″.2454; q = 16″.2802 e = 0.750479 μ = 0.0840472; P = 4283ᵃ.3 T = 1861ᵃ.272

ASTROMETRIC PARAMETERS OF VISUAL DOUBLE STARS DERIVED FROM SIMULATED HIPPARCOS MEASUREMENTS*

H.-H. BERNSTEIN, R. HERING, and H. G. WALTER

Astronomisches Rechen-Institut Heidelberg, F.R.G.

Summary. In the course of the reduction of *Hipparcos* data the astrometric parameter determination of visual binaries is reduced to the parameter determination of single stars. This approach assumes that the off-set of the binary components with respect to the conventional light point directly measured by the Hipparcos telescope are known on the individual scanning circles. The computer algorithms are outlined, preliminary results based on simulated measurements are discussed.

1. Introduction and Objectives

In the presence of visual binaries there are normally two simultaneous lightpoints in the instantaneous field of view of the *Hipparcos* telescope. This fact requires a special treatment of the light curves which has consequences to the determination of astrometric parameters. At the present state of development we start out from the assumption that the analysis of the photoncounts of a double star provides the field coordinates and, eventually, the abscissas on scanning circles for the conventional point of the system, and the offsets of the primary and secondary components with reference to the conventional point defined by the weighted phase of the first and second harmonic of the light curve.

In dealing with parameter determination of double stars it seems economical to pursue along the lines set up for single stars (Walter *et al.*, 1985) and to extend to an enlarged set of parameters representative of a double star system. Consequently, in the presence of double stars the following five parameters for the primary component have to be determined: the ecliptic longitude λ_p, the ecliptic latitude β_p, the proper motion in longitude $\mu_{\lambda p}$, the proper motion in latitude $\mu_{\beta p}$ and the parallax π_p. Furthermore, in the case of wide pairs one should be able to determine the same parameters for the secondary component. For secondary components, however, having separations of about $0''.2$ to $5''$ it will be attempted to solve for the position angle θ and the separation ρ, or to treat the secondary component like a single star if measurements permit it.

2. Method of Parameter Adjustment

The spherical coordinates of the secondary component with respect to the primary component are

$$\lambda_s = \lambda_p + \rho \sin\theta/\cos\beta_p\,, \qquad \beta_s = \beta_p + \rho\cos\theta\,.$$

* Presented by H.-H. Bernstein.

Astrophysics and Space Science **142** (1988) 161–165.

In order to connect these coordinates with the spherical coordinates ψ, ζ in the system of reference great circles, the same relationships are valid as for single stars. The respective mathematical relations and the establishment of the partial derivatives needed for a least squares process are detailed in a paper by Walter (1987). It seems to be expedient to distinguish between three cases of adjustment strategy.

CASE 1

From the observations of the primary component the five usual astrometric parameters are determined. This corresponds to the least squares adjustment for single stars described by Walter *et al.* (1985). The observations of the secondary component will be used to compute the position angle θ and the separation ρ. Under some circumstances one is able to determine $\dot{\theta}$ and $\dot{\rho}$ additionally. In general the adjustment has the form

$$\bar{l} + \bar{v} = \bar{A}\,\bar{x}\,,$$

where $\bar{l}$ designates the observed quantities, $\bar{v}$ is the vector of improvement, and $\bar{x}$ the solution vector. The design matrix $\bar{A}$ is of the form

$$\bar{A}_1 = \left|\begin{array}{ccccccccc} \frac{\partial\psi_{p_1}}{\partial\lambda_p} & \frac{\partial\psi_{p_1}}{\partial\beta_p} & \frac{\partial\psi_{p_1}}{\partial\mu\lambda_p} & \frac{\partial\psi_{p_1}}{\partial\mu\beta_p} & \frac{\partial\psi_{p_1}}{\partial\pi_p} & 0 & 0 & (0 & 0) \\ \hdashline 0 & 0 & 0 & 0 & 0 & \frac{\partial\psi_{s_1}}{\partial\theta} & \frac{\partial\psi_{s_1}}{\partial\rho} & \left(\frac{\partial\psi_{s_1}}{\partial\dot{\theta}}\right. & \left.\frac{\partial\psi_{s_1}}{\partial\dot{\rho}}\right) \end{array}\right|$$

The residual vector (observed minus computed abscissas) contains the following elements:

$$\bar{l} = \left|\begin{array}{c} \psi_{\text{obs},\,p_1} - \psi_{\text{comp},\,p_1} \\ \psi_{\text{obs},\,p_n} - \psi_{\text{comp},\,p_n} \\ \hdashline \psi_{\text{obs},\,s_1} - \psi_{\text{comp},\,s_1} \\ \psi_{\text{obs},\,s_n} - \psi_{\text{comp},\,s_n} \end{array}\right|$$

The parameters of the secondary component are determined following the principle of the treatment of single stars augmented by the additional parameter θ and ρ. Here, no relationship between primary and secondary components are taken into account. One can expect a numerical robust solution. In order to consider possible connections between the components, one can apply Case 2 of our adjustment strategy.

CASE 2

Since the observations of the secondary component are related to the parameters of the primary component, it is reasonable to introduce the partial derivatives of these observations with respect to the astrometric parameters of the primary component into the adjustment process. The design matrix contains the following elements:

$$\bar{A}_2 = \left| \begin{array}{ccccccccc} \frac{\partial\psi_{p_1}}{\partial\lambda_p} & \frac{\partial\psi_{p_1}}{\partial\beta_p} & \frac{\partial\psi_{p_1}}{\partial\mu\lambda_p} & \frac{\partial\psi_{p_1}}{\partial\mu\beta_p} & \frac{\partial\psi_{p_1}}{\partial\pi_p} & 0 & 0 & (0 & 0) \\ \hdashline \frac{\partial\psi_{s_1}}{\partial\lambda_p} & \frac{\partial\psi_{s_1}}{\partial\beta_p} & \frac{\partial\psi_{s_1}}{\partial\mu_p\lambda_p} & \frac{\partial\psi_{s_1}}{\partial\mu\beta_p} & \frac{\partial\psi_{s_1}}{\partial\pi_p} & \frac{\partial\psi_{s_1}}{\partial\theta} & \frac{\partial\psi_{s_1}}{\partial\rho} & \left(\frac{\partial\psi_{s_1}}{\partial\dot{\theta}}\right. & \left.\frac{\partial\psi_{s_1}}{\partial\dot{\rho}}\right) \end{array} \right| .$$

CASE 3

This case is a method which treats both components as single stars. The structure of the design matrix is

$$\bar{A}_3 = \left| \begin{array}{cccccccccc} \frac{\partial\psi_{p_1}}{\partial\lambda_p} & \frac{\partial\psi_{p_1}}{\partial\beta_p} & \frac{\partial\psi_{p_1}}{\partial\mu\lambda_p} & \frac{\partial\psi_{p_1}}{\partial\mu\beta_p} & \frac{\partial\psi_{p_1}}{\partial\pi_p} & 0 & 0 & 0 & 0 & 0 \\ \hdashline 0 & 0 & 0 & 0 & 0 & \frac{\partial\psi_{s_1}}{\partial\lambda_s} & \frac{\partial\psi_{s_1}}{\partial\beta_s} & \frac{\partial\psi_{s_1}}{\partial\mu\lambda_s} & \frac{\partial\psi_{s_1}}{\partial\mu\beta_s} & \frac{\partial\psi_{s_1}}{\partial\pi_s} \end{array} \right| .$$

This case allows to test whether the two stars exhibit double star properties. In the case of a double star one expects the same values for proper motions and parallaxes.

For the lack of real observations simulated data have been employed to test the algorithms. For the primary component we used the simulated observations from CERGA (Falin *et al.*, 1986). A secondary component was chosen by fixing arbitrarily a position angle of 10° and a separation of 1″. According to the single star treatment approximate positions, the abscissas and ordinates are obtainable. The simulated observations of the secondary component are derived from the corresponding computed values by adding the same noise as for single stars.

3. Discussion of the Results

For the primary component Table I shows the average difference between the corresponding single star solution (named T_{6000}) and the results of the three cases. Table II gives the average r.m.s. errors of these parameters. The average difference between input and output data for the secondary component are listed in Table III.

TABLE I

Average differences for the primary component

	T_{6000} – Case 1	T_{6000} – Case 2	T_{6000} – Case 3
$\overline{\Delta\lambda \cos \bar{\beta}}$	1.7 mas	1.7 mas	1.7 mas
$\overline{\Delta\beta}$	– 0.3 mas	– 0.2 mas	– 0.3 mas
$\overline{\Delta\mu_\lambda \cos \bar{\beta}}$	0.8 mas yr^{-1}	0.7 mas yr^{-1}	0.8 mas yr^{-1}
$\overline{\Delta\mu_\beta}$	– 0.02 mass yr^{-1}	– 0.01 mas yr^{-1}	– 0.02 mas yr^{-1}
$\overline{\Delta\pi}$	+ 0.05 mas	+ 0.04 mas	+ 0.05 mas

TABLE II

Average r.m.s. errors of the astrometric parameters of the primary component

	T_{6000}	Case 1	Case 2	Case 3
$\overline{m\lambda}$	3.1 mas	2.9 mas	2.8 mas	3.0 mas
$\overline{m\beta}$	1.9 mas	1.8 mas	1.7 mas	1.9 mas
$\overline{m\mu_\lambda}$	3.4 mas yr^{-1}	3.2 mas yr^{-1}	2.2 mas yr^{-1}	3.3 mas yr^{-1}
$\overline{m\mu_\beta}$	2.7 mas yr^{-1}	2.5 mas yr^{-1}	1.8 mas yr^{-1}	2.6 mas yr^{-1}
$\overline{m\pi}$	3.0 mas	2.8 mas	2.0 mas	2.9 mas

TABLE III

Average input-output differences and r.m.s. errors of the secondary component

	Case 1	Case 2
$\overline{\Delta\theta}$	+ 8.0 mas	0 mas
$\overline{\Delta\rho}$	3.0 mas	-9×10^{-6} mas
$\overline{m\theta}$	1.0 mas	1.4 mas
$\overline{m\rho}$	1.6 mas	2.3 mas
$r_{\theta\rho}$	0.08	0.17

Additionally, Table III contains the average r.m.s. errors of θ and ρ, and the correlation coefficient between θ and ρ, $r_{\theta\rho}$.

In all cases the differences between the parameters of the primary component and those derived from single star treatment are less than the respective r.m.s. errors. Therefore, the solution vector did not change significantly. For the primary component the results obtained from the three cases are nearby identical. Small differences exist among the r.m.s. errors. In general the r.m.s. errors are less than the corresponding values for the single stars.

A comparison between input and output values for θ and ρ shows no relevant differences in Case 2. However, the discrepancies in θ and ρ of 8 and 3 mas, respectively are remarkable in Case 1. For Case 2 the r.m.s. errors of these parameters are about a factor 1.4 greater than the r.m.s. errors obtained in Case 1. Since for the simulation of the measurement errors of the secondary component an intensity model is not introduced, presumably too favourable r.m.s. errors are obtained.

In both cases the correlation coefficient between θ and ρ is negligible. Due to the simultaneous solution the correlation coefficients between λ and ρ as well as β and ρ are remarkably large in Case 2. Both correlation coefficients are functions of θ and ρ. For $\theta = 10°$ and $\rho = 1''$ one obtains -0.67 and -0.68. Other correlations are not important.

The results arrived at so far are preliminary and reflect the state of a pilot study aiming at testing the capability of the computer software. Some of the quantities show a still unexplained behaviour which is now under investigation.

References

Falin, J. L., Froeschlé, M., and Mignard, F.: 1986, 'Simulation des données pour la tâche "Astrometric Parameter Determination"', *Hipparcos Note Technique*, No. 13, Version 3.

Walter, H. G.: 1987, in P. L. Bernacca and J. Kovalevsky (eds.), 'Astrometric Treatment of Double Star Abscissae', *Processing of Scientific Data from the ESA Astrometry Satellite Hipparcos*, Proceedings of the Third Thinkshop of the FAST Consortium held at Bari, 3 to 6 November 1986, ERGA, Grasse, p. 223.

Walter, H. G., Hering, R., Bastian, U., and Bernstein, H. H.: 1985, in J. Kovalevsky (ed.), 'Astrometric Parameter Determination: Methods, Algorithms, and Program Implementation', *Processing of Scientific Data from the ESA Astrometry Satellite Hipparcos*, Proceedings of the Second Thinkshop of the FAST Consortium held at Marseille, 21 to 25 January 1985, CERGA, Grasse, p. 281.

POSTER SESSION

Chairman: E. VAN DESSEL

TENTATIVE COMMENTS ABOUT THE MEETING

J. Dommanget
President of the Scientific Organizing Committee

I have the feeling that an exceptional opportunity appears for me now as chairman of the Scientific Committee of the present Colloquium, to make a few personal comments on some discussions that took place during this first part of the meeting, in connection with what I said in my introductory talk. You will remember that I very much insisted on the contribution that in my opinion, visual double star astronomy should bring to what may be considered as two of the most important "pillars" in modern astronomical research : stellar evolution and galactic dynamics and structure. At the start of this poster session, I would like to come back for a while to both of these aspects of double star astronomy because it seems to me that we have now reached a point of our discussions where each of these two subjects finds or will find a good illustration.

First of all, you may have been surprised that when I so much insisted on having all participants presenting papers concerning wide systems only, I personally come with a poster which deals fundamentally with the closest visual pairs. Actually if I - unfairly (?) - accepted this exception for my-self, it was because the subject of the poster should be considered in my mind as an illustration of the possible contribution of visual systems to galactic research. We may come back to that subject when discussing this poster. Secondly, the noteworthy discussion that took place yesterday afternoon (fourth session) between many of the participants on stellar evolution with mass-loss appeared to me as the best illustration I could expect for the contribution of visual double stars in this other field of research.

Concerning this last point, may I take the opportunity to recall here that twenty-five years ago, I had started with a research [1],[2], on the period-eccentricity correlation, of which the existence was very much contested at that time. Doing so, I came to the conclusion that if

(1) Annales de l'Obs. R. de Belgique, 3e. Série, IX, 5, 1963;
(2) Communication Obs. R. de Belgique, 232, 1964;

Astrophysics and Space Science 142(1988), 169–170.

one accepts that wide pairs have been formed by orbital evolution of close pairs under the effect of a low but marked secular isotropic mass-loss of their components, the period-eccentricity correlation is easily explained. This research was made by considering only classical celestial mechanics and by using statistical results on data of known orbital pairs. Some interesting diagrammes have been established on that occasion as well as at the end the following relation between total mass and eccentricity [1] :

$$e^{2,8}\mathfrak{M}_{AB} = 3,60$$

which is the expression of a limit to the domain of the possible values of these parameters that no selection effect seems to be able to explain.

If thus such a secular mass-loss really exists for all stars - red dwarfs included, as the yesterday's discussion seems to confirm - the concomitant evolution of the orbital parameters of any visual binary will appear as the "conducting wire" that should be taken into account to establish the sequence of the astrophysical evolutionary steps of their components. Any result in this domain will have an important impact on our stellar evolution knowledge. I must confess that it was for me a great pleasure to assist yesterday to the above mentioned discussion, the first one I think on the subject at a visual double star meeting. This is why I felt it necessary to reiterate what I said on the first day, just when I have to present the poster about the space distribution of visual orbital planes that you have seen accompanied by a model that I hope to discuss now with you, as the illustration of the visual double star contribution in this other field of research : galactic dynamics and structure.

(1) Astrophysics and Space Science Library, 98, 1982, p. 119.

THE SPACE ARRANGEMENT OF THE ORBITAL PLANES OF THE VISUAL DOUBLE STARS

J.Dommanget
Royal Observatory, Belgium

ABSTRACT. *A more extended statistical material based on the "Second Catalogue d'Ephémérides de Vitesses radiales relatives" by J.Dommanget and O.Nys (1982) has made possible a new study on the distribution of the orbital poles of visual orbital pairs.*

It is confirmed that orbital planes do not show any tendancy to be parallel to the galactic plane : at the contrary they show a preference to be perpendicular to it.

The most important result is: "Space seems to be divided into elements of the order of 10 to 30 parsecs where visual systems show similar orbital plane orientation".

This confirms a result found in a preceding research (1967) concerning some ten systems surrounding the solar system.

I) EARLY RESEARCHES :

During the first half of the century, various researches have been conducted on the distribution of the orbital poles of the binaries, but no significant results have been found, probably because of the following reasons :

a) insufficiently extended statistical material;

b) use of catalogues of visual double star orbits without taking care of the fact that no uniform convention had been adopted for the geometrical elements, especially for the inclination (i) and for the ascending node (Ω);

c) researches generally conducted with a preconceived conception of what, at that time, was expected to be found, that is to say a tendency of the orbital planes to be parallel to the galactic plane.

II) FIRST UCCLE RESEARCH (1967) :

Before starting a new investigation on the subject, it thus appeared to us that the first precautions to take were : a) to establish valuable statistical material, e.g. a list of binaries for which sufficiently

Astrophysics and Space Science **142**(1988), 171–176.

Figure 1.-

a) Photograph of our model taken from the direction: L = 100°; B = -15°, showing the few systems at less than 10 parsecs from the Sun (white disks) with orbital planes nearly parallel to each other.

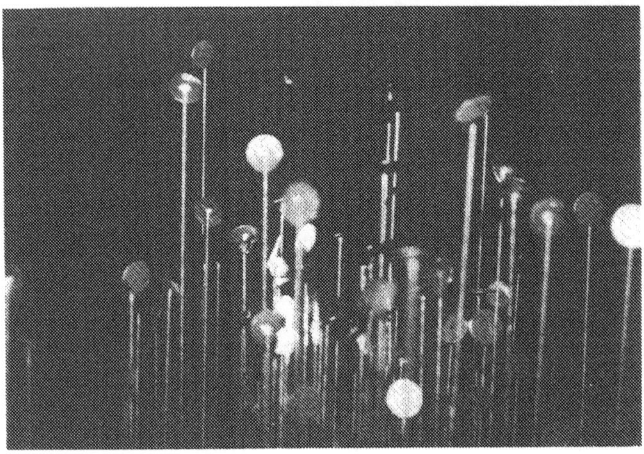

b) Photograph of the same model taken from the direction: L = 280°; B = +15° where the same systems are seen side-face.

Other groupings of this kind may be seen on the model but cannot be clearly shown here.

accurate and coherent radial velocity observations were known for solving the ambiguity on their orbital ascending nodes and b) to use a uniform convention (the one adopted by IAU Commission 26) for the elements of all such orbits. On the other hand, the small extent of the material at our disposal at that time conducted us to invite radial velocity observers to make more observations of visual pairs. Also, in view to easily permit these observers to establish their observation programmes, we started with the realization of a catalogue of ephemerides of relative radial velocities for all known orbital systems. To fulfill our project, we undertook with our colleague O. NYS, a very extensive bibliographic research on radial velocity observations on the orbital pairs. This gave us the opportunity to solve the ambiguity on the ascending node for many systems and to establish the best possible statistical material for our research on the orbital poles distribution. (J. DOMMANGET-O. NYS, 1967).

Taking also into account the systems having an orbital inclination near 0° or 180° and near 90°, we finally had at our disposal a material of 70 systems. The observed features were : (J. DOMMANGET, 1968)

a) no systematic parallelism of the orbital planes with the galactic plane;

b) for the systems nearer than 25 parsecs (nearly all 70 systems) a preferential distribution of the orbital poles is observed with some dispersion around a plane of which the pole is : L = +35°, B = +60° (or : L = +215°, B = -60°);

c) for the systems nearer than 10 parsecs (10 systems) the poles are assembled around a position near : L = +280°, B = +15° (or : L = +100°, B = -15°).

III) SECOND UCCLE RESEARCH (1982/1987) :

Some twenty years later, numerous new relative radial velocity observations as well as new orbits were known for visual binaries. On the other hand, as the first catalogue of ephemerides of relative radial velocities covered the period 1955-1975 only, the need appeared for a new one, covering another twenty years interval. This catalogue, established for the period 1970-1990 was published somewhat late (J. DOMMANGET-O. NYS, 1982). Taking then again into account also the binaries having orbital inclination near 0°, 90° or 180° we had at our disposal a new list of some 145 visual pairs having correctly defined poles.

A new research was thus possible. It confirmed the previously discovered features a) and c) but did not show so well what is expressed under item b). As in 1967, a space model has been constructed to facilitate the study of the characteristics of the statistical material. At that time, the scarcity of the material increasing with the distance to the solar system, did not permit to recognize any other group of systems such as that of the ten systems surrounding the Sun. Now, with the more extended material, the new space model permits the recognition of a few groups of stars inciting us to conclude that :

> "Space seems to be divided into elements of the order of 10 to 30 parsecs where visual systems show similar orbital plane orientation".

IV) ONGOING RESEARCH :

First of all, it should be stressed that only systematic programmes of observations of radial velocity of orbital visual binaries may increase and improve the statistical material needed to assure the features mentioned here concerning the arrangement of their orbital planes and the existence of space elements containing visual double stars showing common characteristics.

However, some researches may already be conducted on the groups that seem to exist, by considering the space motions not only of the binaries but also of all single stars with the hope to find some connection between them.

In that respect, the astrometric mission of the satellite HIPPARCOS may be of some importance by the parallaxes and proper motions that will be determinate with an exceptional accuracy for the great majority of the nearby stars.

Acknowledgment : This research has been made in the frame of a research intitled "Contribution des étoiles de masses et de compositions diverses, à l'évolution chimique des galaxies" supported by the belgian Fonds de la Recherche Fondamentale Collective under ref. n° 2.9009.79.

J. Dommanget, 1968. - La cinématique des couples stellaires visuels dans le voisinage du Soleil,
Bull. Astr., de l'Obs. R. de Belg. 6, pp. 236-249;

J. Dommanget-O. Nys, 1967. - Catalogue d'éphémérides des vitesses radiales relatives des composantes des étoiles doubles visuelles dont l'orbite est connue.
Comm. de l'Obs. R. de Belg. Série B, n° 15;

J. Dommanget-O. Nys, 1982. - Second catalogue d'éphémérides des vitesses radiales relatives des composantes des étoiles doubles visuelles dont l'orbite est connue,
Comm. de l'Obs. R. de Belg. Série B, n° 124.

KOPAL - Do the systems exhibiting parallelism of orbital planes constitute "moving clusters" (possibly interpenetrating each other) ?

DOMMANGET - To give an answer to that question, one should consider the space velocities of all star members of a same group. This has not been done systematically till now for each possible group because the model has been finished only last week and the recognition of the star's membership to a same group needs some careful examination and discussion. All what I can say to-day is that from computations made already in 1967 for all ten systems of the group surrounding the Sun to less than 10 parsecs, it appears that the "mean" vector representing the space velocity of these stars is practically at 90° from the direction of their mean orbital pole.

ABT - Your evidence for regions of space that are roughly the dimensions of open clusters and that have aligned orbital axes may have a bearing on a problem about stellar rotational velocities in open clusters. There is old evidence that some clusters (Pleiades, alpha Persei) have large rotational velocities and some (NGC 6475, IC 4665) have low rotational velocities. Are these differences due to an aspect effect (sin i) rather than to different mean rotational velocities (V_{rot}) ? Your evidence suggests that aspect effects may be important in individual clusters, even though the evidence is against alignments of rotational and orbital axes in visual binaries.

DOMMANGET - The observed organisation of the orbital planes of visual binaries should of course be considered in connection with our knowledge on open clusters. But it is now probably too early to make inference from what is shown here. A specific study should be first conducted on all dynamical characteristics of the celestial bodies belonging to the space elements defined by the groups that may be recognised on the basis of orbital axis orientation.
On the other hand I am not sure that there is no correlation at all between the rotational axis of the visual double stars components and their orbital axis.

WORLEY - How many of the orbits in the model have ascending nodes which are unambiguously determined ?
Comments :
1. I suppose everyone has noticed the lack of orbits for southern pairs, which is due to their poorer observational history.
2. I believe there is considerable observational selection introduced because pairs of high inclination (and also e) are less likely to have an orbit.

DOMMANGET - There are some 140 orbits considered in this research. I would estimate to less than one third those with some ambiguous determination of the ascending node. But even so, half of these only should be considered as uncorrect, because only two possibilities are acceptable in each case.
The fact that orbits are lacking in the southern hemisphere is not important : it does not change the situation in the regions where visual double stars appear as members of a same group.
Finally, I do not think that high orbital inclination may introduce any important selection effect that could explain the present observed distribution of the orbital poles. If their real distribution would be uniform and isotropic, any such high inclination effect would also be uniform and isotropic and would lead to another uniform and isotropic apparent distribution of the poles around the Sun, which does not seem to be the case. Concerning the eccentricity, the situation could be somewhat different.

KOPAL - In order to render your diagnosis fully significant, I think the orientation of the planes must also be looked at for close binaries.

VAN DESSEL - I agree with Prof. Kopal insofar as you remember that there will be different mechanisms of double star formation in action. It has not become clear yet if there is a well defined border line between, say fission, separate development of protostars or caption of components. So which of those mechanisms does Dommanget's statistics refer to ?

DOMMANGET - This is of course an important question, but a priori, it appears impossible to conduct any research on the pole distribution of the orbital planes of the spectroscopic binaries because node and inclination are unknown. Only indirect methods may give partial answers to such questions and I would like to mention here a research made by A. PEDOUSSAUT (*) on the distribution of the spectroscopic pairs on the sky. Such distribution in galactic longitude shows a gap that could be interpreted as the result of a preferential orientation of the orbital planes : in the direction of this gap, the orbital inclination would have a tendency to be small, hampering the discovery of binaries.

WEINBERG - I would like to comment on Dr. Abt's observation that the correlation scale is roughly that of an open cluster. If the cluster were old, one might expect the orbital angular momenta to be rather mixed. A young cluster or association would still retain its initial angular momentum distribution. Thus, it would be interesting to estimate the age of stars in Dr. Dommanget's sample.
I would also like to ask Dr. Dommanget what the distribution of periods is in the sample.

DOMMANGET - Again, this research is a tentative one that needs some extension as for instance, as you say, by introducing the concept of the ages of the systems. Unfortunately, the numbers of stars per group are small and if one starts to consider categories inside each group, there will be only one or two systems per category, which of course is insufficient for further significant research.
This is why an appeal should be repeated here for more radial velocity observations of the components of orbital systems (see : ref. : J. Dommanget-O. Nys - 1982).
About the periods, I can mention that all systems are orbital visual and astrometric ones. The periods are thus for 90 % between a few years and a few hundreds years.

(*) Sur la répartition galactique des binaires spectroscopiques.- Comptes Rendus de l'Académie des Sciences, Série b, Paris 1968, 266, p. 1341.

A POSSIBLE CONNECTION BETWEEN WIDE BINARIES AND OPEN CLUSTERS

L. O. LODÉN
Astronomical Observatory
Uppsala
Sweden

Summary. There are certain indications that some wide binaries might represent an intermediate stage between multiple system and open cluster.

The initial project was a survey of the southern Milky Way starting with a visual inspection of objective prism plates secured at the Boyden Observatory in South Africa between 1956 and 1960. The idea was to select interesting objects for subsequent detailed investigations. At this inspection were also discovered a large amount of "coincidence cases", i.e. two or more spectra with practically identical appearance and situated with so short angular separation on the plate that they formed a highly conspicuous configuration for the inspecting eye. A rough statistics showed that spectral type around 'A' was highly predominant among these objects and also that the observed (and estimated) number of them exceeded the calculated random one with respect to the total number of stars within the relevant magnitude and spectral ranges.

As a possible explanation of the phenomenon we considered a number of different physical circumstances:

a. A projection effect (near conjunction) of a nearby low-luminosity star and a distant one of higher luminosity.

b. A temporary encounter between two identical stars of different origin.

c. A projection of two identical stars belonging to one and the same cluster or association but not necessarily situated close together.

d. A verv small cluster ('minicluster' or 'microcluster') consisting of very few stars but not classifiable as an ordinary multiple star system; maybe some sort of an cluster remnant.

Astrophysics and Space Science **142** (1988), 177–179.

e. A binary system of the type that has been described by Stock and Wroblewski (1972). In such a system the equality of the components is a consequence of the creation of the system by cleavage of a single star.

f. A multiple system with extremely wide separation and with accidentally coinciding spectral type of the components.

g. A spurious phenomenon that is caused by a combination of various technical circumstances.

The basic concept was also that the apparent near identity between the components was the necessary condition for the detection, while the identity itself should - in a first approximation - need no explanation. This in turn would imply that the real physical phenomenon might be a highly common one although it could be detected only in cases of near coincidence in spectral type. The majority of the objects would therefore escape detection.

In addition, the working hypothesis considered the probability that there is no completely unique explanation of the phenomenon, so that more than one of the explanations indicated above might very well be relevant.

In order to test the 'cluster model' the surrounding stellar field of a large number of candidate objects has been thoroughly investigated in attempt to reveal additional members (by means of spectroscopy and photometry). In many cases such members have really been found, thus indicating that the candidate object was a pair of cluster members which happened to be of almost identical spectral type so that they revealed the loose cluster by their conspicuous appearance (Lodén 1980).

In other cases, however, the original candidate object has turned out to be a solitary phenomenon. A second working hypothesis then has been that these objects might (at least partly) constitute the ultimate remnant of an open cluster subject to dynamical disintegration, i.e. a 'hard binary' (Lodén 1984). An indication in that direction could be found in the condition that relatively massive stars seem to be predominating among the candidate objects. We might therefore consider the possibility that we are here facing some kind of a transitional stage between open cluster and multiple star.

One of the most serious problems connected with the study of these objects arises from the difficulty to estimate the number of binaries (or higher multiples) which have less coinciding spectral type of the components and therefore escape detection.

REFERENCES: Lodén, L.O.:1980, Astron. Astrophys. Suppl. Ser. 41, 173.
Lodén, L.O.:1984, Astron. Astrophys. Suppl. Ser. 58, 431.
Stock, J., Wroblewski, H.:1972, Astron. Astrophys. 18, 341.

WEINBERG - Dr. Wasserman and I have also found evidence for clustering around wide CPM pairs in the Yale Bright Star Catalog.

LODÉN - I am not surprised.

ABT - I gather that you are suggesting that the wide pairs are members of disrupted clusters.

LODÉN - Yes, at least a considerable fraction of them.

NEW ORBITS OF FOUR VISUAL DOUBLE STARS BY THE METHOD OF APPARENT MOTION PARAMETERS (AMP)

A.A. Kisselev and O.V. Kiyaeva

Main Astronomical Observatory
196140 Leningrad, USSR.

Summary. The orbits of four binaries ADS48, 9031, 14636 (61 Cyg) and 11632 were determined by the AMP-method from the short arc photographic observations made in Pulkovo with the 26-inch refractor.

The AMP-method was proposed and investigated at Pulkovo for a determination of the orbit elements of visual double stars with long periods on the basis of short arc observations (Kisselev and Kiyaeva, 1980). The following three sets of data are necessary for a determination of the AMP-orbit :

1. Five apparent motion parameters for the given moment t_0, corresponding to an approximate mid of the internal of short arc observations. These parameters are as follows :
   - ρ, θ - polar coordinates of the component B relative to A,
   - μ - the angular velocity of the apparent motion B relative to A,
   - ψ - the position angle of the apparent motion direction,
   - ρ_c - the radius of the curvature of the apparent orbit at a given point corresponding to t_0.

   These parameters may be derived from dense sets of accurate photographic observations made with long-foci astrographs for about 20-30 years.
2. Trigonometrical parallax π_t. We use the data from Woolley's catalogue.
3. The radial velocity of the component B relative to the main component A, Vr(*). In order to determine this parameter within the error box of 0.2 km/s, we organized special spectral observations with the 6-m telescope at SAO.

Thus, it is necessary to receive highly precise information on a double star from different specific observations for a comparatively short period. These data are a basis for a calculation of a reliable dynamical orbit, but there is one unknown parameter - the sum of masses of the components $\mathfrak{M}_{AB}$. To determine $\mathfrak{M}_{AB}$, it is necessary to use some ancient observations of the star made about 100-150 years ago. The ancient

(*) $Vr = VR_B - VR_A$

Astrophysics and Space Science **142** (1988), 181–183.

TABLE 1

N ADS	48	9031	14636	11632
plates	79	49	132	
ΔT	1960-85	1962-79	1958-74	1910-35
t_0	1973.0	1972.0	1966.5	1922.5
ρ	5".889	3.288	28.211	16.950
	± 3	± 5	± 5	± 1
θ	170°.39	152.69	143.12	153.178
	± 3	± 6	± 1	± 5
μ	0".0473	0.0502	0.133	0.0648
	± 5	± 6	± 2	± 1
ψ	243°.7	214.6	195.5	262.52
	± 6	± 9	± 6	± 5
ρ_c	3".3	1.3	25.9	3.1
	± 4	± 1	± 1.0	± 0.1
π_t	0.094	0.075	0.296	0.284
	± 4	± 6	± 4	± 3
Vr	-2.8	+2.9	+1.1	+2.6
	± 2	± 1	± 1	± 2
$\mathfrak{M}_{AB}$	2.0	1.4	1.1	0.68

observations may be not accurate. Our principle is as follows : not to mix ancient observations and new observations. Of a great amount of AMP-orbits calculated only with the basic data, we choose the orbit in a better agreement with the ancient observations than other orbits. The criterion for the agreement is discussed by Kiyaeva (1983). If it's impossible to choose $\mathfrak{M}_{AB}$ for an AMP-orbit agreeing with ancient observations, we have to assume, that the orbit is not satisfactory. Thus we can check our own results. It is an important merit of our method.

The results obtained by the AMP-method for four visual double stars are presented in table 1 and 2. The initial data for computing the AMP-orbit are presented in table 1. The AMP-orbit elements, computed on the basic observations being made during the time ΔT, are given in table 2. The errors of the AMP-orbit elements depend on the errors of the initial data. All our orbits agree with all observations (ancient and new). For a comparison the results of other authors are presented. The basic arc observations for all the stars, except ADS 11632, are observations obtained with the 26-inch refractor at Pulkovo. The basic arc for ADS 11632 is a set of 1910-1935 observations which is published by Hopmann (1955). This part of the apparent orbit is more convenient for determining the radius of curvature. The Pulkovo observations (1961-1982) are used for this star as the ancient observations. The values of ΔVr were derived from spectral observations for ADS 14636 (by Dr. E.L. Chentzov) and for ADS 48 (by Dr. A.A. Tokovinin). For other stars the values of Vr, as the values of $\mathfrak{M}_{AB}$ were determined by means of comparing with ancient observations. For ADS 48 we obtain $\mathfrak{M}_{AB} = 2.0\ \mathfrak{M}_{\odot}$. It is twice as much as the value $\mathfrak{M}_{AB}$ according to the "mass-luminosity" relation. Brosche and Güntzel-Lingner received the same high value of $\mathfrak{M}_{AB}$. This interesting result needs explanations. The investigation shows that for

TABLE 2

a	P	e	ω	i	Ω	T_0	ΔT	Author
				ADS 48				
6".10	370	0.50	279°	61°	23°	1717	25	AMP
± 47	± 43	± 5	± 8	± 2	± 3	± 40		
5.71	327	0.52	280	60	17	1733	75	Brosche
6.18	362	0.52	276	62	19	1710	77	Güntzel-L
11.70	1507	0.50	350	57	174	1969	85	Hopmann
				ADS 9031				
2.45	158	0.44	197	47	157	1916	17	AMP
± 13	± 13	± 9	± 47	± 8	± 15	± 20		
".42	155	0.44	200.9	46.8	155.1	1916.9	124	Strand
				ADS 14636 /61 Cygni				
24.9	734	0.40	158.6	52.3	172.1	1687	16	AMP
± 8	±34	± 4	± 4.8	±1.8	± 1.7	± 20		
24.6	722	0.40	158.0	51.8	172.3	1690	146	Josties
				ADS 11632				
17.47	585	0.35	264	70	142	1778	25	AMP
± 2.17	±112	± 4	±18	±1	± 1	± 14		
44.44	2547	0.63	342	74	146	1866	121	Hopmann
11.83	346	0.55	210	61	140	1767	125	Rabe
13.14	352	0.61	231	65	136	1789	122	Güntzel-L
15.66	453	0.53	248	69	137	1783	134	Heintz
10.50	295	0.70	203	52	139	1778	134	Baize

ADS 11632 we can obtain reliably only e, i and Ω. Therefore it is not surprising that all the authors obtained different orbits.

In conclusion we want to draw attention to the necessity of improvement of spectral observations so, that the relative radial velocity of the visual double star components could be determined with an error less than 0.1-0.2 km/s.

References

Kisselev A.A., Kiyaeva O.V. - 1980 - Astr. J. USSR, 57, 6, 1227.
Kiyaeva O.V. - 1983 - Astr. J. USSR, 60, 6, 1208.
Hopmann J. - 1955 - Mitt. Univ. Sternwarte Wien, 7, 3, 67.

Are Double Star Observations the Most Suspicious Ones in the Paris Astrolabe Programme ?

M.L. BOUGEARD (1)(2)
(1)Université ParisX, I.U.T. "Techniques aérospatiales"
Ville d'Avray,France
(2)Observatoire de Paris
Paris,France

Summary : By use of "influential diagnostic" statistics, we identify star observations that appear to have a disproportionate influence on the estimated model of the usual Paris Astrolabe reductions. We prove that these statistics may reveal the severity of remaining observational errors and detect some difficulties related to double stars and high magnitude stars.

1. INTRODUCTION

Since 1956 , Paris Astrolabe data have been regularly used for latitude and Universal Time (UT) work. For the purpose of future revision of this long astrometrical series for proper motion determinations, a careful analysis of the star observations appears to be necessary.

2. THE DATA : CHOICE OF TWO TEST-PERIODS

The Paris star observations are partitioned in groups g=1,..12, with $n(g,t) \leq 30$ stars observed at the mean date t, following their sidereal time (TS) of passage. The related "equations of condition" are solved -at the mean date t - by an ordinary least-squares fit (OLS regression) in which the dependent variable contains standard corrections such as a refraction correction,diurnal correction, internal correction (CI), curvature of parallel and amulcantar, etc (see:Débarbat,Guinot,1970) . We write it as DHC (resp. DH) when the "internal corrections" of the stars are (resp. are not) included.

The data considered here were obtained between 1956 and 1982. They are computed in the IAU systems of constants in use before 1984. The data are not homogeneous since we have had a variation of the observer team and two instrumental periods: 1956-70 the "Danjon's prism astrolabe (APD) with an optical prism in transmission and 1970-82 a "full pupil" astrolabe (APP) with an optical system in reflection .

(1) permanent address

Astrophysics and Space Science **142**(1988), 185–189.

Two test periods have been chosen: 1977-79 (APP), 1965-67 (APD) : through "factor analyses of correspondences", they have been proved to be homogeneous enough in terms of group and observer effects on the error variance of the usual group reductions (regressions). The groups observed in summer (Nr 11, 12, 1) tend to have higher error variances (smaller "weights" or "quality") as it has been detected.

3. DETECTING INFLUENTIAL OBSERVATIONS AND OUTLIERS

To detect the extent to which the presence of some remaining "outlier" -further mentioned as "atypical observation" - could have affected the estimated coefficients, standard errors and test statistics in the group reductions (regressions), the following "influential diagnostic" statistics have been used in their application to the reduction No 3017, group 10 :

- RSTUDENT: OLS residual divided by an estimate of its standard error calculated through a regression without the considered star,
- COVRATIO: a measure of the improvement or non improvement of the precision,
- DFITTS and Cook's D : measures of the change in the predicted value of the ith star observation obtained by deleting the ith equation,
- DFBETAS : a measure of the change in each parameter estimate (obtained by deleting the ith equation i=1,..n(g,t)).

In the reduction No 3017, a double star (rank k=15 in the group 10) and a high magnitude star (rank 13) have been seen to cause problems and to be very influential.

4. ATYPICAL STARS IN THE OBSERVATIONAL PROGRAMME

For the purpose of detecting "atypical" (influential, with "ill-behaved" residuals) stars, the above statistics have been performed for each DH, DHC usual OLS group regression for the test-periods 1977-79 (17258 observations) and 1965-67 (21465 observations).

To our surprise, 9 stars (among the 340 of the programme) appear as (systematically) atypical : their list is given in the Table below.

No instrumental (they have not the same azimuth) or observer effect has been detected; so, these stars show some lack of fit of the usual Astrolabe model (observation of a photocenter, elliptic trajectories in the case of double systems...).

Moreover, these 9 stars have all a high visual magnitude (mv) and the highest internal correction , CI, in the observational group to which they belong. 8 atypical stars belong to the FK4 Suppl. Catalogue (not so accurate as the FK4) and the star Nr 738 (FK4) is a double star which also has a high internal correction in the group Nr 23 of the Uccle Astrolabe (Paquet and De Rop, 1975).

TABLE: *List and characteristics of the atypical stars*

group g	rank in g	n°FK4 FK4supp.	CI	NCI	Z azimuth	mv	B-V	D	SV	N(77-79)	W
1	12	3656	0.318"	0.453"	144°	5.6	.22	.	.	75(90)	1
1	18	3722	0.259	0.430	85	4.9	1.08	.	.	79(90)	1
2	3	2319	0.287	0.467	257	4.9	-.03	.	1	62(69)	-1
7	26	2889	-0.463	-0.655	45	5.7	.06	1	.	35(44)	1
9	24	3185	-0.257	-0.478	46	4.9	1.50	.	.	31(47)	1
10	13	3594	0.301	0.552	295	5.5	-.06	.	.	65(77)	-1
10	28	3733	-0.422	-0.485	248	4.7	-.12	.	1	66(77)	-1
11	11	3320	0.215	0.397	123	5.3	1.08	.	.	72(78)	1
12	27	738	0.196	0.321	111	4.6	.38	1	.	80(91)	1

CI=internal correction (1964-84), NCI=new internal correction (1983-86)
D=1 (resp. SV=1) stands for a double star (resp. a "variable star") ; N=number of observations (in brackets: group observations of the test period), mv= visual magnitude, B-V= B-V colour index, W=passage (+1=west, -1=east).

5. THE PARTICULAR CASE OF THE DOUBLE STAR NR 2889 (FK4 supp) IN GROUP 7

In this section, we have considered the case of the star Nr 2889 (FK4 supp.), with rank 26 in group 7 : it is a high-magnitude double star and it has the highest internal correction in the Paris programme. We have proved that this star appears as atypical for the usual model (with explanatory variables - sin Z, cos Z - the sine and cosine of the azimuth Z , counted from the North, of the star) as well as for models with more explanatory variables (such as B-V colour index, sidereal time of passage of the star, visual magnitude mv) . Its strong influence on the reduction coefficients , error variance, nonnormality of the residual distribution, has been discussed through the introduction of a dummy variable related to the observations of this star in the reductions of group 7 . This dummy variable has proved to be highly significant.

Finally, the effect of this star in the reductions of group 7 has been examined through a principal component factor analysis with:

- active variables : sin Z, cosZ, B-V colour index, visual magnitude and sidereal time of passage of the star,
- supplementary variables : DH, DHC and internal correction CI .

This star significantly appears only on the 2nd axis of the factor pattern with a strong effect in colour index in the group.

6. FINAL NOTES

The exploration , over two homogeneous test periods, of the OLS residuals of the Paris Astrolabe data, through "influence statistics", has shown that 9 stars of the observational programme are "atypical": the most "suspicious" one being the high magnitude "double star" Nr 2889.

Astronomers have to pay attention to these "atypical" stars : when present in a reduction, they may destroy the (gaussian) normality of the residual distribution (consequently, destroy usual tests) and they have a "perverse effect" on the usual time and latitude determinations and of course on the usual determinations (see Débarbat and Guinot, 1970) of Astrolabe catalogues.

For details, the reader is referred to the following references.

ACKNOWLEDGEMENTS : The author gratefully acknowledges financial support from the Paris Observatory (laboratories B.I.H. and DANOF), from C.N.R.S. (UA 1125), C.N.F.A. and Center C.I.R.C.E (Orsay, France).

REFERENCES

Bougeard M.L.;1987a, 'Détection statistique d'effets perturbateurs dans les observations', Astron. Astroph., 173, 191-203

Bougeard M.L.;1987b, 'Statistical problems about the OLS use in astrometry', Astron. Astroph., 183, 156-166

Bougeard M.L.;1987c, 'Non-normality of the errors and suspicious observations' (submitted to Astron. Astroph.)

Bougeard M.L.;1987d, 'Time-magnitude-colour equation in visual astrometrical data' (in preparation)

Bougeard M.L.;1987e, 'Paris full-pupil astrolabe and Danjon's prism astrolabe: a statistical comparison in terms of time-magnitude-colour', Coll. in Mapping the sky, Paris (in press).

Débarbat S. and Guinot B., 1970, La méthode des hauteurs égales en astronomie, Gordon and Breach.

Paquet P. and W. De Rop, 1975, Bull. Obs. Belgique, 8, Nr 4.

WORLEY - Question 1. Are these bright stars ?
Question 2. Where did you check for duplicity ?
Comment : There are many newly discovered speckle pairs (as well as other visual doubles). I would be happy to check WDS for the pairs you list.

BOUGEARD - Les étoiles observées à l'astrolabe de Paris sont presque toutes de magnitude (visuelle) inférieure à 6, donc considérées comme "bright stars"; néanmoins à partir de la magnitude 4, elles peuvent être considérées comme des étoiles "faibles" pour l'astrolabe visuel. De 1956 à 1983, les équations de condition des étoiles observées à l'astrolabe de Paris ont été évaluées à partir des données du catalogue FK4 et du FK4 sup. et des constantes UAI d'avant 1984 (ceci en particulier pour les deux périodes testées ici). En ce qui concerne les indices de couleur B-V utilisés dans ce travail, nous les avons obtenus à partir de la base SIMBAD du C.D.S. de Strasbourg (France). Nous avons choisi les indices (B-V) de Nicolet (1978) et en leur absence les indices de Mermilliod (1986), ou à défaut la moyenne des deux derniers indices publiés fournis par la base de données du CDS. Nous remercions à ce sujet Dr Anita Stephanovitch-Gomez (Observatoire de Paris-Meudon) pour de nombreuses discussions à ce sujet. Les renseignements concernant les étoiles doubles "atypiques" proviennent eux aussi d'indications de la base CDS (catalogues IDS, ADS ...).
A noter que les "étoiles atypiques" relevées ici n'ont pas été observées en "double passage" : il est donc quasi-impossible à priori mathématiquement de séparer les erreurs sur les différentes composantes et les mouvements propres et ceci encore plus dans le cas d'étoiles doubles. Il est certain que tout renseignement complémentaire sera précieux et le bien-venu (pour le groupe Astrolabe de Paris) s'il permet d'améliorer l'équation de condition associée à chacune de ces étoiles. Il pourra être en particulier important de rectifier les équations de condition des deux étoiles atypiques du groupe 1 (bien que ce ne soit pas à priori des étoiles doubles) car elles ont vraisemblablement perturbé les estimations sur la position optique de la radio-étoile "Algol" observée dans ce groupe comme "étoile (multiple) de catalogue".

VISUAL DOUBLE STARS IN PROFESSIONAL SPANISH ASTRONOMY: PAST, PRESENT AND FUTURE PROJECTS

J. A. Docobo and J. F. Ling
Observatorio Astronómico "R. M. Aller"
Santiago de Compostela
Spain

Summary. This poster describes the contribution of professional Spanish astronomy to all aspects of the study of visual double stars (observation, calculation of orbits and development of new methods of calculation), together with projected programmes of visual and interferometer observations.

The first micrometer measurements taken in Spain seem to have been carried out in 1897 in the Observatori Catalá in San Feliú de Guixols, Barcelona by J. COMAS and R. PATXOT (1898: Astronomische Nachrichten, Band 146, No. 3497, 309-312), using a 20 cm Mailhat astrophotographic refractor. J.COMAS subsequently published further series of observations made with a 38 cm Mailhat and a 15 cm Grubb.

The true founder of the study of double stars in Spain may nevertheless be considered to be the clergyman R. M. ALLER, who besides obtaining micrometer measurements calculated several orbits and created a school of students of the subject.

Dr. Ramón M. ALLER (1878 - 1966)
Director of the Observatory of the University of Santiago de Compostela

Dr. J. COMAS (1868 -1937)
Director of the Fabra Observatory, Barcelona

Astrophysics and Space Science **142** (1988), 191–193.

R. M. ALLER	DISCIPLES OF R. M. ALLER	
Micrometer measurements published in Astronomische Nachrichten and Spanish journals; calculated double star orbits published in Astronomische Nachrichten and Spanish journals; supervisor of doctoral theses on double stars.	Prof. E. VIDAL	Theoretical studies of the calculation of orbits. Designer of the orbigraph an instrument facilitating the calculation of orbits. Calculation of orbits.
	Prof. R. CID	Analytical methods for the calculation of orbits. Methods for improving orbits. Celestial mechanichs.
	Dr. A. FERRIN	Micrometer measurements.
	Dr. J. A. ZAERA	Calculation of orbits.
	J. M. COSTA	Micrometer measurements. Calculation of orbits.

Currently, the Astronomical Observatory "R. M. Aller" of the University of Santiago de Compostela continues to work on various facets of the study of multiple stars in collaboration with the Department of Astronomy of the University of Zaragoza.

A) OBSERVATION	Nice Observatory (R 50, R 74), Pic du Midi Observatory (S 200): J. F. LING under the direction of Dr. P. COUTEAU Fabra Observatory, Barcelona (R 38): J.A. DOCOBO and J.M. COSTA (1985,86,87)
B) CALCULATION OF ORBITS	Double stars: J.A. DOCOBO, J.M. COSTA, J.F. LING, J.A. ZAERA Multiple stars: J.A. DOCOBO, A. ABAD
C) DESIGN OF ANALYTIC METHODS FOR THE CALCULATION OF ORBITS	Visual and spectroscopic double stars: J.A. DOCOBO, J.F. LING, A. ABAD, A. ELIPE Multiple stars: J.A. DOCOBO, A. ABAD, J.F. LING

At our request, the workshops of the Nice Observatory are currently building a wire micrometer suitable for use in several of the large instruments now installed in Spain at the Hispano-German Astronomical Centre at Calar Alto, Almería and the Roque de los Muchachos Observatory in the Canary Islands. In response to Dr. COUTEAU's kind offer, it is likewise hoped to intensify other kinds of collaboration with the Nice Observatory.

Finally, a group of astronomers at the Institute of Astrophysics in the Canary Islands (J. J. FUENSALIDA, F. J. FUENTES and R. NAVARRO) have announced their intention of applying speckle interferometry to the observation of multiple star systems.

A METHOD FOR CALCULATING THE ORBITS OF INTERFEROMETER/SPECTROSCOPIC BINARIES FROM COMBINED DATA

J. A. Docobo[1], A. Elipe[2] and A. Abad[2]
(1) Observatorio Astronómico "Ramón M. Aller",
Santiago de Compostela, Spain
(2) Dept. of Astronomy, Zaragoza University, Zaragoza, Spain

ABSTRACT. An extension of Docobo's analytical method for calculating the orbits of visual double stars allows all orbital elements to be obtained from three interferometer observations $(\theta,\rho;t)$ and three spectroscopic observations $(v;t)$.

Based on Docobo's analytical method for calculating orbits of visual double stars (Celestial Mechanics, 36, 143-153), the present procedure has been designed for use when measurements of radial velocities are also available, and in principle allows preliminary orbits to be calculated from combined data for just three epochs.

As in the original method, the starting point is a mapping

$$x \longrightarrow (P,T,e,a'',i,\Omega,\omega) \qquad (*)$$

of the interval $(0,2\pi)$ or $(0,+\infty)$ into the set of periodic Keplerian orbits whose apparent orbits pass through three given points $(\theta,\rho;t)_i$ $(i = 1,2,3)$, x being the difference between the excentric anomalies for epochs 3 and 1. In the present procedure, however, the control data, instead of also being of the form $(\theta,\rho;\ t)$ as in the original method, are radial velocities $(v;t)$ (usually referring to the main star). In principle it is sufficient to know just three such radial velocities, some or all of whose epochs t_4, t_5 and t_6 may coincide with t_1,t_2 and t_3. The nine measurements $(\theta,\rho;t)_i$ $(i = 1,2,3)$ and $(v;t)_j$ $(j = 4,5,6)$ suffice to determine the seven elements of the visual orbit together with V_o and K.

PROCEDURE

Let

$$Y_i = e\cos\omega + \cos(\omega+f)$$

so that

$$v_i = V_o + KY_i \qquad (**)$$

For each orbit determined by the mapping (*) the values Y_4, Y_5 and Y_6

Astrophysics and Space Science **142** (1988) 195-197.

are calculated for the epochs t_4, t_5 and t_6 respectively. Equation (**) now determines two systems, one for the pair (4,5) and another for the pair (4,6). Thus each value of x determines the values $(V_o)_{45}$, K_{45}, $(V_o)_{46}$ and K_{46}. The solution, the orbit that exactly fits the observed data, is given by the value x_o of x for which

$$\Delta V_o = (V_o)_{46} - (V_o)_{45} = 0 \quad , \quad \Delta K = K_{46} - K_{45} = 0$$

and then

$$d = \sqrt{(\Delta V_o)^2 + (\Delta K)^2} = 0$$

In practice, observational errors may mean that there is no value of x for which ΔV_o and ΔK are both exactly zero, but approximate solutions will be determined by values of x for which ΔV_o and ΔK are both close to zero. Naturally, the more control data are available, the less arbitrary is the choice of a solution.

EXAMPLE

HR 5747 - 15278 N 2906

Data:

$t_1 = 1976{,}370 \longrightarrow \theta = 502^{\circ}7$, $\rho = 0''295$ $\quad t_4 = 1969{,}688$ $v = -27{,}91$ Km/sec

$t_2 = 1980{,}470 \longrightarrow \theta = 329^{\circ}4$, $\rho = 0''098$ $\quad t_5 = 1976{,}173$ $v = -13{,}90$ Km/sec

$t_3 = 1981{,}470 \longrightarrow \theta = 226^{\circ}2$, $\rho = 0''055$ $\quad t_6 = 1978{,}890$ $v = -18{,}85$ Km/sec

since $|\theta_3 - \theta_1| > 180^{\circ}$, we shall work in the interval $180^{\circ} \leq x \leq 360^{\circ}$

FIRST ITERATION

$x = 180^{\circ}$ — $P = 14^y867$ — $\Delta V_o = -76{,}56$ — $\Delta K = +243{,}93$ — $d = 225{,}7$

$x = 210^{\circ}$ — $P = 10^y869$ — $\Delta V_o = -\ 4{,}49$ — $\Delta K = -\ 0{,}64$ — $d = 4{,}5$

$x = 240^{\circ}$ — $P = 8^y574$ — $\Delta V_o = +34{,}98$ — $\Delta K = -114{,}19$ — $d = 119{,}4$

$x = 270^{\circ}$ — $P = 7^y157$ — $\Delta V_o = -73{,}63$ — $\Delta K = +154{,}02$ — $d = 170{,}7$

SECOND ITERATION

$x = 215^{\circ}$ — $P = 10^y399$ — $\Delta V_o = -\ 0{,}38$ — $\Delta K = -\ 0{,}32$ — $d = 0{,}5$

$x = 220^{\circ}$ — $P = 9^y969$ — $\Delta V_o = +\ 1{,}94$ — $\Delta K = +\ 3{,}09$ — $d = 3{,}6$

$x = 225^{\circ}$ — $P = 9^y575$ — $\Delta V_o = +\ 5{,}75$ — $\Delta K = +\ 24{,}05$ — $d = 24{,}7$

THIRD ITERATION

x = 216º —— P = 10^y310 —— ΔV_o = 0,0 —— ΔK = 0,0 —— d = 0,0

SOLUTION:

P = 10^y310; T = 1980,509; e = 0,525; a = 0"206; i = 110º8

Ω = 148º7 ; ω = 183º1; V_o = -18 Km/sec; K_1 = 9,16 Km/sec

Figures showing the approach to the solution:

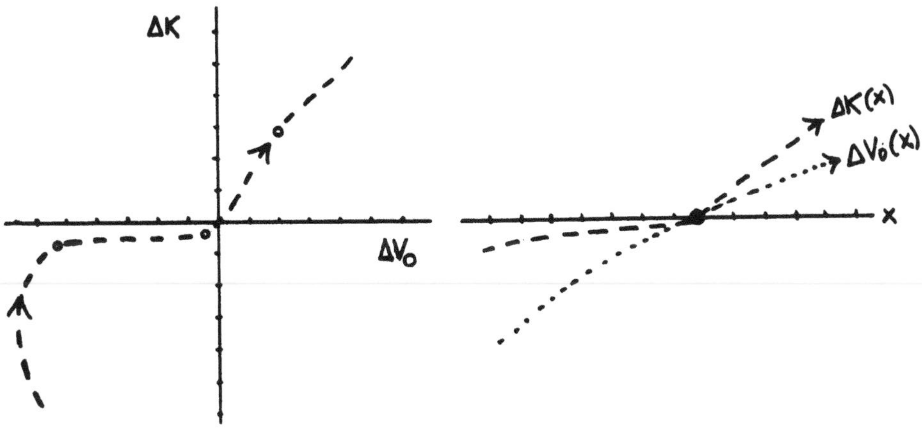

ABT - May I ask Dr. Worley whether this method differs from that being done by McAlister ?

WORLEY - I am under the impression that McAlister and colleagues used a method similar to that developed by Batten/Scarfe (?) I have not had time to compare the two methods but if dissimilar, I think it useful to have more than one method in this instance, where one is combining visual and spectroscopic observations.

THE FREQUENCY OF WIDE BINARIES AMONG THE SOUTHERN BARIUM STARS

Roger B. Culver
Colorado State University
Fort Collins, Colorado, U.S.A.

Philip A. Ianna
The University of Virginia
Charlottesville, Virginia, U.S.A.

Summary. All of the barium and marginal barium stars in the 1972 MacConnell, Frey, and Upgren (1972: MFU) list of such objects which are located south of the celestial equator were examined for visual duplicity. Over 200 objects were observed, and a total of six possible candidates for wide pair visual binaries were found. Of these, four are most likely optical in nature. These data suggest that the incidence of wide binaries among barium stars is of the order of 1%.

The evolutionary stage of the barium stars remains incompletely understood. These G and K giants display enhanced s-process elements for which two scenarios have been suggested as explanation for the peculiar abundances. In one picture the core helium flash and subsequent extensive mixing explains these stars (e.g. Deupree and Cole 1983); the alternative picture makes them all binaries with white dwarf companions that have enriched the original secondary through mass exchange (McClure 1983), although the presence of the white dwarf companions is not confirmed in many cases (Dominy and Lambert 1983).

Earlier studies of the wide binary systems HD 196673 (Culver and Ianna 1976), HD 105902 (Culver, Ianna, and Franz 1977) and HD 126313 (Culver and Ianna 1980) have helped place constraints on the absolute magnitudes and masses of these stars. Additional studies of similar systems would clearly be of help to extend our knowledge of the physical properties of the barium stars. For this reason we have tried to find a few more wide barium star binaries.

Using the 66-cm refractor of the Mt. Stromlo Observatory, 200 southern hemisphere barium and marginal barium stars from the list of MacConnell et al. (1972: MFU) were examined for visual companions. Of the 200 stars examined, a total of six revealed companion stars which could possibly be physical companions. These six stars are listed in Table I along with approximate Δm values, separations, and position angles. Any pair separated by more than 70" and/or possessing a Δm

Astrophysics and Space Science 142 (1988), 199–201.

greater than 3.0 was discarded from the list of potential physical pairs.

Table I. Measurements for Southern Barium Star Doubles.

Star	MFU	Δm	PA	Separation
HD 4084	1	1.0	163°9	6".2
HD 36650	28	2.6	227.7	62.0
HD 107264	80	2.3	342.1	63.6
HD 223586	150	3.0	142.9	56.3
HD 21682	158	1.2	251.0	16.8
HD 31341	165	0.9	129.6	49.3

Each of the candidates listed in Table I was photographed with the 66-cm refractor and the resulting plates were measured and reduced according to standard procedures. Orientations for the plates were determined from stellar trails.

The large Δm values combined with the large separations for several objects in Table I suggest that the 'companion' stars for HD 36650, HD 107264, and HD 223586 may be optical in nature. The true nature of the three remaining stars, HD 4084, HD 21682 and HD 31341 cannot be ascertained in the absence of proper motion or radial velocity data, although low dispersion radial velocity data for both components of HD 31341 (Culver, unpublished) suggest that these two components have differing radial velocities. From data kindly supplied by Nys (1987) and Worley (1987), two of these stars (HD 4084 and HD 21602) have been measured previously. These data are as follows:

HD 4084	= IDS 00374S5620	1877	164°	6.3"
		1953	162°	6.3"
HD 21682	= IDS 03248S4400	1916.69	255.9°	17.28"

The survey was such that any visual pair having a separation larger than 1" and a small to moderate Δm should have been detectable. If we assume an absolute magnitude of +0.2 for the typical barium star (MFU 1972), the limits on the angular separations adopted for this survey (1" - 70") correspond to a range in linear separations of roughly 400 AU to 28,000 AU for systems whose primary is between 8.0 and 9.0 in visual magnitude. The survey also excludes possible companion stars whose absolute magnitudes are fainter than about +3.2, or three magnitudes fainter than the barium star primary. The overall statistical results of this survey strongly suggest that wide binaries are comparatively rare among the barium stars, being on the order of 1% of known objects.

This work has been supported in part by the National Science Foundation under grants AST82-00232 and AST83-18782 and by the Estate of Leander J. McCormick; the support of the Mt. Stromlo and Siding Spring Observatories is also gratefully acknowledged.

REFERENCES

Culver, R.B. and Ianna, P.A.: 1976, Publ. Astron. Soc. Pac. 88, 41.
Culver, R.B., Ianna, P.A., and Franz, O.G.: 1977, Publ. Astron. Soc. Pac. 89, 397.
Culver, R.B. and Ianna, P.A.: 1980, Publ. Astron. Soc. Pac. 92, 829.
Deupree, R.G. and Cole, P.W.: 1983, Astrophys. J. 269, 676.
Dominy, J.F. and Lambert, P.L.: 1983, Astrophys. J. 270, 180.
MacConnell, D.J., Frye, R.L. and Upgren, A.R.: 1972, Astron. J. 77, 384.
McClure, R.D.: 1983, Astrophys. J. 268, 264.
Nys, O.: 1987, Private Communication.
Worley, C.: 1987, Private Communication.

ABT - This is a good example of the use of wide visual binaries because Barium stars infrequently occur in open clusters, so the observations of visual binaries allow the determination of absolute magnitudes.

LODEN - Have you in any case found barium star characteristics also in the companion ?

CULVER - We have not yet examined the spectra of any of the companion stars involved. It is anticipated that this will be done in the near future.

KOPAL - How did the barium get there - and to what extent could the difference between binaries and open clusters be accounted for by the origin of this barium ?

CULVER - The question of how the barium got there would be worth an entire conference in itself. McClure's work at Victoria suggests that the barium enhancement observed in these objects is due to the presence of close companion stars which are now in the white dwarf stage. The observational limits on our survey prevents us from detecting such components.

OBSERVATION ET RESOLUTION DU SYSTEME TRIPLE η Virginis PAR LA METHODE DES OCCULTATIONS

M. FROESCHLE et C. MEYER, CERGA, France

Summary. A high-speed photometer has been used to observe the occultation of η Vir. Analysing the photoelectric record of the signal, it is obvious that a visual companion exists. The data processing using the parameter fitting method drives us to retain a three-component pattern. We provide both the angular separations and flux ratios for this pattern. A detailed comparison of our results and those obtained by other methods is made.

1. INTRODUCTION

L'observation de l'occultation d'une étoile double permet d'accéder à la séparation angulaire projetée sur la direction de déplacement de la Lune et à la différence de magnitude des composantes du couple (Nather et Mc Cants, 1970 ; Froeschlé et Meyer, 1986). On obtient typiquement une précision de ± 2 mas sur la séparation et de ± 0,2 sur la différence de magnitude. Nous avons utilisé le photomètre rapide à comptage de photons TELOC (Froeschlé et al., 1987) pour l'observation de l'occultation de η Vir (SAO 138721) le 19 mai 1986, dans la longueur d'onde λ=0,41 µm avec une bande passante de 0,04 µm . La figure 1 montre la courbe observée brute. On distingue sans ambiguïté la première frange qui précède la disparition de la composante principale, le plateau d'un compagnon et la disparition de ce dernier. Pour l'ensemble des comptages, le rapport signal sur bruit est de l'ordre de 17.

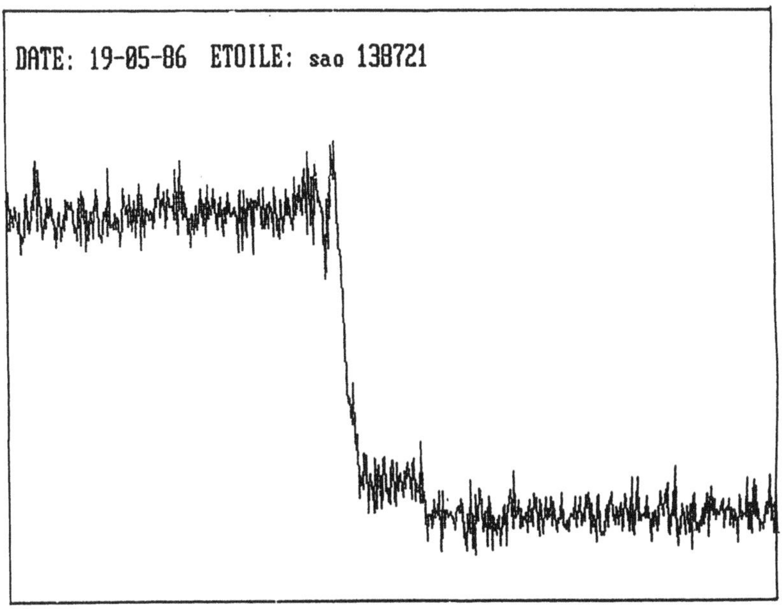

FIGURE 1

Astrophysics and Space Science **142** (1988), 203–206.

2. REDUCTION

Pour réduire notre observation, nous avons utilisé la méthode par ajustement paramètres (Nather et Mc Cants, 1970). L'étoile η Vir est une étoile tri connue (Mc Alister, 1977 ; Harper, 1935). Elle se compose d'un couple vis serré et d'un compagnon spectroscopique. Nous avons successivement utilisé modèle d'étoile double et un modèle d'étoile triple.

a) Hypothèse d'une étoile double. La figure 2 représente, au voisinage de première frange, les points observés, le meilleur ajustement et les résidus O- On remarque que ces derniers ont un aspect systématique significatif avec écart type de 0,12.

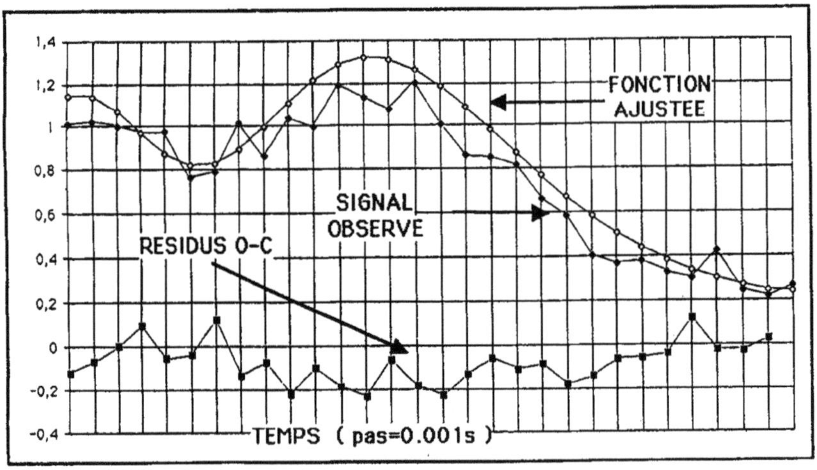

FIGURE 2

Bien que nous n'ayons pas d'observation simultanée dans une autre longu d'onde, il y a de bonnes raisons de penser que cette déformation du sig observé n'est pas due à un accident du relief lunaire au voisinage du po d'occultation ou à un diamètre apparent de la source (Evans, 1970).

b) Hypothèse d'une étoile triple. Dans les mêmes conditions, la figure 3 mon que les résidus n'ont plus d'aspect systématique. L'écart type vaut dans ce 0,07, ce qui est comparable à la valeur 0,06 obtenue pour l'ensemble de l'éch tillon. Nous avons donc retenu ce modèle pour la réduction de l'observation.

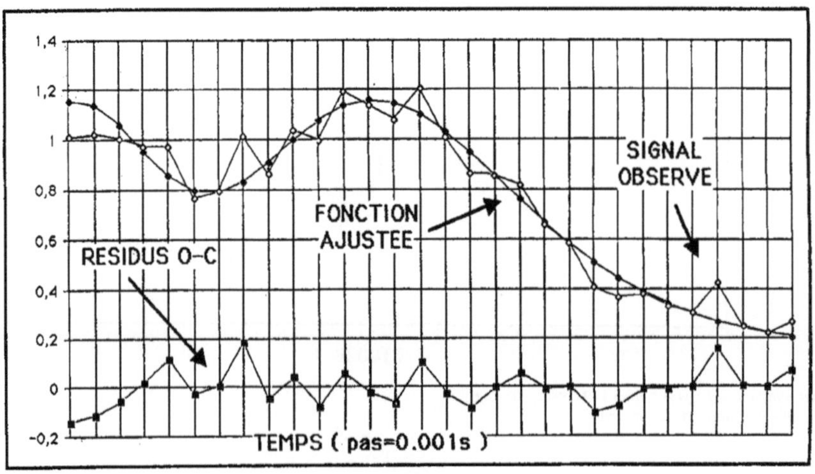

FIGURE 3

3. RESULTATS

Le tableau I donne les valeurs numériques relatives à la séparation, où :

ρ_o est la séparation projetée observée, mesurée à partir de la composante principale,

θ_t et r_t sont respectivement l'angle de position et la séparation du couple visuel, déduits de l'orbite connue de l'étoile double (Balega et al., 1984),

r_o est la séparation réelle, déduite de l'observation, connaissant θ_t.

La valeur de r_o est en bon accord avec celle calculée à partir des éléments de l'orbite obtenus au moyen de mesures interférométriques.

SEPARATION	ρ_o	θ_t	r_o	r_t
DOUBLE VISUELLE	0".028 ±0.002	54°.8 ± 3	0.099 ±0.003	0".094 ±0.007
DOUBLE SPECTRO	0".007 ±0.002	???	???	???

TABLEAU I

DIFFERENCE DE MAGNITUDE	Δm
DOUBLE VISUELLE	2.3±0.2
SYSTEME TRIPLE	
Δm_{01}	1.8±0.2
Δm_{02}	2.1±0.2

TABLEAU II

Le tableau II rassemble les résultats relatifs aux différences de magnitude entre les trois objets. Les écarts Δm_{01} et Δm_{02} s'appliquent respectivement au couple spectroscopique et au couple formé par la composante principale et le compagnon éloigné. La valeur $\Delta m = 2,3 \pm 0,2$ concernant la double visuelle est en bon accord avec celles obtenues par d'autres estimations (Bonneau, 1987 ; Tokovinin, 1984). La différence de magnitude Δm_{01} du système serré est à comparer à la valeur 1,3 obtenue par une méthode spectroscopique (Conti, 1969).

REFERENCES

Balega, Y., Bonneau, D., Foy, R. (1984), A & A Suppl., 57, 31
Bonneau, D. (1987), Communication privée.
Conti, P.S. (1969), Astron. J., 158, 1085
Evans, D.S. (1970), Astron. J., 75, 589
Froeschlé, M., Meyer, C. (1986), IAU Symp. 109, Eichhorn and Leacock eds.
Froeschlé, M., Meyer, C. (1987), Annales de Physique, 12, 39.
Harper, W.E. (1935), Pub. Dom. Ap. Obs., 6, 227.
Mc Alister, H.A. (1977), Ap. J., 215, 159.
Nather, R.E., Mc Cants, M.M. (1970), Astron. J., 75, 963.
Tokovinin, A.A. (1984), Sov. Astr. 10(2).

RAKOS - Attention should be paid to the possible roughness of the moon limb. Spectroscopic observations should be made to prove the results.

MEYER - The best fitting is obtained with the theoretical speed of the fringe system. This means that there is no significant slope near the occultation point. This is confirmed by the fact that a typical 10° slope would produce a difference of about 16 millisecond of arc in the determination of the separation. This is excluded in regard of the close agreement of our measure with the value given by the orbit and also in regard of the method-error wich is of about 0.003".

KOPAL - To what extent could the outcome of your measurements be affected by micro-irregularities of the lunar limb ?

MEYER - The deformation that may arise from micro-irregularities of the lunar limb have been described by Dr. Evans. It appears from this study that this kind of deformation are not similar to those we have observed.

MAZEH - Can you predict what be the amplitude of the radial-velocity modulation of the close pair ?

MEYER - The amplitude of the radial-velocity modulation of the spectroscopic pair is to be deduced from the elements of the orbit given by Harper.

THE RESEARCHES AT TORINO ON WIDE DOUBLE STARS

R. Pannunzio and R. Morbidelli
Astronomical Observatory of Torino
10025 Pino Torinese (TO)
Italy

Summary.. The present poster shows the main researches conduced at the Astronomical Observatory of Torino during this last decade in the field of the wide double stars. From the observational point of view, two different photographic techniques carried out with our 105 cm astrometric reflector over a selected sample of wide binaries are explained. With reference to the Hipparcos mission, we show two aspects of our collaboration with the INCA and FAST Consortiums. From the statistical point of view, a study concerning the systematic and accidental errors detected in visual double star observations is explained. We conclude the poster with the future foreseen researches concerning new observational techniques and new theoretical statistical studies on these wide objects.

1. INTRODUCTION

Since several years the Astronomical Observatory of Torino is engaged in astrometric studies, particularly devoted to the wide double stars, with the main purpose of develop new observational techniques in order to obtain results with an accuracy grater than that attained till now. On the other hands statistical and theoretical studies on these wide double objects are also in progress.

2. OBSERVING TECHNIQUES

Two different observing techniques in this last decade have been developed for double star researches: the photographic multiple exposures method and the photographic method of the trails. The first method, started in 1976, has produced in

Astrophysics and Space Science **142** (1988), 207–210.

this last decade 1760 polar coordinates of some hundreds visual double stars with separations greater than 2" and magnitudes brighter than 13th. The observations were performed with the 105 cm astrometric reflector of our Observatory on Kodak spectroscopic plates of type IIa-O or 103a-O and format 16x16 or 9x12 cm. The observing technique consists in a series of 20 contiguous exposures of the binary followed by an orientating trail. Usually for each binary 3 plated in different nights were carried out. The average error found for a single image was of +/- 0.87", while the average standard deviation per plate was of +/- 0.026" in both coordinates.
The second method consists in a series of two pairs of trails left by a binary star on a plate obtained with the slow motions of the telescope drive. The reduction method consists in the determination of several contiguous separations measured on the two pairs of trails by means of a PDS densitometer. The combination of the average values of the separations obtained for each trail with the angle formed between them, gives the polar coordinates of the binary.
This method has given an internal accuracy greater than that obtained with the multiple exposures method, while the external accuracy is about the same in both methods.

3. SPACE TECHNIQUES

Our group of research has cooperated in these years for the Hipparcos mission. More precisely has cooperated with the INCA Consortium for the accurate determination of the equatorial coordinates of some double and multiple systems that probably should be observed by the astrometric satellite and furthermore with the FAST Consortium for the realisation of the most representative model of double star that the satellite could find during the scanning of the sky.

4. STATISTICAL STUDIES

In this last year our research group is engaged in a statistical study concerning the errors in visual double star observations. The purpose of the work is the detection of the systematic and accidental errors in the measures of visual double stars obtained by means of different techniques (micrometric, photographic and interferometric). At present a first paper has been published concerning the errors found in the micrometric measures of visual double stars performed by the observers of the "old-generation" on a sample of 50 pairs with well-known orbit of grade 1, (Pannunzio et al

:,1986) and as soon as possible a second paper concerning the errors found in the micrometric measures of the "new-generation" observers will be submitted for publication. A third and conclusive paper concerning the errors detected in the photographic and interferometric measures of the same set of binary orbits should be ready before the end of 1987.
The results at present available obtained from the above mentioned studies have shown the existence of a positive increasing trend in the residuals in ρ for normalized separations to a 62 cm telescope lower than 0.2" in both observer-generations and an irregular negative-positive trend in the same residuals for normalized separations greater than 0.75"; while for the residuals in θ there is only a systematic positive trend in both distributions (old and new generation) for normalized separations up to 0.75".
A possible explanation of the systematic positive trend of the (O-C) in ρ for low separations should depend from the resolving power of the assumed 62 cm telescope used for normalization of the measures obtained with different telescope-sizes; while the trend obtained for separations larger than 0.75" should depend from selection effects due to the relative small number of measures in that range of separations. With reference to the accidental errors the study shows that these errors increase either for residuals in ρ or for residuals in θ with the increasing separations and that the standard deviation of the (O-C) in ρ is about double than that of the residuals in θ. However the final purpose of this work is the definition of a weighting procedure for the measures of those binaries without orbit determination or with a preliminary orbit.

5. FUTURE RESEARCHES ON WIDE DOUBLE STARS

The future foreseen researches on wide double stars in our Observatory can be summarized as follows:
- continuation of the observing programme of wide double systems obtained by means of new techniques (for example with a CCD detector)
- dynamical studies concerning the wide double systems by means of the analysis of the available Catalogues (AGK3, Parallaxes and Radial Velocities)
- continuation of the double and multiple star observations for the INCA Consortium
- conclusion of the statistical work concerning the systematic and accidental errors detected in visual double star observations obtained with different techniques over a sample of well-known orbits
- studies concerning the evolution of the close binary sy-

stems towards very large systems.

ACKNOWLEDGEMENTS

The authors of the poster would like to thank the colleague L. Armanelli, L. Bacchelli, A. Delgrosso, M. Lattanzi, G. Massone, F. Siciliano, M. Scardia and V. Zappalà which have cooperated for the researches mentioned in the poster.

REFERENCES

Pannunzio, R., Zappalà, V., Massone, G., Morbidelli, R.: 1986, Astron. and Astrophys., 166, 337

WORLEY - I think that comparing the visual observations with the existing visual orbits of high quality is perhaps not the best solution for determining systematic errors, because these very observers are partly (or largely) responsible for the orbits. A better solution is to use the independent speckle observations to adjust the semi-major axes, a, and then to make the comparison. For example, one of the grade I orbits has recently been shown by Hartkopf to require a correction of 0".01 in its computed a = 0".10. Although a small amount, this has a 30 % effect on the masses.

PRECISE RADIAL VELOCITY MEASUREMENTS OF VISUAL BINARY STARS

S. Weichinger and K. D. Rakos
Institute for Astronomy
Vienna, Austria

Program :

The radial velocity of a selected number of components of visual binaries have been measured. Attention is paid to systems suspected to have third (invisible) companions. A second aim of observation are selected couples from the "Catalog of radial velocities of binary stars" by Dommanget, for which the decision between different orbits is possible by the knowledge of a single radial velocity measurement.

Until now a total of 110 spectra of 31 visual binaries have been obtained. This program is very well suited for the equipment of the observatory and the small number of clear nights we have in Vienna. It will be continued for at least a few years.

Four spectra of the brighter component of ADS 9737 (HR 5834) are presented for illustration. The region of the H Alpha line shows narrow atmospheric lines suitable for exact calibration of the wavelength scale. However the presented examples have been calibrated using a neon comparison source.

Instrumentation :

The Echelle spectrograph of the FOA Vienna Observatory (Weiss et al. 1981, P.A.S.P. 93, 787) attached to the 1.5 m telescope gives a reciprocal linear dispersion ranging from 6 A/mm at 4000 A to 9 A/mm at 7000 A. The linear-silicon-diode array camera, cooled by liquid nitrogen, is equipped with a 936/1872 pixel Reticon array. An accuracy of 0.8 km/s can be achieved using conventional comparison spectra of iron or neon. This accuracy can be increased to better than 0.5 km/s using comparison lines of telluric oxygen and sophisticated data reduction software.

Astrophysics and Space Science **142** (1988), 211–212.

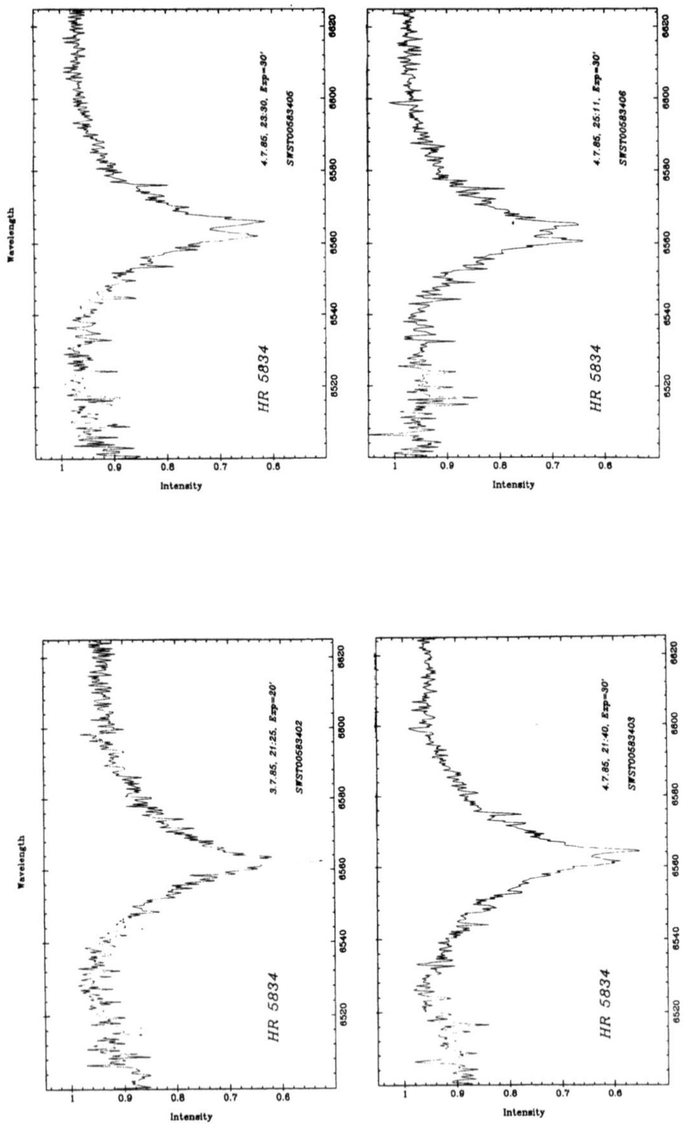
Wavelength
Intensity
HR 5834
4.7.85, 23:30, Exp=30'
SWST00583406
HR 5834
4.7.85, 25:11, Exp=30'
SWST00583406
HR 5834
3.7.85, 21:25, Exp=20'
SWST00583402
HR 5834
4.7.85, 21:40, Exp=30'
SWST00583403

MESURES PHOTOGRAPHIQUES D'ETOILES DOUBLES VISUELLES LARGES

M. SCARDIA
Osservatorio Astronomico di Brera
MERATE
Italy

Etant disponible, depuis 1984, à l'Observatoire de Brera-Merate l'astrographe ZEN de 20 cm. d'ouverture et 2077 mm. de focale, on a commencé un programme d'observation photographique d'étoiles doubles visuelles larges, en general donc negligées par les observateurs, par la methode des expositions multiples.

Le programme de travail a été établi en considerant la focale courte de l'instrument et toutes les causes possibles d'erreurs systématiques, qui sont désormais bien connus d'après le travail fondamental d'Hertzsprung (1920). On a donc choisi des couples de séparation plus grande que 17", dans lesquels la difference de magnitude entre les composantes était, en general, plus petite que 1.5 magnitudes.

La procedure d'observation est traditionelle: on a pris sur la même plaque en moyenne 16 images du couple en étude, en decalant chaque fois la plaque de quelque millimètre; après, le mouvement horaire étant arreté, on a obtenu la trace qui materialise la direction E-O. On a toujours utilisé des plaques KODAK IIaD avec un filtre jaune.

La mesure des plaques a été faite par le comparateur digitisé KORISTKA de l'Observatoire de Brera-Merate, dont les éléments mesureurs sont deux resaux de HEIDENHAIM, leur précision étant meilleure que 0.5 micromètres.

On a corrigé l'angle de position et la séparation même pour l'effect de la réfraction atmosphérique.

Dans la table 1 on montre un exemple des résultats obtenus; en particulier, pour chaque couple, on donne la date d'observation, séparation et angle de position avec leur incertitude. Dans la dernière ligne on donne les moyennes ponderées avec les erreurs relatifs.

REFERENCES

BOTTGER,G.,1962, Publ. Astrophys. Obs. Potsdam n° 102
HERTZSPRUNG,E.,1920, Publ. Astrophys. Obs. Potsdam n° 75
KRANJC,A.,1951, Contr. Oss. Astr. Milano-Merate n° 31
REUIJL,D.,1931,"Photographic Measures of Close Double Stars", Den Boer Ed., Utrecht
SCARDIA,M.,PANNUNZIO,R.,1982, Astron. Astrophys. 107,362

Astrophysics and Space Science **142** (1988), 213–214.

Tableau 1

ADS 639 - STF 1 App. I

1986.920	47".052	0".103	47°.53	0°.126	
1986.975	46".927	0".093	47°.55	0°.113	
1986.977	46".942	0".130	47°.64	0°.158	
1986.991	47".043	0".076	47°.58	0°.093	
1987.008	47".158	0".064	47°.45	0°.077	
1986.974	47".057	0".044	47°.53	0°.032	5 n

ADS 988 - STF 98

1986.920	19".514	0".086	248°.81	0°.253	
1986.975	19".525	0".100	248°.69	0°.273	
1986.977	19".555	0".141	248°.62	0°.342	
1986.991	19".470	0".083	248°.49	0°.215	
1987.008	19".647	0".094	248°.42	0°.266	
1986.974	19".536	0".031	248°.60	0°.072	5 n

ADS 996 - BU 1029

1986.978	22".949	0".129	62°.33	0°.311	
1986.991	22".644	0".108	63°.07	0°.303	
1987.008	22".983	0".080	62°.33	0°.224	
1987.010	22".846	0".082	62°.71	0°.237	
1986.997	22".869	0".071	62°.58	0°.165	4 n

ADS 1982 - STF 5 App. I

1986.978	37".967	0".102	274°.29	0°.133	
1986.989	38".138	0".107	274°.40	0°.190	
1986.991	37".983	0".078	274°.43	0°.085	
1987.008	38".201	0".076	274°.53	0°.131	
1987.010	38".072	0".064	274°.13	0°.054	
1986.995	38".076	0".043	274°.26	0°.077	5 n

POSITIONS PRECISES D'ETOILES DOUBLES VISUELLES LARGES

M. SCARDIA
Osservatorio Astronomico di Brera
MERATE
Italy

Depuis 1984 on a commencé à l'Observatoire de Brera-Merate un programme d'observation de petites-planetes et cometes, à l'aide de l'astrographe ZEN de 20 cm. d'ouverture et 2077 mm. de focale. On utilise habituellement des plaques KODAK IIaD avec un filtre jaune. Le champ utile de chaque plaque mesure environ 2° X 2°.

Sur chaque image, outre les étoiles de référence, la petite-planete ou la comete, on a cherché et mesuré même un certain nombre d'étoiles doubles visuelles larges.

Le but final de ce travail est de donner les positions precises, ascension droite et declinaison, des composantes chaque couple et les coordonnées polaires ρ et θ , en contribuant à la formation du "Catalogue of Components of Double and Multiple Stars" (C.C.D.M.).

La mesure des plaques a été faite par le comparateur digitisé, à deux axes, KORISTKA de l'Observatoire de Brera-Merate, dont les éléments mesureurs sont deux resaux de HEIDENHAIM, leur précision étant meilleure que 0.5 micromètres. La réduction des mesures a été faite en utilisant un groupe de programmes écrits pour cette application spécifique (Ascione et al., 1985). Les étoiles de référence ont été tirées d'une version sur bande magnétique du catalogue S.A.O.

On a utilisé en moyenne 13 étoiles de référence pour chaque plaque et des formules à six constantes de plaque; le residu moyen typique, pour les étoiles de référence, a été de 0".33 en $\alpha \cos \delta$ et de 0".30 en δ , ces valeurs étant la moyenne de plus que 130 plaques reduites jusqu'aujourd'hui. On a fait même une estimation de la magnitude des composantes chaque couple.

Pour toutes les étoiles doubles mesurées on a tempté l'individualisation sur le catalogue B.D. et sur l'I.D.S..

Dans le tableau 1 on donne un example des résultats obtenus. En particulier, pour chaque couple, on donne la date de l'observation en années et fraction, ascension droite et declinaison (1950.0), respectivement de l'étoile principale et du compagnon, séparation et angle de position, calculés à partir de $\Delta \alpha \cos \delta$ et $\Delta \delta$, magnitudes estimées. Dans la dernière ligne on donne les moyennes avec les incertitudes relatives.

Astrophysics and Space Science **142** (1988), 215–216.

REFERENCES

PODOBED,V.V.,1965, "Fundamental Astrometry", The University of Chicago Press, Chicago and London

ASCIONE,A.,BERGAMINI,U.,KRANJC,A. and SCARDIA,M.,1985, Rapporto Interno Osservatorio Astronomico di Brera n° 1/85

Tableau 1

IDS 00038N5904 - ARG 1 - BD +58°.0004 - SAO 21134

1985.559	$00^h06^m31^s.24$	$59^\circ21'12''.6$	$00^h06^m29^s.38$	$59^\circ21'33''.3$	$25''.11$	$325^\circ.5$	9.2 - 9.2
1985.561	00 06 31.24	59 21 12.5	00 06 29.42	59 21 33.4	25.11	326.3	9.2 - 9.2
1985.561	00 06 31.18	59 21 12.4	00 06 29.44	59 21 33.8	25.20	328.1	9.2 - 9.2
1985.561	$00^h06^m31^s.22$	$59^\circ21'12''.5$	$00^h06^m29^s.41$	$59^\circ21'33''.5$	$25''.14$	$326^\circ.7$	9.2 - 9.2
	.02	.1	.02	.2	.03	.8	

\- SCA 1 -

1985.559	$00^h07^m07^s.34$	$58^\circ42'56''.3$	$00^h07^m05^s.41$	$58^\circ42'55''.4$	$15''.06$	$266^\circ.6$	9.5 - 10.5
1985.561	00 07 07.35	58 42 56.1	00 07 05.50	58 42 56.0	14.41	269.6	-
1985.561	00 07 07.34	58 42 56.0	00 07 05.49	58 42 56.0	14.41	270.0	9.5 - 10.5
1985.561	$00^h07^m07^s.34$	$58^\circ42'56''.1$	$00^h07^m05^s.47$	$58^\circ42'55''.8$	$14''.63$	$268^\circ.7$	9.5 - 10.5
	.00	.1	.03	.2	.22	1.1	

IDS 00047N5814 - STI 1287 - BD +57°.0012

1985.559	$00^h07^m26^s.86$	$58^\circ30'03''.3$	$00^h07^m25^s.42$	$58^\circ30'11''.9$	$14''.19$	$307^\circ.3$	8.5 - 10.5
1985.561	00 07 26.84	58 30 02.5	00 07 25.47	58 30 12.0	14.34	311.5	-
1985.561	00 07 26.81	58 30 02.9	00 07 25.49	58 30 11.6	13.52	310.1	8.5 - 10.5
1985.561	$00^h07^m26^s.84$	$58^\circ30'02''.9$	$00^h07^m25^s.46$	$58^\circ30'11''.8$	$14''.01$	$309^\circ.6$	8.5 - 10.5
	.01	.2	.02	.1	.25	1.2	

IDS 00425N1220 - STT 8 - BD +12°.00089 - SAO 92088

1985.520	$00^h45^m04^s.29$	$12^\circ36'24''.7$	$00^h45^m06^s.73$	$12^\circ35'58''.9$	$44''.06$	$125^\circ.8$	8.8 - 8.9
1985.520	00 45 04.25	12 36 24.7	00 45 06.74	12 35 59.0	44.60	125.2	8.8 - 8.9
1985.520	00 45 04.26	12 36 24.7	00 45 06.75	12 35 58.7	44.77	125.5	8.8 - 8.9
1985.520	00 45 04.27	12 36 24.3	00 45 06.73	12 35 59.0	44.01	125.1	8.8 - 8.9
1985.520	00 45 04.26	12 36 24.6	00 45 06.73	12 35 58.8	44.42	125.5	8.8 - 8.9
1985.520	00 45 04.27	12 36 24.7	00 45 06.75	12 35 59.0	44.48	125.3	8.8 - 8.9
1985.520	$00^h45^m04^s.27$	$12^\circ36'24''.6$	$00^h45^m06^s.74$	$12^\circ35'58''.9$	$44''.39$	$125^\circ.4$	8.8 - 8.9
	.01	.1	.00	.1	.12	.1	

SOFTWARE FOR CALCULATING ORBIT AND MASS OF VISUAL BINARIES

S.Siregar
Bosscha Observatory and Department of Astronomy
Institute of Technology,Bandung,Indonesia

Abstract: This paper presents an overview on the software that will be used at the Bosscha Observatory of Indonesia for calculating the orbit and mass of visual binary system. The programme is based on the Thiele-van den Bos methods. The algorithm is designed for elliptic motions and applicable to personal computer.

1.Introduction

Since the Bosscha Observatory was established in 1923 researches on visual binary stars play an important role in astronomical researches in Indonesia. Van Albada (1957) proposed a method to determine the orbit according to three base points of observation for moderate length of secondary trajectories. Several computational techniques proposed by Arend et al(1964) or algorithm by Dommanget(1968) give us a possibility to determine the element orbits. The works which were done by Nys(1983) show that the proposals of Arend and Dommanget are usable for the modern computers of types IBM 1620,IBM 360, SIEMENS 4004 and UNIVAX 1100.

We do not have any computer like those types, so a very simple algorithm must be developed. In this work the Thiele-van den Bos methods is used, that is employing three complete observations of the form (t,θ,ρ) together with the double areal constant which must be obtained from additional data. To avoid a trivial solution a sufficient and/or necessary condition for the existence of a periodic orbit is taken from several authors.

2.Notation

P, T, e, a, i, ω, Ω have their usual meaning for orbits

$u = 2\Pi/P$: mean annual motion

Astrophysics and Space Science **142** (1988), 217–221.

t_i : time at which the secondary star occupies its i-th position

E_i : eccentric anomaly at time t_i

$t_{ij} = t_j - t_i$

$E_{ij} = E_j - E_i$

$V = E_{13}$; $U = E_{23}$ and $W = E_{12}$

θ_i , ρ_i : coordinates of the secondary star at time t_i

$\Delta_{ij} = \rho_i \rho_j \sin(\theta_j - \theta_i)$

C : the double areal constant of the apparent orbit

We define several supplement variables;

$S_1 = (t_{23} - \Delta_{23}/C)$; $S_2 = (t_{13} - \Delta_{13}/C)$; $S_3 = (t_{12} - \Delta_{12}/C)$

$R = \Delta_{12}/\Delta_{13}$; $S = \Delta_{12}/\Delta_{23}$

$M = R(t_{12} - St_{23})/(Rt_{13} - St_{23})$

$N = S(t_{12} - Rt_{13})/(Rt_{13} - St_{23})$

$Q = (RS + R - S)^2/2RS$

3. Fundamental equations

If, for the three pairs $(t_1, t_2), (t_2, t_3)$ and (t_3, t_1) the Thiele Innes equations (vide; Aitken ,1935) are written in Kepler' equation, the result is;

$$
\begin{aligned}
f_1 &= U - \mathrm{Sin}U - uS_1 \\
f_2 &= V - \mathrm{Cos}V - uS_2 \\
f_3 &= W - \mathrm{Cos}W - uS_3 \\
f_4 &= U + W - V
\end{aligned}
\qquad (3.1)
$$

The three elliptical segments S_1, S_2 and S_3 must be positive. I case of an elliptical orbit, the equation(3.1) will lead to solution for a mean annual motion only when (Dommanget, 1960)

$$f = S_1^{1/3} - S_2^{1/3} + S_3^{1/3} > 0 \qquad (3.2)$$

It should be remembered that the function f always increases wit increasing C. Hence a slight increase of C, may manage a sufficient increase of f, so that an elliptical solution may finally appear.

By choosing mean annual motion u and regarding the condition U+W = V , the values of U,V and W are determined by iteration method of Newton(see for example, Burden et al,1981). To start an iterative process, as initial value u must be taken around a real mean annual motion. Arend(1965) has shown that the best choice is to take;

$$u = 77.5(-f/(S_1-S_2+S_3))^{3/2}\, e^{-1.9138f} \qquad (3.3)$$

In case the convergence of the mean annual motion is not reached,i.e, its mean $V \neq U+W$, the cycle of calculation must be repeated with $u+\delta u$, $U+\delta U$, $V+\delta V$ and $W+\delta W$ where δu, δU, δV and δW are solution of linear systems

$$\begin{bmatrix} -f_1 \\ -f_2 \\ -f_3 \\ -f_4 \end{bmatrix} = \begin{bmatrix} \frac{\partial f_1}{\partial U} & \frac{\partial f_1}{\partial V} & \frac{\partial f_1}{\partial W} & \frac{\partial f_1}{\partial u} \\ \frac{\partial f_2}{\partial U} & \frac{\partial f_2}{\partial V} & \frac{\partial f_2}{\partial W} & \frac{\partial f_2}{\partial u} \\ \frac{\partial f_3}{\partial U} & \frac{\partial f_3}{\partial V} & \frac{\partial f_3}{\partial W} & \frac{\partial f_3}{\partial u} \\ \frac{\partial f_4}{\partial U} & \frac{\partial f_4}{\partial V} & \frac{\partial f_4}{\partial W} & \frac{\partial f_4}{\partial u} \end{bmatrix} \begin{bmatrix} \delta U \\ \delta V \\ \delta W \\ \delta u \end{bmatrix} \qquad (3.4)$$

The iterative process is stopped when $(\delta U, \delta W, \delta V, \delta u) < \varepsilon$. In this work the tolerance factor ε is less than 10^{-4} radian .Once the relationship between U,V and W has been obtained, the well-known expression

$$e^2 = 1 + \frac{S(1 - \cos U) + 1 - \cos W - R(1 - \cos V)}{Q} \qquad (3.5)$$

define e as a function of U,V and W . The behaviours of this function and various types of solutions have been published by Dommanget(1981) and Docabo(1985). For calculating the orbital parameters from U,V,W,u and e is used the classical method such as;

$$(U,V,W,u,e) \rightarrow (\text{Thiele-Innes elements}) \rightarrow (a, i, \omega, \Omega)$$

Through mass-luminosity relation, primary and secondary masses are determined. For detail explanation see Heintz(1978).
In case of an elliptical motion this mapping is guaranteed, but for non periodic orbits the Thiele-Innes elements must be modified(Dommanget,1983)

4.Application of the programmes

As an example we shall consider the case of the binary 02158 : 1814 (ADS 1733=BDS1179=Hasting I=Lalande 4219). The orbit fo this double star has previously been calculated by Choreshi(1957 and by Siregar(1976). In order to make the example a comprehensive as possible, we shall start from three fundamenta points and the apparent areal constant C with the same data whic has been used by Siregar. The following table gives the thre base points and the relative positions of the secondary star o ADS 1733 which will be used to determine the mean annual motion

Table 1. Three base points of ADS 1733 which will be used to determine the mean annual motion.

No	t	θ	ρ	Δ_{ij}	t_{ij}	θ_{ij}
1	1906	355°.22	2".171	2.250	30	32°.33
2	1936	387 .55	1 .937	2.129	30	37 .45
3	1966	435 .00	1 .491	3.185	60	79 .78

The orbital elements of visual binary ADS 1733 computed b our programmes is presented on the following table. A compariso is made with other results.

Table 2. Orbital elements of ADS 1733 according to our software and other authors

Element orbits	Our Software	Horeschi	Siregar
u	2°.324	2°.135	2°.139
P	$154^{y}.875$	$168^{y}.6$	$168^{y}.303$
T	1993.347	1987	$1995^{y}.733$
e	0.493	0.48	0.428
i	57°.296	31°.25	30°.972
ω	65°.317	71°.72	57°.521
Ω	125°.013	103°.13	127°.335

The system has masses M_1=0.49 and M_2=0.38(spectral type KO apparent magnitude 8.3 and 9.3). All these BASIC programmes ar at the disposal of any user who would be interested in.

Acknowledgments

I would like to thank the Director of the Leids Kerkhoven Bossch Fonds for financial support to attend the I.A.U colloquium No.97 I appreciate the Institute of Technology, Bandung for making th travel possible.

References

Aitken,R.G	1935	*The Binary Stars*(Mc Graw Hill Book Company) p.89
Arend,R.G and Freitas Mourao ,R.R	1964	*Publ.Scientif et Techn.du Minestere de l'air,Paris.,*157,83
Arend,R.G	1965	*Comm.Obs.Roy.Belgique* No.243
Burden,L.R. Douglas,F.J and Reynolds,A.C	1981	*Numerical Analysis*(Prindle,Weber and Schmidt.Inc)p.439
Docabo,J.A	1985	*Celestial Mechanics.,*36,143
Dommanget,J	1960	*Comm.Obs.Roy.Belgique* No.204
ibid	1968	*Astron.Astrophys.,*68,315
ibid	1981	*Astron.Astrophys.,*94,45
ibid	1983	*Comm.Obs.Roy.Belgique* No.128
Heintz,W.D	1978	*Double Stars*(D.Reidel.Publ.Co.Dordrecht Holland) p.31
Horeschi,E	1957	*A.N.,*284,57
Nys,O	1983	*Comm.Obs.Roy.Belgique* No.128
Siregar,S	1976	*Master Thesis*(Unpublished),Bandung Institute of technology,1976
Van Albada,G,B	1957	*Contr.Bosscha.Obs.*No.4,Vol I

ABT - Is this a new method or a computerized form of a well-known method ?

SIREGAR - Computerized form from the merging of several well-known methods.

KOPAL - Does your computational method furnish not only the most probable value of the respective elements, but also the uncertainty within which such elements are defined by available observational data ?

SIREGAR - Yes, the most uncertainty in orbital elements comes from observational data.

VRI PHOTOMETRY OF LATE DWARF COMMON PROPER MOTION STARS

Edward W. Weis
Van Vleck Observatory, Wesleyan University
Middletown, Connecticut 06457, U.S.A.

Summary. From the photometric observations of a sample of late dwarf common proper motion pairs it may be inferred that about 30% - 50% of such systems contain one or more additional stellar components. A small fraction of such systems have separations in excess of 5000 A.U.

Observations on the V system of Johnson and the R,I system of Kron,et al have been made for about 200 common proper motion systems composed of late-type dwarf stars. Over half of the observations are new while the remainder have been made during the course of other observing programs (Weis, 1984,1986,1987). From the photometry and previously determined color - absolute magnitude relations (Weis 1984), distance moduli for the component stars in each pair may be computed as well as the difference in distance moduli between the stars in each pair. For eight of the pairs the difference in distance moduli exceeds 1.3 magnitudes, and because of this, the physical reality of these pairs must be regarded as very doubtful.

The differences between true distance moduli and distance moduli determined from mean color - absolute magnitude relations typically have a variance on the order of 0.4 - 0.5 magnitudes for late dwarfs. A significant portion of this scatter arises as a result of differing chemical compositions and ages among the stars. It is assumed, however, that nearly all of the common proper motion pairs have been formed in star clusters since the conditions for their formation are so much more favorable there than in the general field. Thus, for a given pair, we may expect that chemical differences or age differences between the two stars will not be significant and can not be responsible to a significant degree for the differences in distance moduli of the two stars being much different than zero. Aside from the fact that a few of the pairs may not be physical, there appear to be two other factors which will contribute to a modulus difference being significantly different than zero: 1) observational and calibration errors, and 2) the presence of additional stellar components.

Errors in the modulus differences resulting from inappropriate

Astrophysics and Space Science **142** (1988), 223–225.

color - absolute magnitude calibrations may be reduced to the point of relative insignificance by considering only those pairs with small color differences between the two stars. In this case only observational errors, typically on the order of 0.15 magnitude in the modulus difference, will be significant. From a subset of 49 of the observed pairs with color differences $\Delta(V-I) \leq 0.12$ magnitude one finds that 41% have a modulus difference greater than 0.2 magnitude. Since this difference is larger than the expected error one may conclude, with due regard for Poisson statistics, that about 30% - 50% of these common proper motion pairs have one or more additional stellar components.

Abt (1986) has suggested that among pairs of solar age and older separations exceeding 5000 A.U. are not found. However, about 11% of the pairs in the present sample have separations larger than this. While the physical reality of some of these wide pairs may be questioned either because of large modulus differences or questionable common motion, not all may be so explained. Since we do not know the ages of these late pairs it can not be said with certainty that old pairs with separations well in excess of 5000 A.U. exist, but it would seem likely to be the case for at least some of them.

All of the observations and further details will be published elsewhere.

REFERENCES

Abt, H. A. 1986. Astrophys. J. **304**, 688.
Weis, E. W. 1984. Astrophys. J. Suppl. **55**, 289.
Weis, E. W. 1986. Astron. J. **91**, 626.
Weis, E. W. 1987. Astron. J. **93**, 451.

POVEDA - Can you tell us what is your distribution of separations expressed in Astronomical Units, what is your largest separation ?

WEIS - About 10 % of my sample have projected separations in excess of 5000 A.U. The pair in my sample with the largest separation (G30-15/G29-61) has a separation of $\sim$ 85000 A.U., but the physical reality of this pair may be questioned.

RAKOS - What was the criterion for the existence of third companion, pure photometric measurements or the spectroscopy ?

WEIS - Purely by photometry (which is very crude spectroscopy).

LODEN - How large is the difference in distance modulus that corresponds to the uncertainty in your effective temperature determination ?

WEIS - I estimate that the error in differential modulus due to errors in differential photometry alone are on the order of $0^{m}_{\cdot}15$ to $0^{m}_{\cdot}20$. Comparison with the photometry of Eggen and Weistrop for stars in common is consistent with this estimate.

LODEN - I find that surprisingly good !

ABT - This is an important paper in giving good data regarding the frequency of comoving pairs. Previous papers seem to differ as to whether comoving pairs are infrequent, as Upgren implied, or frequent, as Luyten implied. If I may make a prediction, it is that when you have high-quality astrometric measures the members of comoving pairs will have similar, but not identical, proper motions.

WEIS - You may be right.

GLIESE - You set a limit of about 0.2 magnitude in distance moduli. Exceeding this limit is most likely due to the presence of additional stellar components. By my experience the distance moduli of components of a common proper motion pair sometimes deviate by 0.5 magnitude and even more from each other.

WEIS - My notion is that modulus differences for physically associated pairs which are this large are due to an additional stellar component. Remember that the stars in each pair are presumed to have the same age and chemical composition. The errors in the modulus difference resulting from errors in the photometry enter differentially, and differential photometry is usually more accurate than absolute photometry.

CCD SUITABLE FOR DOUBLE STAR ASTROMETRY. (*)

D. Sinachopoulos(1), H. Nicklas(2), M. Geffert(1)
(1) Sternwarte der Universitaet Bonn, D-5300 Bonn, F.R.G.
(2) Landessternwarte Heidelberg-Koenigstuhl, Heidelberg, F.R.G.

Summary

Astrometric measurements of 18 visual double stars carried out with a CCD are given.

The present observations are a test of the propriety of the CCD technique for astrometric observations of visual double stars.

The duration of the observations was 75 minutes. The precision of the CCD measures corresponds the precision of the photographic method, so the advantages of CCD's are the great loss of reduction work and the corresponding time saving, inclusive shortering the observing time.

CCD-frames of 18 visual double stars and 3 "astrometric standard" visual double stars have been taken with the ESO 2.2m reflector and the CCD camera attached to the Cassegrain focus without corrector, which results in a focal legth of 17.62m. The CCD was a RCA SID 501 EX chip with 320X512 pixel (each 30X30 m) and respectively a field size of 1.9X3 arcmin . The chip was not preflashed, but dark and flat-field exposures, the later inside the dome with dimmed sunlight, were made.

The read out noise of the chip is given with 45 electrons. An U filter with central wavelength of 356nm and a bandpass (FWHM) of 51nm was used for the exposures.

The frames were reduced with the software packet STARLINK at the VAX 11/780 of the astronomical institutes of Bonn. The present observations are a test on the performance of CCD's for astrometric observations of visual binaries.

One CCD-frame was made for each double star. A series of 4 contiguous exposures of the binary were taken on the same CCD-frame setting a telescope offset of 20 arcseconds between the exposures in alpha direction. As the test objects were relative

(*) Based on observations made at E.S.O., La Silla, Chile.

Astrophysics and Space Science 142 (1988), 227–230.

bright (usually less than 9 magnitudes) the duration of the exposures was between 4 and 8 seconds each. About four minutes were needed to locate each binary, to observe it, to read out the CCD and to store the frame. We used the three "astrometric standard" binaries of table 1 to calibrate the CCD. These are visual double stars for which the relative coordinates and the relative proper motion of the components are calculated according to P. Brosche and D. Sinachopoulos, (1986).

The brighter component of the double star was often overexposured. For this raison we made the astrometry but not the photometry too. The seeing during the observations was constant and it is estimated about 2". The signal to noise ratio was usually better than 100 for the not overexposured companion.

The position of each star was determined using a projection of the two dimensional profile on the x- and y-axis of the CCD and fitting a gaussian separately for each profile created by the projections. This is too the standard technique for estimating the photocenter of stellar profiles on photographic plates scanned with PDS and other maschines. And there the overexposured stellar images are centered accuracely using the same technique. In the CCD's moreover is the geometrical dimensions' constancy guaranteed by the construction of the chip and can be not affected from mechanical or electronical factors as in the PDS and other types of scanner maschines.
The mean precision - s - of these fits for all projections of all stars on all frames is estimated: s = 0.07 [pixels], or by a pixel size of this CCD-chip of 30X30 micron per pixel: s = 2.1 [micron].

The determination of the relative polar coordinates and their weighted average values have been performed following the standard method described by E. Hertzsprung, (1920). For the position angle the definition of R.G. Aitken, (1935) is used. The mean standard deviation of the mean errors of the position angle for all binaries observed is estimated to σ_θ=0.30 [degrees] and the same for the angular separation is σ_ρ=0.12 [pixels] or σ_ρ=3.6 [micron]. From our "astrometric standards" we determined the scale - m - of our optical system to m = 0.351 [arcsec/pixel]. Using this scale we get for the angular separation a mean standard deviation of σ_ρ=0.042 [arcseconds]. This result agrees good with the accuracy estimared by A.N. Argue (1984) using also the CCD technique. Unfortunately, the accuracy of the scale of our telescope obtained from our "standard" double stars is very small: σ_m= 0.002 [arcsec/pixel].In the future more accurate "standards" will be used to obtaine the accuracy permited by the used optical system. The value of the is responsible for the large errors of the angular separations for pairs with great distances of the components.

The results of our observations are presented in table 2. The equatorial coordinates are given for 1987.0 epoch, the name is the BD (Bonner Durchmusterung), CoD (Cordoba Durchmusterung) or CpD (Cape Durchmusterung) number of the primary. The photographic accuracy given in the literature is about the same with the accuracy obtained by these CCD observations. The duration of

Table 1. List of the used "astrometric standard" binaries.

INDEX (1900)	α_{2000} h m	δ_{2000} o ,	Co/CpD	RELATIVE COORDINATES (1989.0) $\Delta\alpha \cos\delta$ [arcsec]	$\Delta\delta$ [arcsec]	REL. PROPER MOTION $\mu\cdot\cos\delta$ [arcsec / century]	μ' [arcsec / century]
03362S4141	03 39.8	-40 22	S40 1027	-4.11 +/- 0.06	6.49 +/- 0.08	0.18 +/- 0.08	-0.27 +/- 0.10
04222S5718	04 24.2	-57 04	S57 659	-5.15 +/- 0.05	-2.35 +/- 0.03	0.25 +/- 0.07	1.19 +/- 0.05
04561S1339	05 00.7	-13 30	S13 1047	5.60 +/- 0.06	-1.69 +/- 0.04	0.23 +/- 0.09	-0.22 +/- 0.06

Table 2. Relative polar coordinates of the measured binaries.

INDEX (1900.)	α 1987. h m	δ 1987. o	BD/CoD/CpD	comp.	epoch [year]	distance ρ ''	σ_ρ ''	pos. angle θ o	σ_θ o	number of used exposures
03365S1207	3 40.6	-11 50	S12 689		1985.977	15.39	0.10	9.58	0.09	3
03593S1201	4 03.4	-11 47	S12 784		1985.977	16.23	0.11	295.23	0.14	3
04343N0941	4 39.0	09 50	N09 621	AB-C	1985.977	33.98	0.23	163.30	0.23	2
04347N0618	4 42.0	06 27	N06 738		1985.977	38.28	0.24	280.22	0.71	4
04444N0647	4 49.1	06 56	N06 762		1985.977	76.56	0.49	160.64	0.27	2
04554N0307	4 59.9	03 14	N03 738		1985.977	22.82	0.15	47.18	0.18	3
04587S0517	5 03.0	-05 10	S05 1138	AB	1985.977	13.93	0.09	73.08	0.13	4
04587S0517	5 03.0	-05 10	S05 1138	AC	1985.977	25.35	0.20	147.22	0.14	2
05086S0447	5 13.3	-04 40	S04 1073		1985.977	33.23	0.21	269.94	0.48	4
05105N0252	5 15.0	02 58	N02 901		1985.977	36.24	0.23	116.29	0.14	4
05130S1048	5 17.1	-10 42	S10 1144		1985.977	31.16	0.20	2.01	0.06	4
05134N0352	5 18.0	03 57	N03 845		1985.977	16.72	0.11	108.67	0.26	4
05179N0111	5 22.3	01 15	N01 992		1985.977	61.72	0.39	10.24	0.14	3
05220N0344	5 26.5	03 48	N03 905		1985.977	7.75	0.05	95.42	0.68	4
05235N0134	5 28.0	01 38	N01 1026		1985.977	26.68	0.17	109.62	0.18	4
05261S0616	5 30.4	-06 12	S06 1204		1985.977	27.62	0.18	226.39	0.11	3
05548N3132	6 00.5	31 32	N31 1165	AB	1985.977	2.56	0.02	193.39	0.08	4
05548N3132	6 00.5	31 32	N31 1165	AC	1985.977	21.81	0.14	211.38	0.14	3

the observations was about seventy five minutes. Under the same observational conditions, no more than five photographic plates could be taken in the same time.

It is not necessary to use a big reflector like the ESO 2.2m. to obtain the above astrometric accuracy with a CCD camera. Only the focal length of the telescope has to be the biggest available.

So, the use of CCD's for observing visual binaries has a good astrometric accuracy and a very important advantage with respect to photographic techniques: observational and measuring speed.

References.

AITKEN,R.G.: 1935, in " The binary stars " (Mc Graw-Hill, New Yo and London), p. 41.
ARGUE,A.N.: 1984, Mon. Not. R. Astron. Soc. Vol. 211, No. 3, p 713.
BROSCHE, P., SINACHOPOULOS, D.: 1986, Astron. Astrophys. Suppl. Ser. 65, p. 189.
HERTZSPRUNG, E.: 1920, Publ. Astron. Observ. Potsdam 24, p. 75.

ABT - 1. At what effective wavelength were these measures made ?
2. Do you plan to obtain photometry also with this equipment ?

SINACHOPOULOS - 1. We used the Johnson U filter.
2. We plan to obtain photometry too but with other telescopes like the ESO 1.5 m Danish, always using the CCD-technique.

INFRARED COMPANIONS TO PRE-MAIN SEQUENCE STARS *

H. Zinnecker[1] A. Chelli[2] L. Carrasco[2] I. Cruz-Gonzalez[2] C. Perrier[3]

[1]Royal Observatory, Edinburgh, Scotland

[2]Instituto de Astronomia, Mexico City, Mexico

[3]Observatoire de Lyon, Saint-Genis Laval, France

SUMMARY

We report 1.2, 1.6, 2.2 and 3.6 μm slit scan observations of the young low-luminosity stars Elias 22 (also known as GSS 31) in the Rho Ophiuchus dark cloud and Glass I in the Chamaeleon dark cloud. We discovered an infrared companion to Elias 22 at a projected separation of 2.0 arcsec (320 AU) and an optical/infrared companion to Glass I at a projected separation of 2.9 arcsec (410 AU). The companion of Elias 22 is a very cool optically almost invisible object (~1000 K) less luminous than the primary, while the companion to Glass I is an optically visible late-type star with an infrared excess so large that it is actually bolometrically brighter than the primary. In both cases the infrared excess is likely to be due to hot circumstellar dust grains heated by a central young star (much more obscured in the case of Elias 22 than in the case of Glass I). We outline a new method to determine the mass ratio of such systems assuming that both components lie on an isochrone in the H-R diagram. If the companions are in a bound orbit, the estimated specific angular momentum exceeds or is of the order of 10^{21} cm^2 s^{-1}, only one order of magnitude larger than that of the Sun-Jupiter system.

A short version of this work is published in IAU Symposium 122 "Circumstellar Matter" (eds. Appenzeller & Jordan, Reidel; p.117) while the full paper is currently prepared for submission to Astron. Ap. (Chelli et al., 1988).

* Based on observations collected at the European Southern Observatory at La Silla, Chile.

Astrophysics and Space Science **142** (1988), 231.

DYNAMICAL EVOLUTION OF TRIPLE SYSTEMS
(Scientific film) (*)

J.P. Anosova, V.V. Orlov
Astronomical Observatory Leningrad State University
Leningrad, U.S.S.R.

The study of the classical three-body problem lasts nearly 300 years. But some concrete examples have been obtained only by computers in the general three-body problem. For the first time, such works were begun at the Astronomical Observatory of Leningrad State University in 1965 by T.A. Agekian and J.P. Anosova. Later V.V. Orlov has joined them. During 20 years the extensive studies have been carried on - more than 30 000 runs have been considered by the computer simulations.

The basic result of the studies is that evolution of a triple system is finished with an escape : a binary is formed, and a distant body is escaped from it on a hyperbolic orbit. An average life-time is near 10^6 years for typical triple stars and near 10^{11} years for triple galaxies.

In the Astronomical Observatory, a film has been made which shows some main features of motions in triple systems. In this film, the typical trajectories are represented. The numbers show the time moments in dynamical units τ. For triple stars $1\tau = 10^4$ years, for triple galaxies $1\tau = 10^9$ years.

The film shows how in a triple system, a number of double and triple encounters, ejections on elliptic orbits take place. Evolution is completed by a close triple approach that results in escape of one component from the triple system. Such motions are being demonstrated in the first part of the film. In the second part of it, the trajectories of bodies are represented during the close triple approaches of the "fly-by" and "exchange"-types. The third part of the film shows the motions of bodies with different masses : a) an escape of the smaller mass body; b) an escape of the intermediate mass body; c) an escape of the maximum mass body.

(*) Presented by J. Dommanget - Full details on the subject have been given in : Soviet Astronomy, 30, 1986, pp. 380-390.

Astrophysics and Space Science **142** (1988), 233.

FIFTH SESSION

STATISTICAL STUDIES ON WIDE PAIRS

Chairman: S. VERESHCHAGIN

STATISTICAL STUDIES ON WIDE PAIRS

Introductory Lecture

J.-L. HALBWACHS
Observatoire de Strasbourg, France

Summary. Star pairs with wide separations need to be approached from a statistical point of view if the three following problems are to be dealt with; firstly, the selection of physical systems, secondly, the statistical properties of wide binaries, and thirdly, the derivation of single star properties when wide binaries are considered as two stars of the same age and observed from the same distance.

1. Selection of Physical Systems

The distinction between random pairs of stars and true physical systems may be based on many criteria: the most important is the separation between the components, but additional conditions such as the compatibility of proper motions, parallaxes, etc., may also be considered.

1.1. Selection based upon the apparent separation

From a historical point of view, the first study on this question dealt with the discovery of binary stars. In 1767, Michell (1768) argued that the probability of observing two stars by chance closer than a certain apparent separation, was related to the area of a disk with a radius equal to that limit. The probability $p(\rho_{\max})$ that two given stars are closer than the separation $\rho_{\max}$ is given by

$$p(\rho_{\max}) = \pi\rho_{\max}^2/(4\pi) \tag{1}$$

($\rho_{\max}$ is assumed to be small and expressed in radians). The probability of these stars having a separation wider than $\rho_{\max}$ is then $1 - p(\rho_{\max})$. If we now consider N stars randomly distributed in the sky, the probability that no pair is closer than $\rho_{\max}$ is $((1 - p(\rho_{\max}))^N)^N$. Michell applied this relation in order to calculate the probability that no pair like β Capricorni could exist in the sky. The separation between the components of β Capricorni was $3\frac{1}{3}'$, and he estimated that 230 stars were at least as bright as these stars. He calculated that the probability of finding no star with a companion at least as close and at least as bright as the secondary component of β Capricorni was 1–1/80. Then, the probability that a system like β Capricorni could appear by chance was $\frac{1}{80}$, and Michell concluded that β Capricorni was a system of stars bound by gravitation, a hypothesis that has yet to be refuted.

Struve (1827, 1852) calculated $n(\rho_{\max})$, the number of optical pairs with an apparent separation less than a certain limit $\rho_{\max}$ that should occur among N stars counted in a given area, A: i.e.,

$$n(\rho_{\max}) = N(N - 1)\pi\rho_{\max}^2/(2A)\,. \tag{2}$$

Astrophysics and Space Science **142** (1988) 237–244.

Kubikowski *et al.* (1959) considered the probability of finding a field star with magnitude m in the neighborhood of a given star given by

$$P(\rho_{\max}) = N(m)\pi\rho_{\max}^2/A\,, \tag{3}$$

where $N(m)$ is the number of stars with magnitude m. They assigned a given level (1%) to P, and they derived $\rho_{\max}$ as a function of the magnitude of the secondary components. For example, they found that 1% of stars with magnitude 10 should have an optical bright companion with a separation closer than 1′.6. This 1%-level is certainly not the proportion of optical systems among double stars, but only a means of deriving the contribution of optical pairs in the total number of binaries. A similar procedure for rejecting optical companion from the IDS was used by Poveda *et al.* (1982). They estimated that the 'filtered IDS' contained on average 1000 optical systems among 51,201 stars.

Van Albada (1968) and Bahcall and Soneira (1981) used another method, one based upon the distribution of the nearest neighbours. The probability that the nearest optical companion of a star is at a separation s is stated by a Poisson's law

$$P(\rho \pm \mathrm{d}\rho/2) = 2\pi\rho N/A \exp(-\pi\rho^2 N/A)\,\mathrm{d}\rho\,. \tag{4}$$

They compared the histogram of the separation of the observed stars to this theoretical distribution and found that both were similar with regard to the larger separations in their samples. For close separations, they obtained an excess of systems due to physical binaries.

Bahcall *et al.* (1986) preferred the 'near neighbours' method. They considered the separations between stars and all their neighbours closer than a given limit. It follows from Equation (3) that the number of optical systems with separations $\rho \pm \mathrm{d}\rho/2$ is then a linear function of ρ.

Another formulation of this method is the so-called 'two point correlation function' (Bahcall and Soneira, 1981). It consists of calculating $w(\rho)$, which is the ratio between physical and optical pairs with the separation ρ.

1.2. Improvement due to additional criteria

The number of stars that may produce optical systems may be reduced by adding a condition that should be satisfied by the components of physical binaries.

Struve (1852) suggested considering only pairs with compatible proper motions. These binaries are the 'common proper motion' (CPM) stars. When proper motions are greater than 0.1″ yr^{-1}, the number of stars that may produce optical systems is so limited that Luyten (1967) estimated that only 3% of all pairs with components brighter than 14.5 mag and separations closer than 200″ could be optical. Halbwachs (1986b) has applied the near neighbour method to a sample of CPM stars. Since the probability of obtaining an optical pair with a given separation is much lower for stars with large proper motions than it is for others, the ratio between the apparent separation and the proper motion was considered instead of the separation. In fact, this ratio refers to the time during which the displacement of a binary is equal to the separation between the components. It appears that the proportion of optical pairs among CPM stars extracted from the AGK2/3 catalogue is only 1.3% when this time is less than 1000 years.

Other criteria may also be used. Bahcall and Soneira (1981) considered that the components of physical systems should have the same photometric parallaxes. Lindroos (1985) used photometric data and spectral types. Latham *et al.* (1984) verified that the components have similar radial velocities. Dommanget (1955, 1960) investigated the selection of long-period binaries showing evidence of orbital motion originating from proper motions or from radial velocities.

1.3. The case of the 'moving pairs'

The spatial velocities of stars were used to find pairs with very wide separations, or the so-called 'moving pairs' (Vandervoort, 1968; Lü and Upgren, 1973; Beavers *et al.*, 1983). These star pairs differed from usual binaries in that their separations exceeded a few parsecs and in that they were not assumed to be bound by gravitation. Branch (1974) and Upgren and Chabotte (1983) have pointed out that the distribution of the velocities of stars was not as homogeneous as these studies assumed and that these moving pairs could in fact be produced by random combination.

2. The Statistical Properties of Wide Binaries

Three fundamental questions are related to the study of the origin of wide binaries. These questions are: (a) What is the distribution of the mass ratios of these systems? (b) What is their distribution of semi-major axes? (c) What is the frequency of these systems among stars?

In practice, the mass ratios of visual binaries are derived from the difference between the magnitudes of the components Δm and from the mass-luminosity relation. For that reason, question (a) may be restated as: What is the distribution of Δm of dwarfs binaries?

Since wide binaries have very long periods, it is not possible to measure the semi-major axes of their orbits. If the distances of the systems are known, the projections of the separations between the components s may be computed. Kuiper (1935), Couteau (1960), van Albada (1968), and Halbwachs (1983) have shown that this parameter is nearly equivalent to the semi-major axis. Moreover, in the case of CPM stars, proper motions may be used instead of the parallaxes (Luyten, 1939).

2.1. The selection effects

Because they are affected by selection effects, the distribution of the observed binaries cannot be considered as the distribution of all binaries in space. Some of these selection effects were introduced in order to avoid optical systems, but physical wide binaries may also be rejected. Other selection effects are the failure to detect systems with faint secondaries and the failure to detect binaries with close apparent separations; this last point depends on whether we are dealing with visual binaries or with binaries detected on photographic plates.

The detectability of visual binaries was investigated by Öpik (1923, 1924), Kuiper (1935), Heintz (1969), and Halbwachs (1983). They all found that the probability of visually detecting a close secondary components depends only on Δm and ρ. Photographic binaries, especially CPM stars, were studied by Luyten (1939). For these stars, the possibility of detecting a close companion depended on the size of the images of the stars, and consequently on their magnitudes: the brighter the primary the larger the minimum value of the separation. Luyten also considered that the proper motions of bright primaries are difficult to estimate and this intervened also in the detection of bright CPM pairs.

For the derivation of the properties of binaries in space, the selection effects may be taken into account in two ways. The first method, initiated by Öpik (1923, 1924) and Kuiper (1935), considers systems for the purposes of the problem being treated, as representative of all binaries. For example, binaries with bright primaries and separations neither close nor wide are convenient for deriving the distribution of Δm, but these binaries are rather rare and the distribution so derived would be unaccurate. For this reason, binaries with faint primaries and close separations are also used, except for small Δm. The second method (Halbwachs, 1983) is to generate a sample of synthetic binaries that are assumed to have the same properties as all binaries in the sky. This synthetic sample is filtered in order to select systems that ought to appear in a sample of true observed binaries. If the sample of synthetic 'observed binaries' has the same properties as the true binaries, then the distributions of Δm and of ρ used in the generation of the synthetic sample are possible solutions.

2.2. Intrinsic distribution of ΔM of dwarf binaries

The intrinsic distribution of Δm of dwarf binaries was derived from visual binaries by Öpik (19224), Kuiper (1935), Heintz (1969), Abt and Levy (1976), and Halbwachs (1986). CPM stars were used for the same purpose by Luyten (1939), Abt and Levy (1976), Trimble (1987), and Halbwachs (1987). These studies all found a nearly constant or increasing distribution of Δm, and a decreasing distribution of mass ratios. It is generally acknowledged that wide binaries have a distribution of mass ratios compatible with the hypothesis that they were constituted from stars of the same age that were randomly associated (Öpik, 1924; Abt and Levy, 1976). For B-type dwarf binaries, however, van Albada (1968b) has demonstrated that they could be produced by a dislocation of small clusters.

2.3. Intrinsic distribution of $\log s$

This distribution is more difficult to derive than the distribution of Δm, because the selection effects do not depend directly on s but on the apparent separation ρ. This is one explanation for the discrepancies found between the various distributions. Another particularity is that selection effects inhibit detection for close and for wide separations. For this reason, the distribution of $\log s$ of a sample of binaries appears as a nearly gaussian curve (Luyten, 1967; Halbwachs, 1986).

Two kinds of distribution of $\log s$ were found: a nearly constant distribution (Öpik,

1924; Luyten, 1939; Dommanget, 1970; Halbwachs, 1986) and the decreasing part of a bell-shaped curve with its center around $\log s(\mathrm{AU}) = 1$ (Kuiper, 1935; Luyten, 1939; Heintz, 1969; Abt and Levy, 1976). The fact that Luyten found these two distributions from two different samples while working on the same study illustrates the difficulties encountered in this derivation. In both cases, Dommanget (1971, 1984) estimated that the distribution of $\log s$ cannot be produced by a random combination of stars. He concluded that wide binaries initially had separations closer than their present ones. This increase may be a consequence of secular mass loss (Dommanget, 1963, 1971, 1981). Zinnecker (1984) have mentioned different theories of the formation of binaries and provided many references on this subject.

The shape of the curve for very wide separations was investigated by Bücher and Laques (1974), who found a sharp decrease for separations wider than 12000 AU. Retterer and King (1982) found a similar decrease, but for separations wider than 20000 AU. This limit is a touchstone for the study of the disruption of binaries due to gravitational perturbations of field stars.

2.4. The Proportion of Wide Binaries

The derivation of the proportion of wide binaries among stars depends on the distribution of Δm and of $\log s$ that is assumed. Since these distributions are not very well known, it is evident that the results must be considered with caution. Moreover, the range in Δm and $\log s$ considered varies from one study to another, thus making any comparison of the different studies hazardous. We just mention some of the results.

Öpik (1924) estimated that 50% of F-type dwarfs had a companion with absolute magnitude brighter than 13 mag and separation between 220 and 28000 AU. According to Kuiper (1935), the proportion of stars with a companion with a separation wider than around 10 AU is $\frac{4}{10}$. The discrepancy in Öpik's results was obviously a consequence of the difference between the distributions of $\log s$ that were assumed. Dommanget (1970) found a result nearly similar to Öpik's but Abt (1979) found a result nearly similar to Kuiper's. Other references are given in Herczeg (1984).

3. The Wide Binaries and the Properties of Single Stars

The components of a wide binary may be considered as two single stars of the same age, the same initial chemical composition, and the same distance. Since they have a wide separation, their physical properties are not modified by mutual interactions, such as mass exchange. Moreover, when their apparent separation is also wide, spectroscopic or photometric observations may be done without encountering difficulties due to proximity.

Wide binaries were used for drawing the HR diagram. Luyten (1957) and Rakos (1985) obtained the Main-Sequence. Using binaries with dwarf primary components, Luyten (1969) derived the sequences of subdwarfs and white dwarfs. Tolbert (1964) and Franz (1981) investigated the bright sequences (from subgiants up to supergiants).

The components of wide binaries are of the same age and were also used in testing theoretical isochrones (Olsen, 1982; Corbally, 1984; who gave many references on this subject). Scalo *et al.* (1978) used binaries with giant primaries in order to derive the average mass of giant stars. They found that giants had masses between 0.8 and 1.2 solar masses.

Hynek (1967) and Halbwachs (1986) also investigated stellar evolution, but they started from the distributions of Δm of binaries with primaries on different sequences (secondaries were assumed to be dwarfs in all cases). Halbwachs estimated that the progenitors of giants should be F-type dwarfs.

4. The Future

The methods of investigating wide binaries are still based on the same principles that were used by Michell, Struve, or Öpik. Any important progress that can be expected will come from the systematic use of computers and from new and more accurate data.

With computers and data in machine-readable form, the histogram of many thousands of binaries may be obtained in a few seconds. Moreover, the use of a sampling experiment is now a very fast and accurate method for the correction of selection effects.

Due to Luyten's work, the number of CPM pairs is increasing regularly. Up to now, he has discovered 6124 systems (Luyten, 1987) and now that they have been included in the Washington Double Star catalogue (Worley and Douglas, 1984), data on more than 5000 of them are available on magnetic tape. The 6124 systems were also entered at the Centre de Données of Strasbourg (CDS).

New data will also be obtained from space. The wide-field camera of the Hubble Space Telescope (Westphal, 1982) will provide material for research on wide systems, such as the Bahcall and Soneira (1981) study. Hipparcos and Tycho will also provide data on companions. For Hipparcos stars, the separations will be from around 0.1″ to 5″ (Lindegren, 1982; Bacchus, 1982; Dommanget, 1982). Part of the preparation of the Hipparcos mission includes a new catalogue of double stars (Dommanget, 1983, 1985; Nys, 1985). Tycho will detect all companions with separations wider than 1 or 2″ (Høg, 1985). When the positions obtained by Tycho are compared to previous surveys, such as the AGK 2, proper motions with an accuracy of $0.003'' \text{ yr}^{-1}$ in the northern hemisphere and $0.01'' \text{ yr}^{-1}$ in the south will be available for use in the search for CPM stars. Halbwachs' investigation (1986b) could then be processed again and extended to wider separations. The photometric B, V measures of Tycho will also be of great interest in the selection of physical systems or for studies of the HR diagram of wide binaries. Moreover, the selection of some systems will be improved by the availability of the radial velocities that are systematically measured for all the stars in the Hipparcos program (Mayor, 1982; Fehrenbach and Duflot, 1982).

Acknowledgement

I would like to thank Mme C. Holtzer for correcting the English. The travel expenses were supported by the 'Comité National Français d'Astronomie'. The subsistence support provided by the Local Organizing Committee on Belgian funds is gratefully acknowledged.

References

Abt, H. A.: 1979, *Astron. J.* **84**, 1591.
Abt, H. A. and Levy, S. G.: 1976, *Astrophys. J. Suppl. Ser.* **30**, 273.
Bacchus, P.: 1982, *The Scientific Aspects of the Hipparcos Space Astrometry Mission*, ESA SP-177, p. 155.
Bahcall, J. N. and Soneira, R. M.: 1981, *Astrophys. J.* **246**, 122.
Bahcall, J. N., Jones, B. F., and Ratnatunga, K. U.: 1986, *Astrophys. J. Suppl. Ser.* **60**, 939.
Beavers, W. I., Russell, J. L., and Simonson, G. F.: 1983, 'The Nearby Stars and the Stellar Luminosity Function', in A. G. D. Philip and A. R. Upgren (eds.), *IAU Colloq.* **76**, 299.
Branch, D.: 1974, *Observatory* **94**, 17.
Bücher, A. and Laques, P.: 1974, *Astron. Astrophys.* **33**, 15.
Corbally, C. J.: 1984, *Astrophys. J.* **285**, 195.
Couteau, P.: 1960, *J. Obs.* **43**, 41.
Dommanget, J.: 1955, *Bull. Astron.* **XX**, fasc. 1, *Comm. Obs. Roy. Belgique*, No. 91.
Dommanget, J.: 1960, *Bull. Astron.* **XXIII**, fasc. 1, *Comm. Obs. Roy. Belgique*, No. 174.
Dommanget, J.: 1963, *Ann. Obs. Roy. Belgique 3ème Série* **IX**, fasc. 5.
Dommanget, J.: 1970, *Ciel et Terre*, p. 463, *Comm. Obs. Roy. Belgique Ser. B*, No. 56.
Dommanget, J.: 1971, *Sciences* **II**, No. 3, *Comm. Obs. Roy. Belgique Ser. B*, No. 70.
Dommanget, J.: 1981, 'Effects of Mass Loss on Stellar Evolution', in C. Chiosi and R. Stalio (eds.), *IAU Colloq.* **59**, 507.
Dommanget, J.: 1982, *The Scientific Aspects of the Hipparcos Space Astrometry Mission*, ESA SP-177, p. 161.
Dommanget, J.: 1983, *Bull. Inf. C.D.S.* **24**, 83.
Dommanget, J.: 1984, *Astrophys. Space Sci* **99**, 23.
Dommanget, J.: 1985, *Proceeding of a Colloquium on the European Astrometric Satellite Hipparcos – Scientific Aspects of the Input Catalogue Preparation*, ESA SP-234, p. 153.
Fehrenbach, C. and Duflot, M.: 1982, *The Scientific Aspects of the Hipparcos Space Astrometry Mission*, ESA SP-177, p. 161.
Franz, O. G.: 1982, 'Binary and Multiple Stars as Tracers of Stellar Evolution', in Z. Kopal and J. Rahe (eds.), *IAU Colloq.* **69**, 123.
Halbwachs, J. L.: 1983, *Astron. Astrophys.* **128**, 399.
Halbwachs, J. L.: 1986a, *Astron. Astrophys.* **168**, 161.
Halbwachs, J. L.: 1986b, *Astron. Astrophys. Suppl. Ser.* **66**, 131.
Halbwachs, J. L.: 1988, *Astrophys. Space Sci.* **142**, 139 (this issue).
Heintz, W. D.: 1969, *J. Roy. Astron. Soc. Can.* **63**, 275.
Herczeg, T.: 1984, *Astrophys. Space Sci.* **99**, 29.
Høg, E.: 1985, *Proceedings of a Colloquium on the European Astrometric Satellite Hipparcos – Scientific Aspects of the Input Catalogue Preparation*, ESA SP-234, 225.
Hynek, J. A.: 1967, in *Modern Astrophysics*, Gauthier-Villars, Paris, p. 225.
Kubikowski, J., Lukaszewicz, J., and Opolski, A.: 1959, *Acta Astron. Ser. A* **9**, 4.
Kuiper, G. P.: 1935, *Publ. Astron. Soc. Pacific* **47**, 15.
Kuiper, G. P.: 1935, *Publ. Astron. Soc. Pacific* **47**, 121.
Latham, D. W., Tonry, J., Bahcall, J. N., and Soneira, R. M.: 1984, *Astrophys. J.* **281**, L41.
Lindegren, L.: 1982, *The Scientific Aspects of the Hipparcos Space Astrometry Mission*, ESA SP-177, p. 147.
Lindroos, K. P.: 1985, *Astron. Astrophys. Suppl. Ser.* **60**, 183.
Lü, P. K. and Upgren, A. R.: 1973, *Astrophys. J.* **180**, 91.
Luyten, W. J.: 1939, *Publ. Astron. Obs. Univ. Minnesota* **II**, 185.
Luyten, W. J.: 1957, *J. Roy. Astron. Soc. Can.* **51**, 41.
Luyten, W. J.: 1967, 'On the Evolution of Double Stars', in J. Dommanget (ed.), *IAU Colloq.* **17**, 45.
Luyten, W. J.: 1969, *Proper Motion Survey with the 48 inch Schmidt Telescope XVIII*, Univ. of Minnesota, Minnesota.
Luyten, W. J.: 1971, *Astrophys. Space Sci.* **11**, 49.

Luyten, W. J.: 1987, *Proper Motion Survey with the 48 inch Schmidt Telescope LXXI*, Univ. of Minnesota, Minnesota.
Mayor, M.: 1982, *The Scientific Aspects of the Hipparcos Space Astrometry Mission*, ESA SP-177, 217.
Michell, J.: 1768, *Philosophical Transaction, London* **LVII**, part I, 234.
Nys, O.: 1985, *Proceeding of a Colloquium on the European Astrometric Satellite Hipparcos – Scientific Aspects of the Input Catalogue Preparation*, ESA SP-234, p. 161.
Olsen, E. H.: 1982, *Astron. Astrophys.* **110**, 215.
Öpik, E. J.: 1923, *Publ. Obs. Astron. Univ. Tartu* **XXV**, No. 5.
Öpik, E. J.: 1924, *Publ. Obs. Astron. Univ. Tartu* **XXV**, No. 6.
Poveda, A., Allen, G., and Parrao, L.: 1982, *Astrophys. J.* **258**, 589.
Rakos, K. D.: 1985, *Astrophys. Space Sci.* **110**, 143.
Retterer, J. M. and King, I. R.: 1982, *Astrophys. J.* **254**, 214.
Scalo, J. M., Dominy, J. F., and Pumphrey, W. A.: 1978, *Astrophys. J.* **221**, 616.
Struve, F. G. W.: 1827, *Catalogus novus stellarum duplicium et multiplicium Dorpaty.*
Struve, F. G. W.: 1852, *Stellarum fixarum imprimis compositarum positiones mediae pro epocha 1830.0*, Academia Caesareae Petropolitanae.
Tolbert, C. R.: 1964, *Astrophys. J.* **139**, 1105.
Trimble, V.: 1987, *Astron. Nachr.* (in press).
Upgren, A. R. and Chabotte, S. C.: 1983, *Proceedings of 'Statistical Methods in Astronomy' Strasbourg*, ESA SP-201, p. 231.
Van Albada, T. S.: 1968a, *Bull. Astron. Inst. Neth.* **20**, 47.
Van Albada, T. S.: 1968b, *Bull. Astron. Inst. Neth.* **20**, 57.
Vandervoort, P. O.: 1968, *Astrophys. J.* **152**, 895.
Westphal, J. A.: 1982, *The Space Telescope Observatory*, Special Session of Commission 44, IAU 18th General Assembly, NASA CP-2244, p. 28.
Worley, C. E., Douglas, G. G.: 1984, *The Washington Double Star Catalogue*, U.S. Naval Observatory.
Zinnecker, H.: 1984, *Astrophys. Space Sci.* **99**, 41.

GLIESE – Is there anything known about programs under way to observe Luyten's LDS stars? Photometry at least of parts of this large number? I do ask every one among us here.

HALBWACHS – I do not know. LDS stars are usually very faint and, moreover, we do not know exactly what selection criteria were used for their detection; for that reason, a statistical study should be difficult.

WAELKENS – Grenon in Geneva is actually carrying out a large photometric program on Luyten's stars.

SCARDIA – Un procédé lent mais recommandable, comme déjà appliqueé par Luyten et très récemment par Gellera (1987) est de mesurer l'angle de position et la séparation des couples C.P.M. sur des plaques photographiques prises par de grands astrographes ou chambres de Schmidt.

HERCZEG – In order to check the 'reality' of the wide pairs, the radial velocities do, of course, offer the usual possibility. How many cases (approximately) are already covered by radial velocity observers? How much more material can realistically be gained by further observing programs?

MAZEH – I am collaborating with Dr Latham from the Harvard-Smithsonian Center for Astrophysics, where echelle spectrographs for radial-velocity measurements are intensively active. The spectrographs can easily measure radial-velocity of stars brighter than 12 mag to a precision of about 0.8 km s^{-1}. Fainter stars, up to 15–16 mag can also be observed with the MMT, but the observing time on this telescope is difficult to get. Therefore, I think that no large stellar survey of faint objects can be performed in the near future.

STATISTICAL STUDY OF VISUAL BINARIES*

S. VERESHCHAGIN[1], Z. KRAICHEVA[2], E. POPOVA[2], A. TUTUKOV[1], and L. YUNGELSON[1]

Summary. A statistical study of physical pairs from IDS catalogue has been performed. For 2811 binaries with luminosity class V primaries it has been found that their mass function is a Salpeter one; their distribution over semi-major axes of orbits is $dN \propto d \log a$. The rate of duplicity of stars is close to 1.

1. Introduction

This investigation of visual binaries (VB) is a part of a study of physical properties of the whole family of double stars. The aim of our study is to find the distributions of binaries with respect to such generic properties as masses of components, their ratios, axes of orbits.

The possibilities of observations of binaries of different types are limited by a number of selection effects. As the statistical study shows (see Figures 2 and 5(a) below), most of physical pairs among VB have angular separations $0''.1 \lesssim a'' \lesssim 10''$ and visual magnitudes of primaries $8^m \leq m_{v_1} \lesssim 10^m$ (Poveda *et al.*, 1982; Vereschagin *et al.*, 1987a, b). The lower limit $a'' \approx 0''.1$ is set by the aperture $D \lesssim 100$ cm of the telescopes, that are customarily used for the study of VB. Pairs with $a'' > 10''$ and $m_{v_1} > 10^m$ when studied statistically as a rule turn out to be optical ones with $\gtrsim 99\%$ confidence. When $m_{v_1} > 10^m$ the poor quality of stellar images embarasses the discovery of VB (Couteau, 1978). Making use of the mass-luminosity law it is possible in the $\log M_1 - \log a$ diagram (Figure 1, Kraicheva *et al.*, 1985) to separate the regions, occupied by binaries of different types, among them by VB. We indicate in Figure 1 the number of binaries, for which it turned out possible to estimate the physical parameters by using the data of several double star catalogues. We show also in this Figure 1 the upper limit of the semi-major axes of binaries (Kraicheva *et al.*, 1985). The existence of this limit is due to encounters with giant molecular clouds, single field stars, and tidal influence of the Galaxy. It appears that only a small fraction of the most close VB with $a \sim 10^3 R_\odot$, $M_1 \sim M_\odot$ are close binaries in the evolutionary sense, i.e., that mass exchange is possible between their components.

2. The Data and the Method of the Analysis

We have performed a statistical analysis of VB from the *Index Catalogue of Double Stars* (IDS). We have employed the magnetic tape version of it from Strasbourg and Soviet

* Presented by: L. Yungelson. A similar paper has been published in Russian, under the title: 'Statistical Investigation of Visual Binaries in IDS Catalogue', in *Scientific Informations* **63**, 1987, pp. 3–18.

[1] Astronomical Council of the USSR, Academy of Sciences, Moscow, U.S.S.R.
[2] Section of Astronomy, Bulgarian Academy of Sciences, Sofia, Bulgaria.

Astrophysics and Space Science **142** (1988) 245–254.

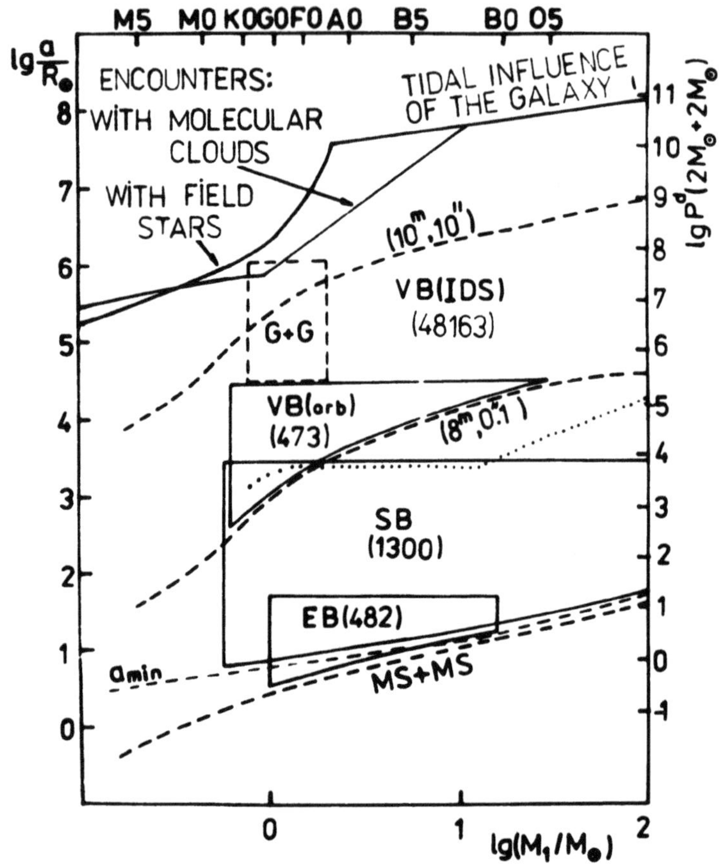

Fig. 1. Position of binary stars of different types in log (primary mass)–log (linear separation) diagram. EB – eclipsing binaries, SB – spectroscopic binaries, VB – visual binaries, VB(orb) – VB with known orbits. The MS + MS line indicates the lowest limit of separations set in by the contact of two homogeneous stars. The line a_{min} shows the minimal separation for new-born stars. Thick lines in the upper part of the diagram indicate the maximum semi-major axes of binaries. The region G + G is occupied by VB with giant components. The dotted line limits the region of evolutionary close binaries. For further details see the text.

Astronomical Data Centers (Worley, 1980). This catalogue contains about 70 000 entries. By rejecting spurious systems and applying the 'one percent filter' as suggested by Poveda *et al.* (1982) we have picked out 48 163 pairs for which the probability to be optical is less than 1% (slightly different from Poveda *et al.* number of stars in the 'filtered' catalogue is due to the use of interpolation formula for the dependence of number of stars on magnitude and galactic coordinates $N(M_2)_{l,b}$ rather than tables). This 'filtered' IDS was merged with the *CSI Catalogue with Selected Data* which contains MK-spectra of stars. Thus we get 2811 VB with luminosity class V primaries that were analysed further. The final number of pairs is low because MK-classification has been performed mainly for bright southern stars ($\delta \lesssim -40°$), but among VB northern stars dominate. We were unable to identify about 25% of IDS stars because they lack DM numbers.

For stars with known MK-spectra we had determined the masses and mass ratios of components making use of Sp – M_v and Sp – $\mathfrak{M}$ relations by Allen (1973). The distances to stars r_i were determined by using the known relation $m - M_v = 5 \log r_i - 5 + 0.0019 r_i$. We had assumed after Couteau (1978) that on the average the true and projected separations are related as $a'' \approx 1.25 d''$.

In addition to the above mentioned sample of VB we have also studied in a similar manner 477 VB with known orbital solutions with luminosity class V primaries from the Finsen and Worley (1970) catalogue (due to technical reasons this sample was 'filtered out' from the IDS).

3. Results and Discussion

The distribution of 48163 presumingly physical pairs from IDS and of 473 pairs with known orbits from Finsen–Worley catalogue is shown in Figure 2. In agreement with Poveda *et al.* (1982) the data may be considered as statistically complete up to $m_{v1} \approx 10^m$. The number of VB depends on m_{v1} as $d \log N / dm_{v1} \approx 0.4$. This reflects the concentration of stars to the galactic plane.

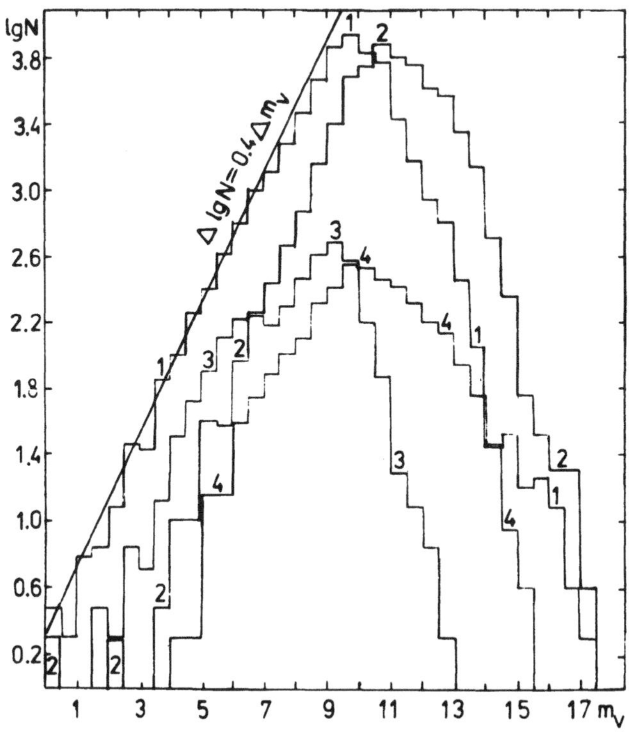

Fig. 2. Distribution of visual binaries from the filtered IDS and Finsen–Worley (FW) catalogue over the visual magnitudes. 1- and 2-primaries and secondaries of filtered IDS, 3- and 4-primaries and secondaries of FW.

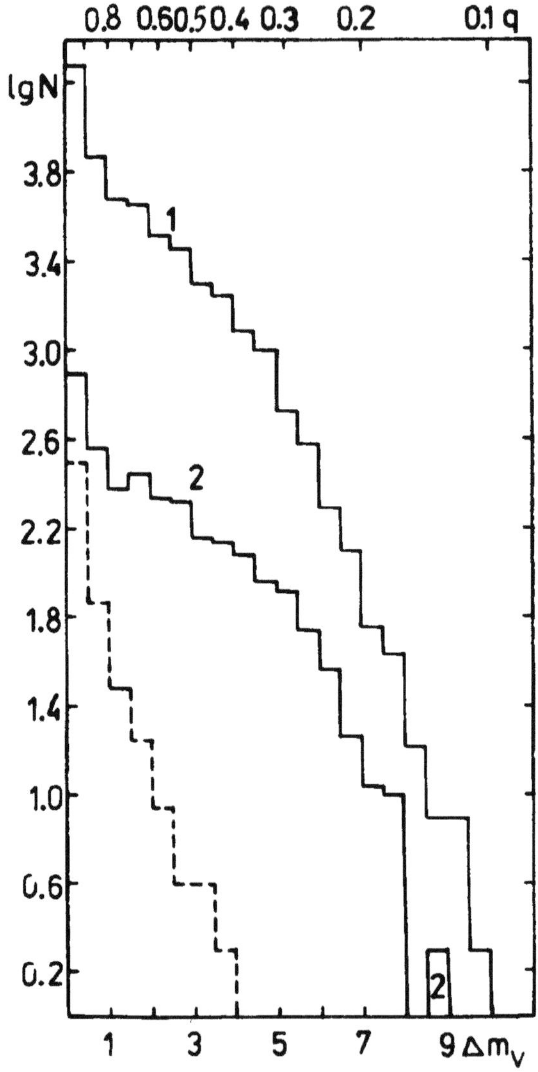

Fig. 3. Distribution of VB over magnitude differences of components. 1 – filtered IDS stars, 2 – VB with V luminosity class primaries. Dashed line – VB with known orbits.

The distribution of VB with respect to the magnitude difference of components Δm_v (Figure 3) reflects their distribution with respect to the mass ratios $q = M_2/M_1$. The companions of VB with the Main-Sequence primaries have to be, as a rule, also Main-Sequence stars. Really, the Main-Sequence lifetime of solar-mass stars is $T_1 \approx 10^{10}\, M_{\odot}^3/M_1^3$ years. The time-scale of gravitational contraction of M_2 star is $3 \times 10^7\, M_2^2/R_2 L_2$ years where R_2 and L_2 stand for radius and luminosity of the secondary. Therefore, T_2 is smaller than T_1 only if $q < 0.15$. But due to the observational selection such systems do not appear among observed VB. On the other hand, there are no white dwarfs among secondaries of observed VB. The luminosity of cooling white dwarf depends on time as $L_{wd}/L_{\odot} \approx 10^9 \times (1\ \text{year}/t)^{7/5}$ (Iben and Tutukov, 1984). The luminosity of (1–3) $M_{\odot}$ Main-Sequence primary depends on mass as $L_{MS}/L_{\odot} \approx (M/M_{\odot})^4$. Therefore $L_{wd}/L_{MS} \approx 10^{-5} M^{1/5}$. For $M \sim M_{\odot}$, the typical magnitude difference would be about 12^m. However there are not VB with $\Delta m_v > 9^m$ among our sample. (Systems like Sirius are discovered due to their proximity to the Sun only.)

Approximating the mass-luminosity relation of Allen (1973) for (1–3) $M_{\odot}$ stars as $\log(M/M_{\odot}) \approx 0.56\text{–}0.12\, M_v$ it is possible to transform Δm_v into q (the upper scale of Figure 3). The apparent concentration of VB to $q \approx 1$ is caused at least by two main effects. The maxima of distributions of VB over visual magnitudes are at $m_{v_1} \approx 10^m$, $m_{v_2} \approx 11^m$ for the sample of all 'physical' VB, and at $m_{v_1} \approx 9\overset{m}{.}5$, $m_{v_2} \approx 10\overset{m}{.}5$ for VB of V luminosity class (Figure 2). Thus, for weak stars which dominate in both samples we

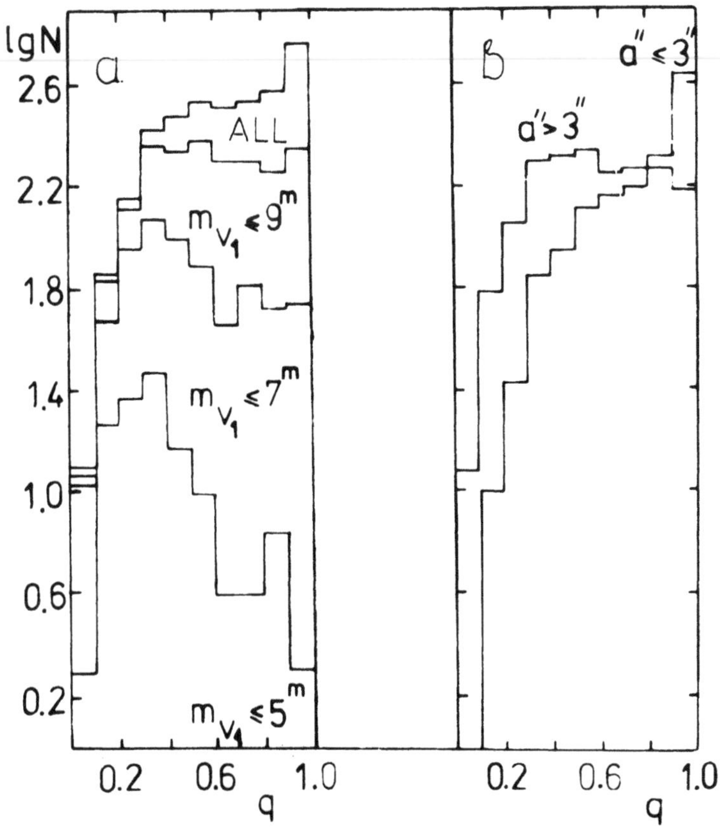

Fig. 4. Distribution of VB with luminosity class V primaries with respect to the mass ratios of components, depending on limiting stellar magnitude of the sample (a) and separation (b).

observe only companions of comparable luminosity. With the increase of the limiting stellar magnitude of the sample the concentration of stars to $q = 1$ decreases (Figure 4(a)). The second selection factor is that in systems with small angular separations ($\sim 1''-2''$) it is difficult to discover weak companions. Thus, only relatively bright and wide systems provide reliable information on the distribution of stars over q.

In Figure 5(a) we show the distribution of VB with respect to the semi-major axis of apparent orbits. This figure clearly indicates that most of the registered VB with $a'' > 10''-20''$ are most probably optical pairs.

The simplest and most obvious selection effects that could be taken into account numerically, are the difference in lifetimes of stars and the difference in volume of space

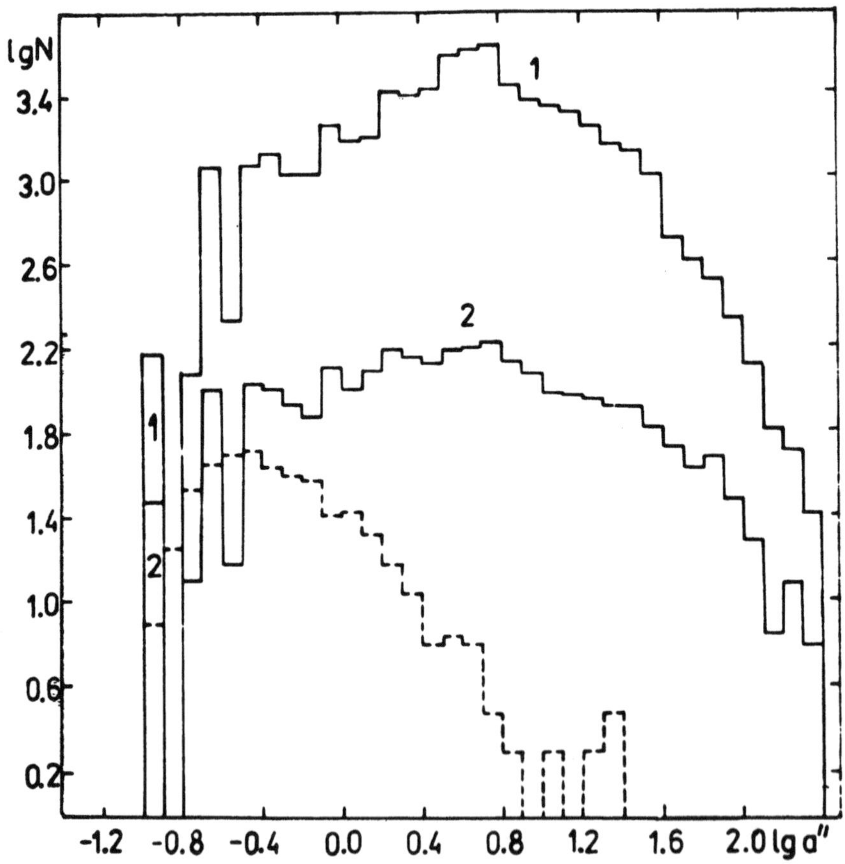

Fig. 5a.

Fig. 5a–b. The distribution of VB over large semi-major axes of apparent orbits (a) and over linear semi-major axes of orbits (b). 1 – all physical pairs, 2– VB with V luminosity class primaries. Dashed line – the VB with known orbits.

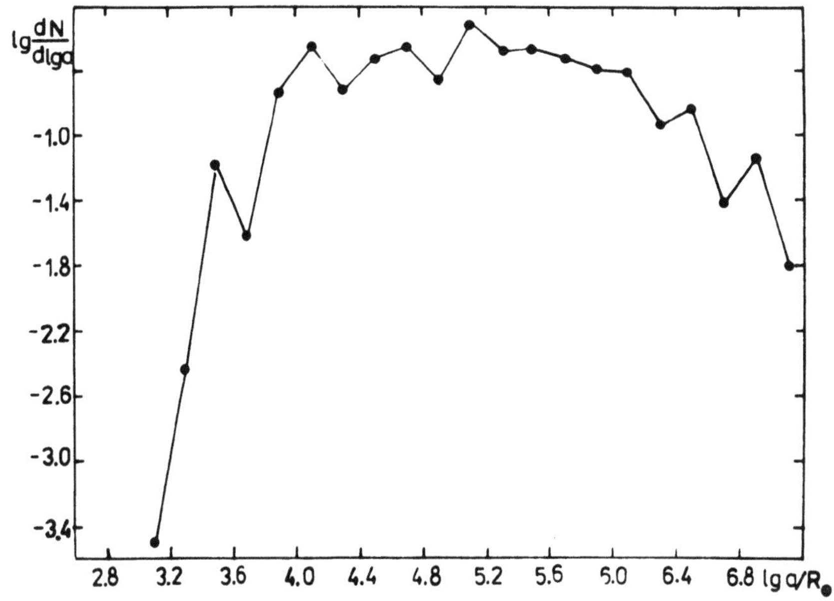

Fig. 5b.

over which they are sampled depending on mass. In order to account for these effects, each entry of the sample of mass M_i was weighted by $\alpha_i = (V_i T_i)^{-1}$, where V_i is the specific volume of space which is occupied by stars of a given mass and visual magnitude, T_i is their lifetime. The volume V_i was specified as $V_i = 4/3\pi r_i^3$ for $r_i \leq \beta$ or $V_i = \frac{1}{3}\pi\beta(3r_i^2 - \beta^2)$ where β is the scale-height for stars of the given spectral class. Lifetimes appropriate for disk population are

$$\log T = \min\{9.83 - 3.65 \log(M/M_\odot) + 1.01 \log(M/M_\odot)^2, 10\}\ .$$

The initial mass function IMF for primaries of VB is then $\xi(M) = 2\,\Sigma_{i=1}^{n} \beta_i \alpha_i(\log M)$, where n is the number of stars within certain mass interval. The IMF is shown in Figure 6. It may be approximated by a power law with $n = -1.55 \pm 0.21$ (for $-0.4 \lesssim \log M/M_\odot \lesssim 1.4$). For stars of luminosity class V and all Main-Sequence stars from the CSI catalogue the slope of IMF is -1.35 ± 0.20. (We assume that 'Main-Sequence stars' are all stars that lie within the hydrogen-burning strip in the Hertzsprung–Russell diagram; these include, apart from class V stars, all O-stars, A- and B-stars of luminosity classes III and IV). Thus, the IMF of VB primaries appears to be a Salpeter one, as well as the IMF of more close spectroscopic binaries (Popova *et al.*, 1982). One may note the apparently non-monotonous behaviour of IMF, which was already noticed earlier for spectroscopic pairs and single stars (Svechnikov, 1969; Piskunov *et al.*, 1979; Giannuzzi, 1981). The reason for this is still unclear. It is probable that it simply reflects some unknown non-monotony of spectrum-mass relation.

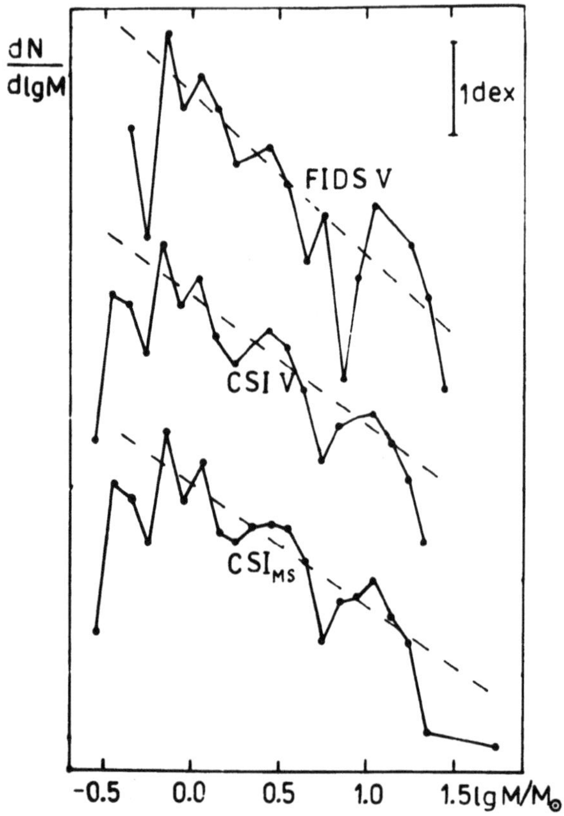

Fig. 6. The initial mass function of VB with V luminosity class primaries (FIDSV), V luminosity class stars from the CSI catalogue (CSIV), and Main-Sequence stars from CSI catalogue (CSI_{MS}).

To account for unequal effects of stars of different masses in the distributions with respect to mass ratios and separations, we took into consideration also the values of IFM for given M (with slope -1.35) when assigning the individual weights to stars. The distribution of VB with luminosity class V primaries, corrected for selection effects is shown in Figure 7. Its dependence on the limiting stellar magnitude of the sample shows that the concentration of stars to $q = 1$ is simply a selection effect. In the range $0.4 \lesssim q \lesssim 0.8$ which is relatively free of selection effects, the distribution is close to that expected in the case of Salpeter's mass function of companions. This means that formation of components of VB is a continuation of the same fragmentation process which forms the common mass spectrum of stars.

Figure 5(b) shows the distribution of VB with primaries of luminosity class V with respect to a, corrected for selection effects. It is flat in the range $4 \lesssim \log a/R_\odot \lesssim 6$, like the distribution of spectroscopic binaries ($\log a/R_\odot \lesssim 3$, Popova *et al.*, 1982), and of VB with known orbits ($3 \lesssim \log a/R_\odot \lesssim 4$, Kraicheva *et al.*, 1985). Also according to Abt (1978), within a factor ~ 2 this distribution is flat for stars with $M \sim M_\odot$ in the

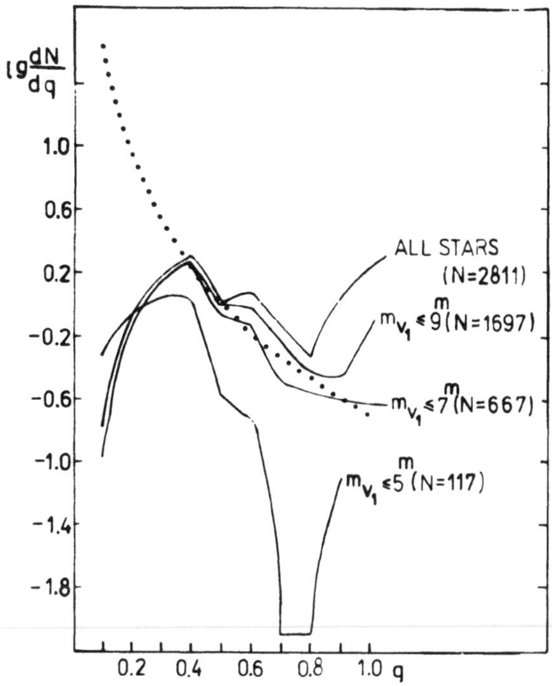

Fig. 7. The distribution of VB with luminosity class V primaries with respect to the mass ratios of components, depending on the limiting stellar magnitude of the sample.

range $1 \lesssim \log a/R_\odot \lesssim 6$. Thus one may suppose that any fixed interval of $\log a$ contains the same proportion of binaries, respectively, to all stars: $dN_d/dN_t = k \, d \log a$. On the other hand, if one neglects the influence of selection effects near the maximum of distribution of VB over $\log a''$ then one finds that the effective width of distribution is 1.1. Thus at any fixed distance to the VB, the observational selection allows to observe stars only in about a unit interval of $\log a$. Thus, the coefficient k has to be equal to the apparent proportion of binaries N_d/N_t among all stars. The data on N_d/N_t given by Allen (1973) allows to estimate that for bright stars $k = 0.2$–0.4. This is roughly equal to the value of k, which we had found previously for spectroscopic binaries. Thus, it appears that practically all stars are binaries with $1 \leq \log(a/R_\odot) \leq 6$.

4. Conclusion

The analysis of VB with primaries of luminosity class V shows that their distribution with respect to the mass is a Salpeter one, the distribution with respect to $\log a$ is flat, as for binaries of other types. The behaviour of distribution with respect to the mass ratios for $q \leq 0.4$ remains unclear because of selection effects.

References

Abt, H.: 1978, in T. Gehrels (ed.), *Protostars and Planets*, Univ. of Arizona Press, Tucson.
Allen, C. W.: 1973, *Astrophysical Quantities*, Athlone Press, London.
Couteau, P.: 1978, *L'Observation des Étoiles Doubles Visuelles*, Flammarion, Paris.
Finsen, W. S. and Worley, C. E.: 1970, *Rep. Obs. Johannesburg, Circ.* **7**, No. 129.
Giannuzzi, M. A.: 1981, *Astron. Astrophys.* **104**, 81.
Iben, I., Jr. and Tutukov, A. V.: 1985, *Astrophys. J.* **282**, 615.
Kraicheva, Z. T., Popova, E. I., Tutukov, A. V., and Yungelson, L. R.: 1985, *Astrofizika* **22**, 105.
Miller, G. E. and Scalo, J. M.: 1979, *Astrophys. J. Suppl. Ser.* **41**, 513.
Piskunov, A. E., Tutukov, A. V., and Yungelson, L. R., 1979, *Pis'ma Astron. Zh.* **5**, 81.
Popova, E. I., Tutukov, A. V., and Yungelson, L. R.: 1982, *Astrophys. Space Sci.* **88**, 55.
Poveda, A., Allen, C., and Parrao, L.: 1982, *Astrophys. J.* **258**, 589.
Svechnikov, M. A.: 1969, *Catalogue of Orbital Elements, Masses and Luminosities of Close Binaries*, Ural University Press, Sverdlovsk.
Vereschagin, S. V., Kraicheva, Z. T., Popova, E. I., Tutukov, A. V., and Yngelson, L. R.: 1987a, *Nauchnyje Informatsii* **63**, 3.
Vereschagin, S. V., Kraicheva, Z. T., Popova, E. I., Tutukov, A. V., and Yungelson, L. R.: 1987b, *Pis'ma Astron. Zh.* **13**, 63.
Woorley, C.: 1980, *Bull. Inform. Centre Donn. Stell.* **18**, 20.

MAZEH – I noticed on your last figure that the distribution of semi-major axes drops at about $\log a = 6$, which is 0.03 pc. Can you comment whether this drop seems real?

YUNGELSON – This drop is most probably not very significant. A real drop starts a little further, at about $\log(a/R_\odot) \approx 6.5$.

WASSERMAN – (1) Have you, or has anyone else, considered the dependence of your results on filtering? (2) What did you use below $\sim 1\,M_\odot$ for the age $T(M)$?

YUNGELSON – (1) We did not analyse the dependence on filtering. We considered the '1% filter' as powerful enough.
(2) We had used 10^{10} yr as the life-time of stars with $M \leq M_\odot$.

POVEDA – I am very glad to see that you verified the validity of Salpeter's Main-Sequence spectrum for the secondaries; can you tell us down to what masses of secondaries Salpeter's spectrum holds?

YUNGELSON – I guess about 0.3–0.4 $M_\odot$. Our data does not extend to lower masses.

THE OPTICAL FRACTION OF A HIPPARCOS SAMPLE OF WIDE BINARIES*

P. BROSCHE and D. SINACHOPOULOS

Sternwarte der Universität Bonn, F.R.G.

Summary. Considerations on the optical fraction of our sample.

It was our task to estimate actual relative positions for 3657 pairs with distances $\rho \geqq 5''$. The pairs were selected by Dr Dommanget and the data were kindly provided by Dr Worley.

In 1173 cases, the data are sufficient to fulfill the task. For 582 of the 1173 pairs, a significant relative proper motion was obtained (Figure 1). An overall figure for the

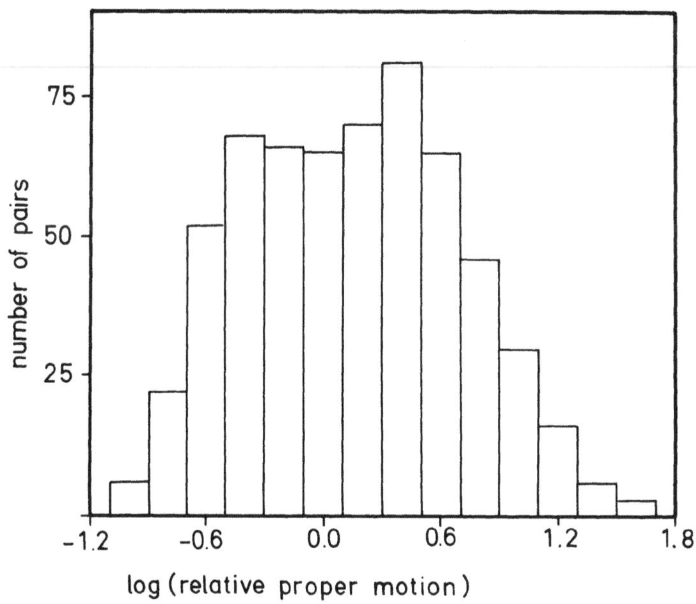

Fig. 1. The histogram of the logarithms of the relative proper motions (measured in seconds of arc per century).

lower limit of a detectable relative motion is $\mu_0 = 0''.1$/century. We estimated the parallax π and mass from the assumption that the primary is a Main-Sequence star. With the additional assumption of a circular orbit with radius $\simeq \rho$ we obtained a guess for

* Presented by D. Sinachopoulos.

Astrophysics and Space Science **142** (1988) 255–258.

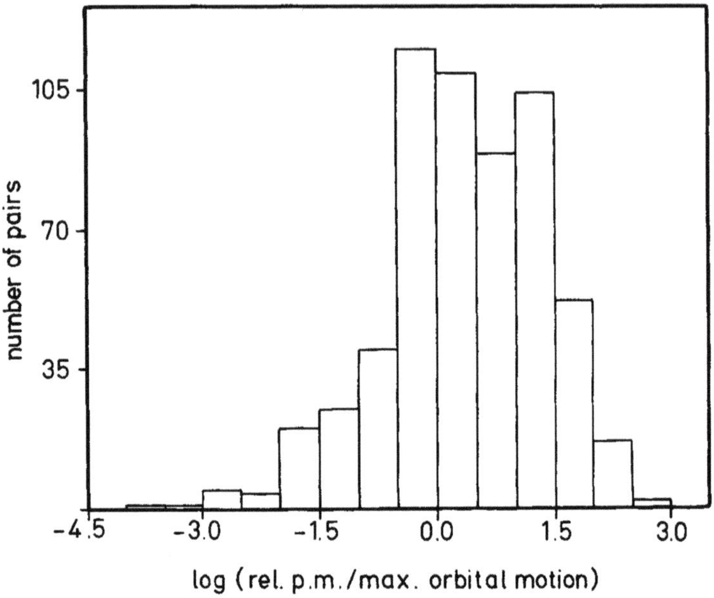

Fig. 2. The histogram of the logarithms of the ratio (relative proper motion/maximum orbital motion). For the estimation of the latter see text.

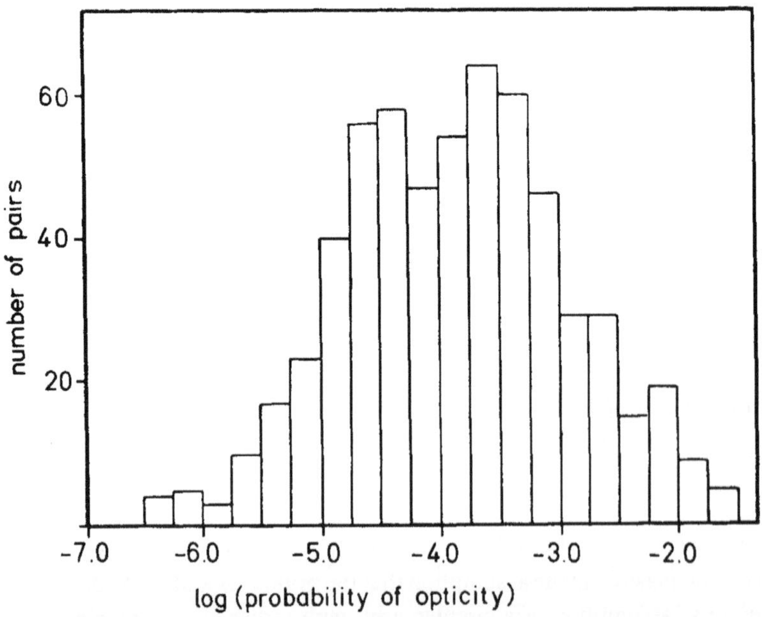

Fig. 3. The histogram of the logarithms of the probability that a given pair is optical only. The values are derived on the basis of mutual distance, magnitude of the secondary, and galactic latitude of the pair.

the orbital velocity V and compared it with the observed relative motion $v = \mu/\pi$ (Figure 2). It is clear that the uncertain fulfillment of our assumptions plus the varying influence of projection factors must lead in case of physical pairs to a large dispersion of the ratio v/V around an average value of the order 1. Nevertheless, (a) if the primary is a giant, the orbital velocity were much smaller (making our adopted value an upper limit) and (b) the relative motion is rather a lower limit for an orbital velocity because of projection. Hence, our ratio $\nu = v/V$ is more likely *smaller* than 1 for physical pairs. So we conclude from Figure 2 that at least one half of our 600 pairs are optical. This is at a first glance surprising since criteria like the one of Aitken (1911) tell us that there is a high probability for a physical nature of all our pairs. Therefore, we took this case as a chance to check the statistical values in a quantitative way. The distinction between optical and physical pairs is also of interest for the evaluation of the HIPPARCOS data: namely, for the choice of the appropriate 'model'.

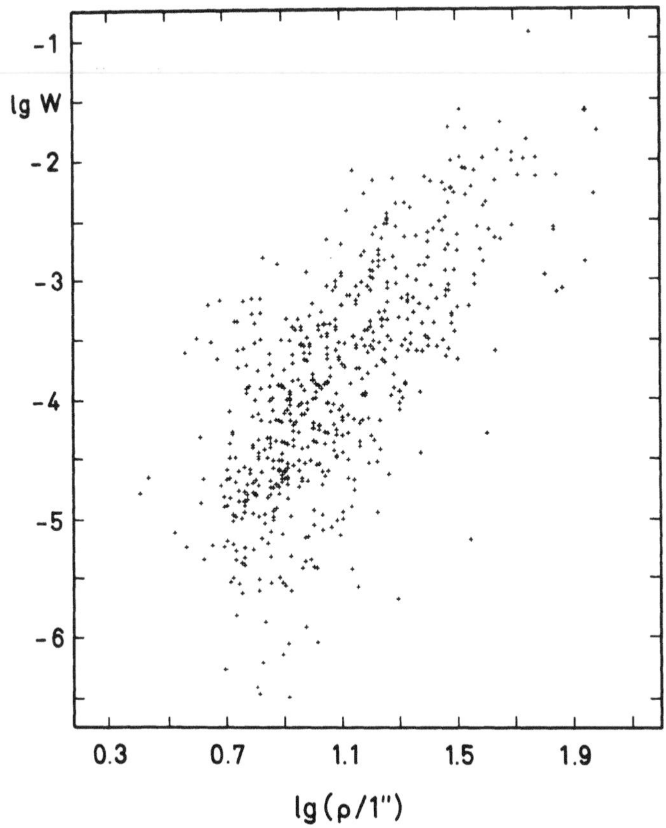

Fig. 4. The probability pairs being optical only as a function of distance (both quantities represented by their logarithms).

As it is well known, the probability W to find by chance a star brighter than m in a circle with radius ρ around a preselected point is given by

$$W = \pi\rho^2 \sigma(m)\,,$$

where $\sigma(m)$ is the surface density of stars brighter than m at the celestial spot under consideration (Kubikowski *et al.*, 1958). We used tables for the stellar densities as a function of apparent photographic magnitude and of galactic latitude (Scheffler and Elsässer, 1965) and computed the probability W that each of our pairs with $\mu \neq 0$ is merely an optical one individually in dependence on the magnitude m of the secondary and the latitude of the pair (Figure 3). One might want to compare our results with the usual formulae by plotting the W for our sample as a function of ρ (Figure 4). The large scatter in W at fixed ρ illustrates that the influence of the other parameters on W is of comparable importance. The average probability for our sample is $\langle W \rangle = 1.23 \times 10^{-3}$. This number has to be multiplied with the number of stars which have been considered as candidates for the *Input Catalogue*, that is with about $N = 2 \times 10^5$, since the known double stars have been found from inspecting approximately such a body of stars (there are more double stars than the ones of our list, but the latter contains the majority of the *wide* binaries, and only those are suspected of being mere optical pairs). If we do so, we expect about 300 optical pairs within our material – in good agreement with the number of those binaries which have a relative proper motion larger than a reasonable orbital motion. Because the numbers agree, it seems that we know already a large part of the optical pairs among the pairs with measurable relative motion.

Acknowledgement

The support of the Federal Ministry of Research and Technology is gratefully acknowledged.

References

Aitken, R. G.: 1911, *Astron. Nachr.* **188**, 281.

Kubikowski, J., Lukaszewicz, J., and Opolski, A.: 1958, *Acta Astron.* **9**, 4.

Scheffler, H. and Elsässer, H.: 1965, *Landolt-Börnstein*, Vol. I, Astronomy and Astrophysics, Springer-Verlag, Berlin, p. 604.

WIDE BINARIES IN A HIPPARCOS SAMPLE CENTERED AROUND $b = 0°$, $l = 300°$*

H. HENSBERGE[1], A. JONGENEELEN, and R. LE POOLE

Sterrewacht Leiden, The Netherlands

Summary. $3' \times 3'$ pictures centered on *Hipparcos Input Catalog Objects* (L3–2 INCA) have been produced in Leiden from measuring 87 plates of the Dutch glass copy of the ESO Quick Blue Survey by use of astroscan. We extract from this material our studied sample, consisting of 19825 regions around INCA objects brighter than 11 mag and outside open clusters. It enables to study in a statistical way wide binaries with apparent angular separations between $0\overset{'}{.}1$ and $1\overset{'}{.}5$ and with components within 3 mag (in the blue spectral region) of the pointed object. We present in this paper our first results on the observed characteristics of this population.

1. Introduction

In the frame of the commitment of Leiden Observatory to the INCA consortium preparing the HIPPARCOS mission, 87 plates of the Dutch glass copy of the ESO Quick Blue Survey have been measured with the ASTROSCAN microdensitometer, providing information on $3' \times 3'$ regions around approximately 22000 INCA stars. An accurate position and a photographic blue magnitude B have been measured for each INCA star and for stars in their vicinity that could influence significantly the signal received from HIPPARCOS. Explicitly, position and magnitude have been determined for all additional stars within $1\overset{'}{.}5$ and brighter than $B + 3$; fainter stars have only been measured progressively with decreasing separation to the pointed star.

Although measured originally for the purpose mentioned above, the sample obviously contains valuable information on wide binaries and deserves attention from that point of view. The available sample, after statistical correction for optical companions, gives information on very wide binaries composed of mostly rather young stars of comparable luminosities (mass). The sample is dominated by young stars, and includes several OB associations in the constellation Scorpius. A typical system, with apparent separation of the order of $1'$, apparent magnitude 10 and spectral types A0 + A0 would correspond to a system with linear separation of roughly 50000 AU and escape velocity of 0.4 km s^{-1}. Such systems have a non-negligible probability of disruption over a stellar life-time, so that age/spectral-type dependencies may be expected. Moreover, the sample is suitable to study wide binary characteristics for association members versus field stars, once the HIPPARCOS programme of the SPECTER group in Leiden will have furnished extensive membership data.

In this contribution, a subsample selected to look for the presence of wide double stars is described, some of its apparent characteristics are pointed out and prospects for further research are discussed.

* Presented by H. Hensberge.

[1] ESA research fellow.

Astrophysics and Space Science **142** (1988) 259–264.

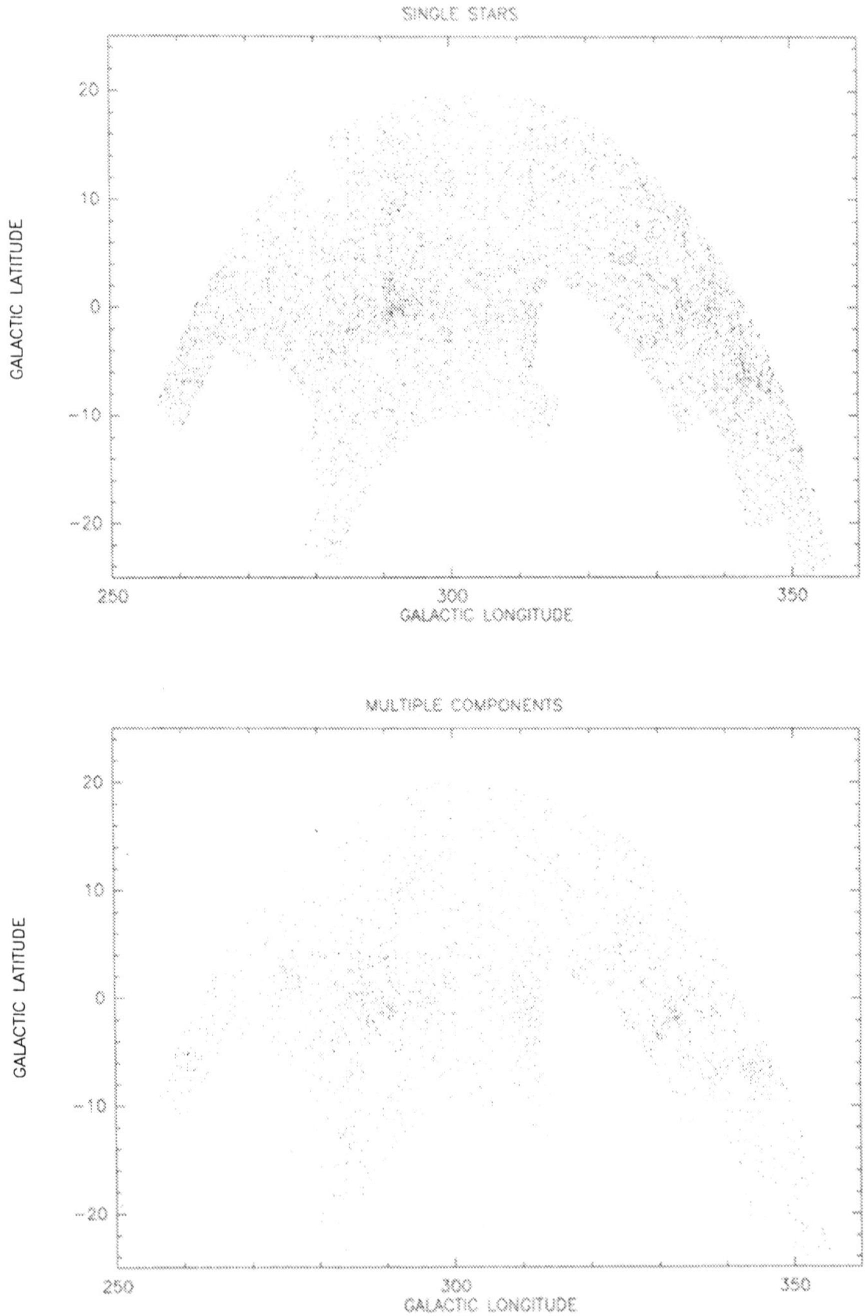

Fig. 1. Selected sample of INCA stars in galactic coordinates.

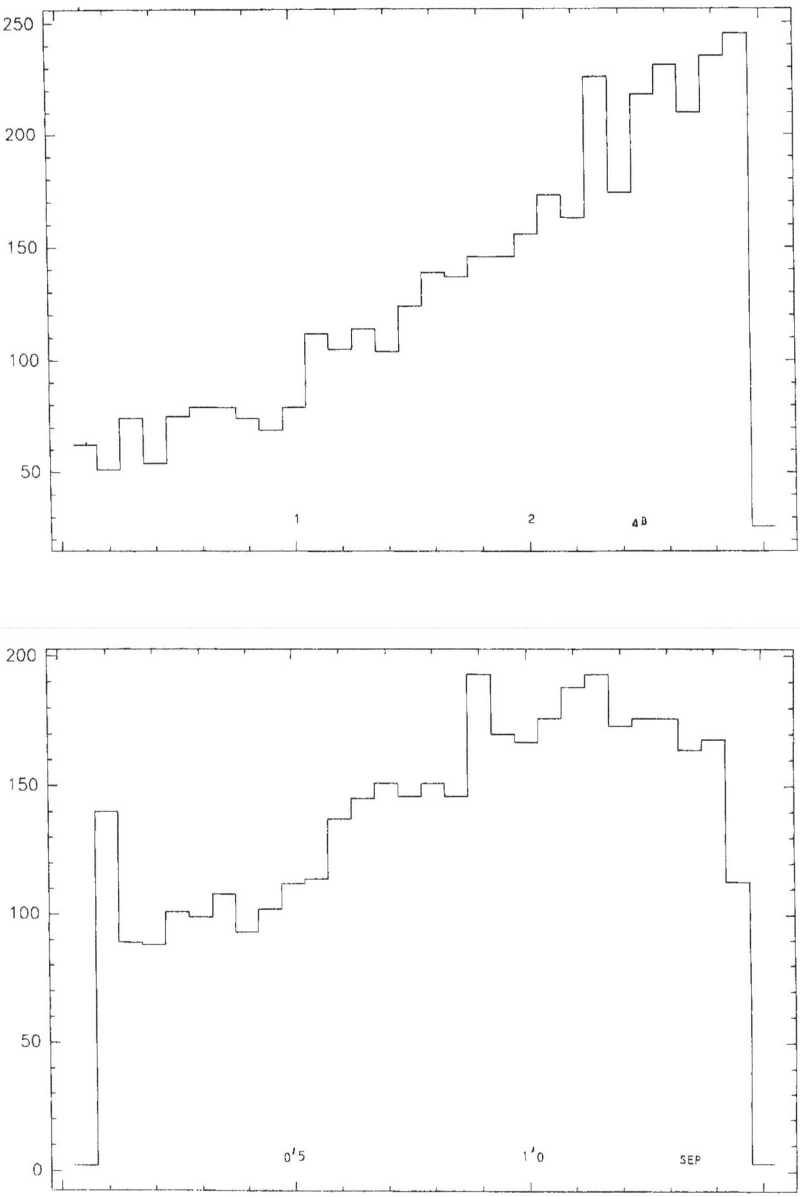

Fig. 2. Frequency (ordinate) per magnitude contrast ΔB bin (above) and per apparent separation bin (below). Apparently our sample is incomplete in the separation range 1'.1 to 1'.5, mainly because faint companions (ΔB near 3) at relative large distance of faint INCA stars (B fainter than 10) have not always been recognized.

2. Description of the Selected Sample

The original sample has been corrected for redundancies (overlap of plates, overlap of $3' \times 3'$ regions) and has been limited to INCA stars brighter than $B = 11$, and to additional stars within magnitude $B + 3$ with respect to the INCA object and within the apparent separation range 0$'$.1 to 1$'$.5. Furthermore, stars in the direction of open clusters have been deleted. The sample then contains 19825 INCA stars and 4752 additional stars, the latter being distributed over 3891 regions around INCA objects. Notice that stars within 0$'$.1 of each other have always been counted as one object in our subsample, as well for INCA as for additional stars. Half of these INCA stars have spectral types A2 or earlier, 25% belong to types G0 and later. Half of the stars have magnitudes in the range 9.0 to 10.4, only 12% are fainter. Their spatial distribution is given in Figure 1. Figure 2 shows the distribution of the apparent separations and the magnitude differences in the sample.

3. Preliminary Discussion

From this selected sample, subsamples have been drawn with limiting conditions with respect to galactic position, spectral type of INCA star, magnitude of INCA star, magnitude contrast ΔB, apparent separation, etc. A straightforward computation using average stellar count rates for low galactic latitudes b shows that the expectation value for additional stars up to $B + 3$ in a 1$'$.5 radius around a 11 mag INCA star is already near to unity. Hence, fainter stars contribute at large separations mainly optical companions. About 3 out of 4 listed additional stars in our sample are expected to be optical companions.

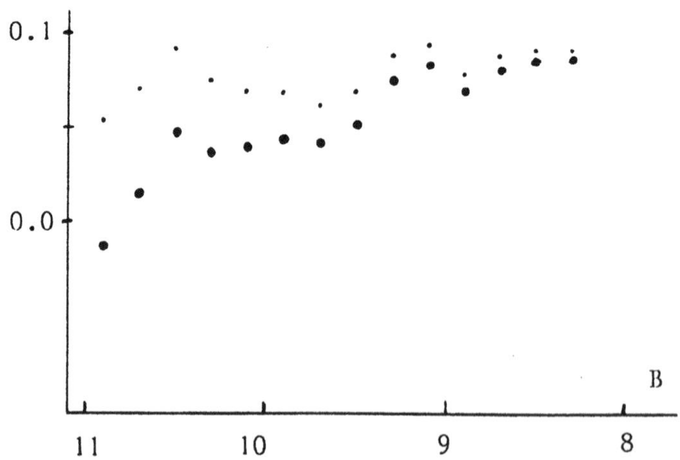

Fig. 3. Frequency of physical companions per star, within the apparent separation range 0$'$.1 to 1$'$.5 as function of magnitude B of the INCA stars. Small dots show the effect of a 10% decrease in the correction for optical companions.

Frequencies of physical companions have been computed for several selected subsamples, taken into account the expected frequency of optical companions for given galactic position, magnitude, magnitude contrast, and apparent separation. In subsamples selected with respect to spectral type, the number of physical companions per star as a function of magnitude shows a tendency to remain constant or decrease slightly towards fainter magnitudes (the apparent decrease may be due to the selection effect at large separations mentioned with Figure 2 or/and to a 10% underestimation of the correction for optical companions). Figure 3 visualizes the situation for B7–A3 stars. If confirmed, however, the decrease should give information on the 'cut-off' separation distance. Notice that our separation field-of-view corresponds very tightly to the spectral type dependent 'cut-off' distance mentioned by Abt at this colloquium.

Better established is the presence, in our sample, of a somewhat higher density of physical companions at lower galactic latitudes for given spectral type. Table I summarizes the results for the subsample of B9–A0 stars, as a function of 'linear' separation assuming their magnitude B to be a distance indicator. The narrowing of the separation intervals for fainter stars eliminates already most optical companions, the result, therefore, being not severely dependent on the adopted correction. The presence of reddening in the galactic plane would enhance the observed difference. Hence, the observed ratio of number of physical companions per INCA star at $|b| < 7.^\circ5$ versus the number at $|b| > 7.^\circ5$ amounts at least to a factor 1.5 in the B9–A0 sample.

TABLE I

Dependence of density of physical companions on galactic latitude. The results refer to a sample of 3118 B9–A0 stars in the magnitude range $B = 8$–11. 'Linear' separations (first column) are given as angular separations at $B = 9.5$. Densities are given in units of number of physical companions per star per surface of $1' \times 1'$ within the given separation range. The results at low b refer to 321 companions of which 60 were treated as optical, these at high b refer to 70 companions of which 10 were treated as optical. The separations covered is roughly 8000 to 25000 AU.

Separation at $B = 9.5$	Density/star/$1' \times 1'$ $|b| < 7.^\circ5$	Density/star/$1' \times 1'$ $|b| > 7.^\circ5$
$0.'25$ to $0.'35$	0.09	0.05
$0.'35$ to $0.'45$	0.08	0.07
$0.'45$ to $0.'55$	0.07	0.04
$0.'55$ to $0.'65$	0.06	0.04
$0.'65$ to $0.'75$	0.07	0.05

Finally, an inspection of about 100 regions with at least three additional components (data in Table II and Figure 4) reveals that they might be explained by optical components (as judged from their distribution with magnitude contrast and separation) for all spectral types except for pictures centered around O and early B stars.

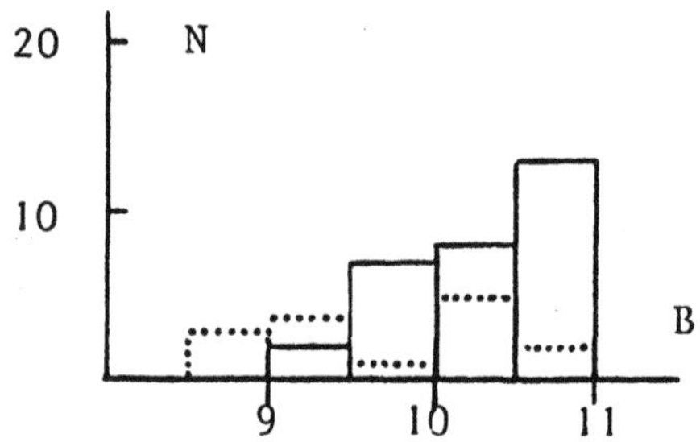

Fig. 4. Distribution with magnitude for the subsamples of INCA stars used to construct Table II. Full line corresponds to the late-B type stars, dotted line to O and early-B stars.

TABLE II

Companions of INCA stars showing 3 or more companions. Three samples are compared: two empirical ones, referring to O and early-B stars (a), respectively, late-B stars (b), and one theoretical distribution corresponding to the case of optical companions (c). The results of (b) and (c) are compatible, those of (a) and (c) not: 38 are expected, 22 are observed below the line crossing (a).

	(a)			(b)			(c)		
Separ. / ΔB	0.′1–0.′5	0.′5–1.′0	1.′0–1.′5	0.′1–0.′5	0.′5–1.′0	1.′0–1.′5	0.′1–0.′5	0.′–1.′0	1.′0–1.′5
0–1	2	3	6	2	2	8	1	3	5
1–2	5	8	5	6	6	13	3	8	14
2–3	6	9	8	5	27	33	7	22	37
	52 companions			101 companions			100 companions		

4. Prospects

With respect to the analysis of HIPPARCOS data, this work draws attention to the fact that measurements may be expected to be disturbed more often by the presence of nearby stars than indicated by relying on double star catalogs that use cut-off conditions based on the probability of physicity for the system.

The discussed apparent characteristics have been derived from preliminary trial samples. Further work on the data is likely to contribute on questions related to the most wide binaries. Progress can be made by looking in more detail to the correction procedure for opticity; by considering additional information on the companions (spectral type if available, or by measuring red plates to select out obviously optical systems); by arranging results with respect to relevant physical parameters as estimates of the binding energy and distance from the galactic plane, etc. Checks on the validity of the correction model for opticity and on the reality of correlations should finally be made by long slit spectroscopy in order to find out in smaller and thus treatable selected samples which fraction of systems shows compatible radial velocities.

SIXTH SESSION

ORIGIN AND EVOLUTION OF WIDE PAIRS

Chairman: J. SMAK

THEORETICAL MODELS FOR WIDE BINARY EVOLUTION

Introductory Lecture

I. WASSERMAN*

Center for Radiophysics and Space Research Cornell University, Ithaca, New York, U.S.A.

Summary. Wide binary evolution under the influence of stars, giant molecular clouds, and the galactic tidal field is reviewed. Recent results show that binary lifetimes at $a \gtrsim 0.1$ pc are determined by an interplay between relatively gentle evolution due to stellar encounters and catastrophic collisions with cloud sub-clumps. It is emphasized that binary semimajor axis distributions may not be understood from lifetimes alone. To determine the theoretical distribution binary 'birthrate functions' must be convolved with probability distributions for the evolution of an ensemble of binaries from given initial separations. Simple models show no sharp breaks below the tidal cut-off imposed by the Galaxy.

1. Introduction

Very widely separated binary star systems are enormously interesting to theoreticians because they may be used to probe the galactic disk environment. Because of their extremely low binding energies, wide binaries are especially susceptible to perturbations by any constituent of the galactic disk. Given sufficient data on the distribution of binary semi-major axes a at large a, one could conceivably discriminate among various models for the birth and subsequent orbital evolution of wide binaries. In this way significant constraints on the various input parameters characterizing different types of perturbers may, at least in principle, be derived.

A 'minimal' model for the orbital evolution of wide binaries should include the influences of passing stars, giant molecular clouds (hereafter GMCs) and the galactic tidal field (Bailey, 1983, 1985; Bahcall *et al.*, 1985; Hut and Tremaine, 1985; Weinberg *et al.*, 1986, 1987). Perturbations due to stars are relatively easy to treat statistically (cf. Weinberg *et al.*, 1987, Section IIb and Section 2 below). Since the densities, masses, velocity dispersions and scale heights for the different types of stars are well known (e.g., Bahcall and Soneira, 1980; Mihalas and Binney, 1983), their perturbative effects on binaries may be accurately specified. Similarly, galactic tidal effects, whose scale is set locally by the Oort (differential) rotation constants, may also be included 'exactly'. However, an exact treatment would necessarily require non-Keplerian orbits for binaries near the tidal limit, $a_T \sim 1$ pc $(M/M_\odot)^{1/3}$ for a binary mass M (Antonov and Latyshev, 1972; Heisler and Tremaine, 1986; Weinberg *et al.*, 1987). For a statistical study of an ensemble of wide binaries it is probably sufficient to retain a Keplerian description of the orbits while modelling tidal disruption by the Galaxy using an upper semi-major axis cut-off at $a = a_T$, beyond which binaries are unbound (e.g., Weinberg *et al.*, 1987; see, however, Torbett and Smoluchowski, 1984). The theoretical situation is less settled for perturbations by GMCs, whose abundance, masses and internal structure are still

* Alfred P. Sloan Foundation Fellow.

Astrophysics and Space Science **142** (1988) 267–276.

controversial (Blitz, 1985; Sanders *et al.*, 1984). We shall discuss below recent calculations that explore the dependences of the characteristic lifetimes of wide binaries ($a \gtrsim 0.01$ pc) on GMC parameters.

Still more problematic are the perturbative effects of the Oort disk dark matter, for which the local mass density may be estimated and the scale height is bounded (Bahcall, 1984a, b). Bahcall *et al.* (1985) have pointed out what wide binary observations may be used to constrain the particle masses M_p of hypothetical low mass dark matter particles, and have suggested that $M_p \lesssim 2\, M_\odot$ based on the Latham *et al.* (1984) study of binaries with $a \lesssim 0.1$ pc in the Weistrop (1972) catalog. (See also Bahcall and Soneira, 1981.) While it is, in fact, premature to use available data to constrain M_p (Wasserman and Weinberg, 1987) significant upper bounds may conceivably be inferred from future observations, at least in the context of simple models for the birth rate and initial semi-major axis distribution for wide binaries.

2. Physics of Binary Evolution

Wide binary evolution is governed by an interplay among the various types of perturbers. Even though the dynamical problem is linear, one cannot understand the results by separately considering the effects of, for example, stars and GMCs. This is because the relatively gentle, diffusive evolution of binary semi-major axis due to stellar perturbations sets the stage for disruptive encounters with cloud subclumps as binaries pass through GMCs.

Field stars cause a steady 'advective diffusion' of wide binaries to large separation (Retterer and King, 1982). Typical random stellar velocities $v_* = 20$ km s^{-1} (Mihalas and Binney, 1983) are much larger than the characteristic orbital speeds $v \approx 0.2\,(M/M_\odot)^{1/2}\,(a/0.1\ \text{pc})^{-1/2}$ km s^{-1} for wide binaries. As a result wide binaries tend to 'heat up', on average, due to their encounters with passing stars. The dominant encounters are those with impact parameters $b < a$, which occur with a frequency (Myr $\equiv 10^6$ yr)

$$\Gamma^*(b < a) = 2\pi a^2 n_* v_* \approx (7.5\ \text{Myr})^{-1} \left(\frac{a}{0.1\ \text{pc}}\right)^2 \left(\frac{n_*}{0.1\ \text{pc}^{-3}}\right) \left(\frac{v_*}{20\ \text{km s}^{-1}}\right). \tag{1}$$

(The factor of 2 arises because there are two stars in a binary.) Such encounters are not disruptive in general. Since the typical change in binding energy per unit mass is $\Delta E \approx 2(GM_*/bv_*)^2$ for an impact parameter $b < a$ and stellar perturber mass M_* (e.g., Weinberg *et al.*, 1987; e.g., Equation (5)) disruptive encounters are those with

$$\frac{b}{a} \lesssim 2\,\frac{M_*}{M}\,\frac{v}{v_*} \approx 0.02\,\frac{M_*}{M}\left(\frac{M}{M_\odot}\right)^{1/2} \left(\frac{a}{0.1\ \text{pc}}\right)^{-1/2} \left(\frac{v_*}{20\ \text{km s}^{-1}}\right)^{-1}, \tag{2}$$

which occur with a frequency

$$\Gamma_{\text{kill}}^{*} \approx (1.9 \times 10^4\ \text{Myr})^{-1} \left(\frac{M_*}{M}\right)^2 \left(\frac{M}{M_\odot}\right)\left(\frac{a}{0.1\ \text{pc}}\right) \times \\ \times \left(\frac{n_*}{0.1\ \text{pc}^{-3}}\right)\left(\frac{v_*}{20\ \text{km s}^{-1}}\right)^{-1} . \tag{3}$$

Equation (3) is also a fair estimate of the binary lifetime against stellar perturbatons when the galactic tidal field is ignored, even though individually disruptive encounters actually contribute far less effectively to binary dissolution than the cumulative perturbations due to relatively distant encounters (Bahcall *et al.*, 1985). However, galactic tidal effects are important at $a \gtrsim 0.1$ pc, and both the numerical coefficient in Equation (3) and the scaling with a are altered when finite a_T is included (Weinberg *et al.*, 1987, especially Figure 7).

Giant molecular clouds also play a central role in the physics of wide binary evolution. Penetrating encounters with GMCs, in which binaries pass through clouds and interact with their subclumps, predominate. Such encounters occur with a frequency

$$\Gamma_{\text{pen}}^{\text{GMC}} = \pi n_{\text{GMC}} R^2 v_b \\ \approx (9.7 \times 10^2\ \text{Myr})^{-1} \left(\frac{n_{\text{GMC}}}{4 \times 10^{-8}\ \text{pc}^{-3}}\right)\left(\frac{R}{20\ \text{pc}}\right)^2 \left(\frac{v_b}{20\ \text{km s}^{-1}}\right), \tag{4}$$

where $v_b \sim v_*$ is the binary center of mass speed, and n_{GMC} is the number density and R the radius of GMCs (Solomon *et al.*, 1984; Blitz, 1985). Comparing Equations (3) and (4) we see that several GMC penetrations are likely to occur within a binary lifetime, provided that wide binaries are not invariably disrupted on passing through a cloud. If we assume N uniformly distributed subclumps, the smallest impact parameter expected during a single GMC penetration is

$$b_{\text{min}}^{\text{SC}} = \left(\frac{8}{3\pi}\right)^{1/2} \frac{R}{\sqrt{N}} = 3.7\ \text{pc}\left(\frac{R}{20\ \text{pc}}\right)\left(\frac{N}{25}\right)^{-1/2} . \tag{5}$$

Since $b_{\text{min}}^{\text{SC}} \gg a$, encounters with GMC subclumps only produce differential tidal effects on the members of a binary. For point-mass subclumps of mass M_{SC}, the resulting mean energy change is $\Delta E \approx (7/3)\,(GM_{\text{SC}}/bv_b)^2\,(a/b)^2$ (Weinberg *et al.*, Equation (10) and Appendix A). Thus the critical impact parameter for disruption is

$$b_{\text{crit}}^{\text{SC}} \approx 1.8\ \text{pc}\left(\frac{M}{M_\odot}\right)^{-1/4}\left(\frac{M_{\text{SC}}}{2 \times 10^4\, M_\odot}\right)^{1/2}\left(\frac{v_b}{20\ \text{km s}^{-1}}\right)^{-1/2}\left(\frac{a}{0.1\ \text{pc}}\right)^{3/4} . \tag{6}$$

Since $b_{\text{min}}^{\text{SC}} \gtrsim b_{\text{crit}}^{\text{SC}}$ for $a \lesssim 0.3$ pc, wide binaries generally survive passages through GMCs as long as their semi-major axes are not too large. This conclusion can only be reinforced by taking into account the finite radii, $R_{\rm clump} \sim 2$–3 pc (Sanders *et al.*, 1984; Blitz, 1985), of the subclumps, thus softening the impact of their gravitational perturbations (Hut and Tremaine, 1985; Weinberg *et al.*, 1987, p. 376 and Figure 5). However, calculations with point mass subclumps only underestimate binary lifetimes by factors $\gtrsim 2$ at $a \lesssim 0.03$ pc, even though $b_{\rm crit}^{\rm SC} \ll R_{\rm clump}$ at such small values of a (see Figure 1 and Weinberg *et al.*, 1987).

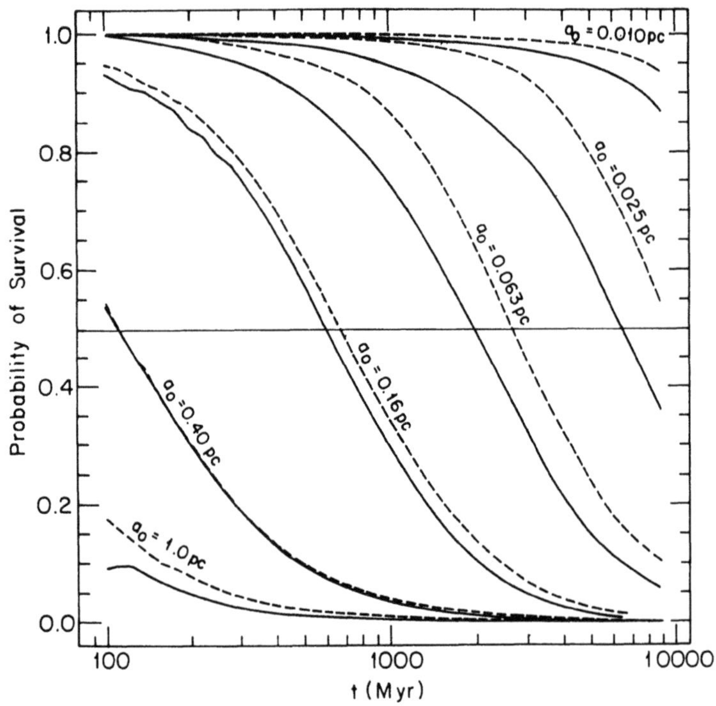

Fig. 1. Probability of survival versus time for selected initial semi-major axes a_0. Solid lines were computed for point-mass subclumps. Dashed lines were computed for $R_{\rm clump} = 4$ pc.

Underlying this apparently counterintuitive result is the dynamical interplay between stellar and cloud perturbations. From Equation (6) it is clear that disruption by subclumps becomes increasingly likely as a increases. As we have seen, stellar perturbations tend to steadily widen loosely bound binaries, thereby promoting dissolution at the hands of GMC subclumps. Even though a binary only becomes unbound when $a > a_T \sim 1$ pc $(M/M_\odot)^{1/3}$, disruption becomes exceedingly probable in a single GMC penetration when $a \gtrsim 0.3$ pc. Thus, binaries with $a \lesssim 0.1$ pc that might survive in the face of perturbations due to stars alone cannot withstand the combined onslaught of stars and GMCs.

The characteristic time-scale for the probability of survival to fall below $\frac{1}{2}$ for a $1\ M_\odot$ binary whose initial semi-major axis is a_0 is well-fitted by the formula (Weinberg *et al.*, 1986)

$$t_{1/2}(a_0) \approx 2.6 \times 10^3 \text{ Myr} \left(\frac{n_{\text{GMC}}}{10^{-8} \text{ pc}^{-3}} \right)^{-0.53} \left(\frac{a_0}{0.1 \text{ pc}} \right)^{-1.4} \tag{7}$$

for $a_0 \gtrsim 0.1$ pc. Note that $t_{1/2}(a_0) \propto n^{-\beta}$ with $\beta < 1$, which reflects the fact that binary dissolution is not an unavoidable consequence of GMC penetration. If perturbations by GMCs are ignored we have the corresponding result for stars alone,

$$t_{1/2}(a_0) \approx 4.3 \times 10^3 \text{ Myr} \left(\frac{a_0}{0.1 \text{ pc}} \right)^{-1.4} . \tag{8}$$

Equation (8) illustrates, on comparison with Equation (3), the destabilizing influence of the tidal cut-off a_T. Setting $t_{1/2}(a_{\text{trit}}) = 10^{10}$ yr in Equation (7) gives $a_{\text{crit}} \approx 0.04 (n_{\text{GMC}}/10^{-8} \text{ pc}^{-3})^{-0.38}$ pc, or $a_{\text{crit}} \approx 0.025 \text{ pc} \approx 5000$ AU for $n_{\text{GMC}} \approx 3.6 \times 10^{-8} \text{ pc}^{-3}$. Although Equation (7) is not strictly speaking valid at such low values of a_0, the derived value of a_{crit} agrees fairly well with the results shown in Figure 1.

3. Binary Distributions

For comparing theory and observations, it is not enough to know the characteristic dissolution time $t_{1/2}(a_0)$ as a function of initial semi-major axis a_0. First, for a given set of parameters describing the various perturbers, $\mathcal{P}(a; a_0, t)\, \mathrm{d}a$, the probability that after time t a binary with original semi-major axis a_0 has a semi-major axis between a and $a + \mathrm{d}a$, must be computed. Next, the probability distribution must be convolved with a 'birthrate function' for wide binaries, which may be a function of a_0, t, spectral type, binary mass, etc. These two steps suffice to calculate the theoretical semi-major axis distribution. Further convolutions are required to take into account observational selection effects, such as magnitude and angular cut-offs (e.g., Wasserman and Weinberg, 1987; Weinberg 1987 (these proceedings)).

Very little is known about the binary birthrate function. To gain some insight, Weinberg *et al.* (1987) considered very simple separable models $\dot{N}(a_0, t) = N(a_0) f(t)$, independent of spectral type and binary mass. (These are admittedly oversimple model assumptions; see, e.g., Abt, 1982 and references given therein.) The models were further specialized by choosing $N(a) \propto a^{-\alpha}$ and either $f(t) = \delta(t)$ (a 'burst' model) or $f(t) = \text{constant} \equiv 1/T$ (a 'continuous creation' model). The binary distribution is then

$$\begin{aligned} N(a, t) &= \int_0^t \mathrm{d}t' \int_0^{a_T} \mathrm{d}a_0 \dot{N}(a_0, t') \mathcal{P}(a; a_0, t - t') \\ &= N(a_0) \mathcal{P}(a; a_0, t) \quad \text{(burst)} \\ &= \int_0^t \frac{\mathrm{d}t''}{T} \int_0^{a_T} \mathrm{d}a_0 N(a_0) \mathcal{P}(a; a_0, t'') \quad \text{(continuous)} . \end{aligned} \tag{9}$$

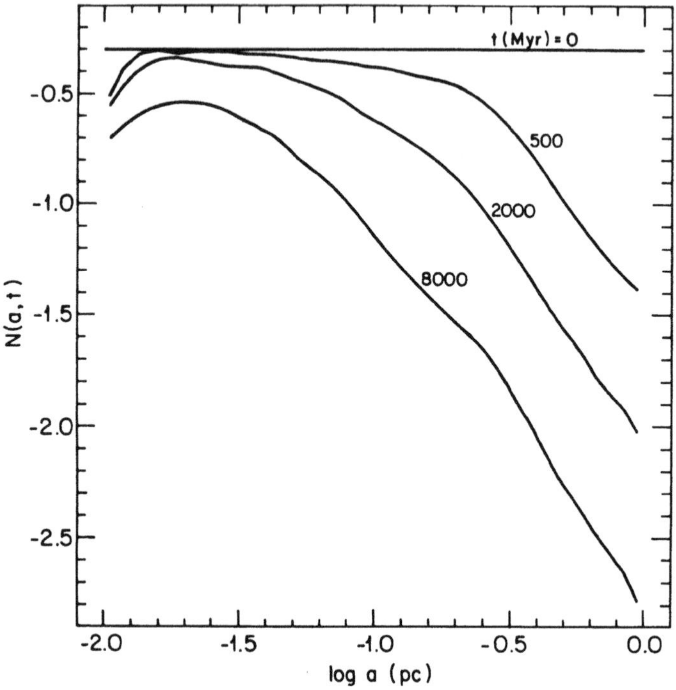

Fig. 2a.

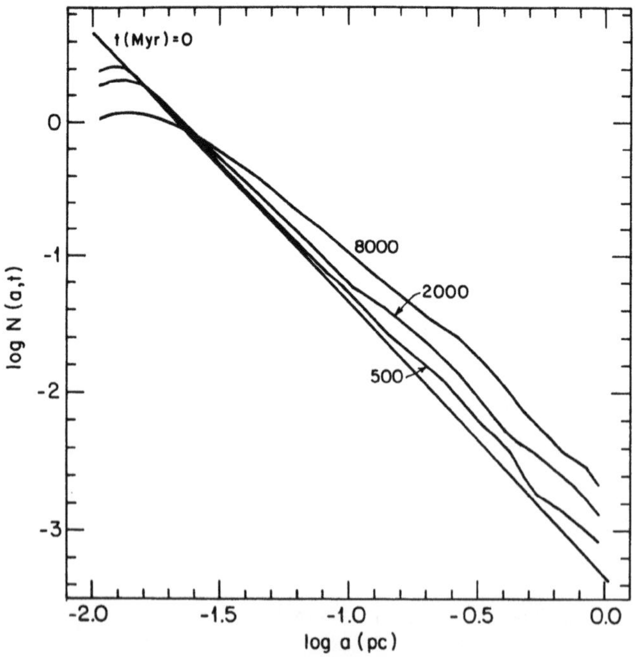

Fig. 2b.

Fig. 2. Semi-major axis distributions at $t = 0$, 500, 2000, and 8000 Myr for burst models with (a) $N(a_0)$ = constant, (b) $N(a_0) \propto a_0^{-2}$.

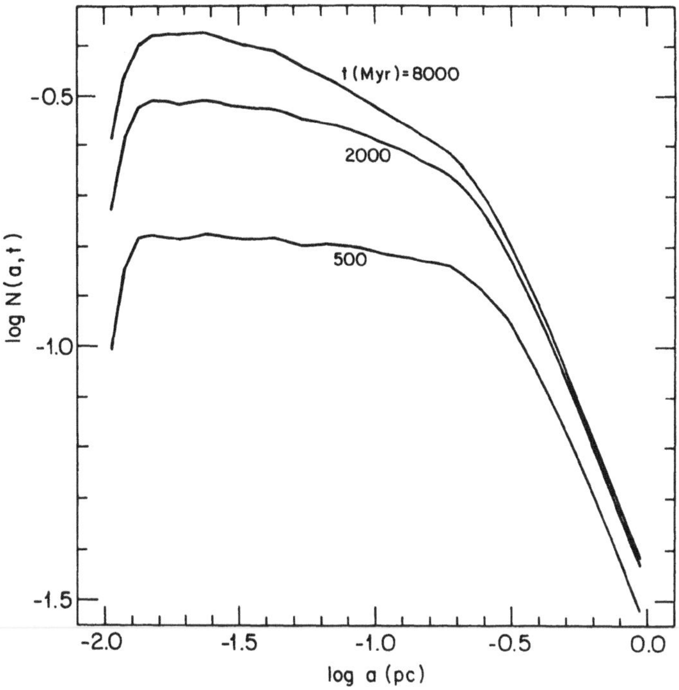

Fig. 3a.

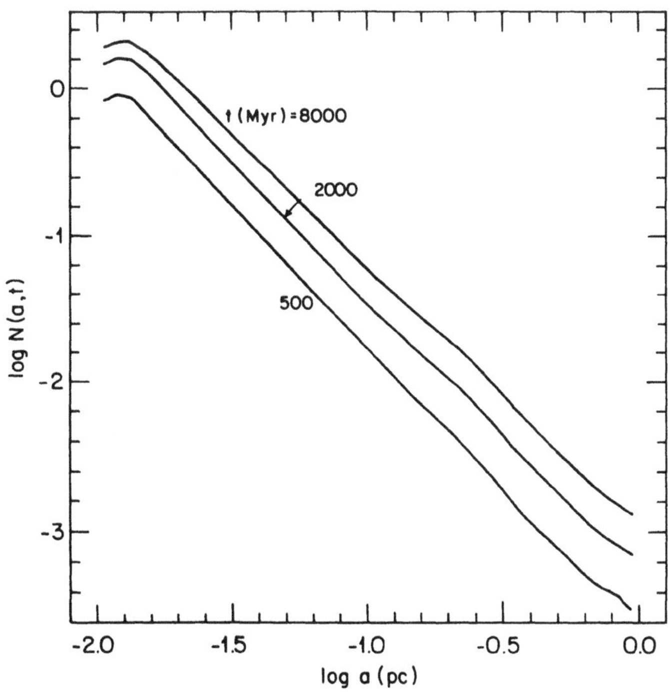

Fig. 3b.

Fig. 3. Semi-major axis distributions at t = 500, 2000, and 8000 Myr for continuous creation models with (a) $N(a_0)$ = constant, (b) $N(a_0) = a_0^{-2}$.

Clearly, the resulting theoretical distribution must reflect a competition between the distribution of binary orbits at birth and subsequent orbital evolution.

This competition is particularly well illustrated for the burst models. For $N(a_0)$ = constant, Figure 2(a) demonstrates that the dissolution of binaries at large a dominates the evolution of the ensemble. The distribution develops a power-law tail $N(a, t) \propto a^{-\lambda}$ as binaries widen towards the tidal boundary a_T. (The dip at $a \approx 0.01$ pc is unphysical since binaries with $a_0 < 0.01$ pc, which have very long dissolution times, were not included.) The powerlaw index in the tail, which eats its way inward with time, is $\lambda \gtrsim 1.5$. At larger values of α, the situation is quite different. Figure 2(b) shows results for $N(a_0) \propto a_0^{-2}$. Now, the high a tail actually grows as time goes on, demonstrating the naiveté of arguments based purely on binary lifetimes. Asymptotically, the distribution appears to seek a shallower power-law tail, once again with $\lambda \gtrsim 1.5$, but $\lambda \leq \alpha = 1$.

Continuous creation models reflect the competition between dissolution at large a and replenishment from smaller a somewhat less starkly. Figure 3(a) shows the distribution that arises when $N(a_0)$ = constant and the suppression at large a seen in the burst models clearly is important for the continuous creation case as well. Less dramatic is the case $N(a_0) \propto a_0^{-2}$, shown in Figure 3(b), where the tail appears to grow almost self-similarly, with only the slightest hint of becoming less steep.

4. Conclusions

Wide binary orbits are easily perturbed by any massive constituent of the galactic disk. From detailed simulations, it appears that binary dissolution ultimately occurs in disruptive encounters with GMC subclumps. However, softer collisions with ordinary stars pave the way for destruction by subclumps, as they tend to increase binary separations, and hence the cross-section for disruptive collisions (see, e.g., Equation (6)). Galactic tidal effects hasten binary disruption as well, and binary lifetimes calculated with $a_T = \infty$ differ dramatically from those calculated with finite a_T for initial semi-major axes $a_0 \lesssim a_T$.

Considerations of binary lifetimes are not, however, sufficient to understand data on wide binaries. The binary distribution at any given time is determined by convolving a 'birthrate function' for wide binaries with the probability distribution for semi-major axis evolution from a_0 to a. Two competing effects are evident when such a convolution is performed. First, as expected, binaries at large a tend to dissolve. Second, however, binaries originally with small semi-major axes tend to widen, thus replacing the disrupted binaries. Which of these effects wins out depends in detail on the birthrate function. For simple power-law birthrate functions no tendency toward sharp cut-offs (except at a_T) is evident in detailed calculations (Figures 2 and 3; Weinberg *et al.*, 1987).

Because wide binaries are so easily perturbed, observational determinations of their semi-major axis distribution may be used to constrain the properties of real or hypothetical constituents of the Galaxy. Most exciting is the possibility of limiting the particle masses of the Oort dark matter in the disk (Bahcall *et al.*, 1985). To do so requires computing theoretical distributions as described in this review (Section 3), and then performing a further convolution with 'effective volumes' for binary detection that include observational selection effects (Wasserman and Weinberg, 1987; Weinberg, 1987 (these proceedings)). While attempts to do this are only just beginning, it would be fitting if the study of binary stars, which has contributed so critically to astrophysics, were to once again promote a fundamental advance in our understanding of the cosmos.

Acknowledgement

This research was supported in part by NSF grant AST 84–15162 and by an Alfred P. Sloan Foundation fellowship.

References

Abt, H. A.: 1982, *Ann. Rev. Astron. Astrophys.* **21**, 343.
Antonov, V. A. and Latyshev, I. N.: 1972, *Soviet Astron.* **15**, 676.
Bahcall, J. N.: 1984a, *Astrophys. J.* **276**, 156.
Bahcall, J. N.: 1984b, *Astrophys. J.* **276**, 169.
Bahcall, J. N. and Soneira, R. M.: 1980, *Astrophys. J. Suppl.* **44**, 73.
Bahcall, J. N. and Soneira, R. M.: 1981, *Astrophys. J.* **246**, 122.
Bahcall, J. N., Hut, P. and Tremaine, S.: 1985, *Astrophys. J.* **290**, 15.
Bailey, M. E.: 1983, *Monthly Notices Roy. Astron. Soc.* **204**, 603.
Bailey, M. E.: 1985, *Monthly Notices Roy. Astron. Soc.* **218**, 1.
Blitz, L.: 1985, private communication.
Heisler, J. and Tremaine, S.: 1986, *Icarus* **65**, 13.
Hut, P. and Tremaine, S.: 1985, *Astron. J.* **90**, 548.
Latham, D. W., Tonry, J. Bahcall, J. N. Soneira, R. M., and Schechter, P. S.: 1984, *Astrophys. J.* **281**, L41.
Mihalas, D. and Binney, J.: 1983, *Galactic Astronomy*, Freeman, San Francisco.
Retterer, J. M. and King, I. R.: 1982, *Astrophys. J.* **254**, 214.
Sanders, D. B., Solomon, P. M., and Scoville, N. Z.: 1984, *Astrophys. J.* **276**, 182.
Torbett, M. V. and Smoluchowski, R.: 1984, *Nature* **311**, 641.
Wasserman, I. and Weinberg, M. D.: 1987, *Astrophys. J.* **312**, 390.
Weinberg, M. D.: 1987, these proceedings.
Weinberg, M. D., Shapiro, S. L., and Wasserman, I.: 1986, *Icarus* **65**, 27.
Weinberg, M. D., Shapiro, S. L., and Wasserman, I.: 1987, *Astrophys. J.* **312**, 367.
Weistrop, D.: 1972, *Astron. J.* **77**, 366.

POVEDA – If you were to divide the binaries by age I think you would find some sort of 'break' in the distribution of separations as we found for the 'old' stars in the solar vicinity.

WASSERMAN – Certainly looking at the distributions at different ages is interesting and one could learn something from the contrast between distributions at large and small age. We haven's looked in detail at a ~ 0.01 pc (≈ 2000 AU) primarily because we focussed on the purported cut-off at $a \sim 0.1$ pc. However, since the characteristic binary age for $a \sim 0.01$ pc should be $\sim 10^{10}$ yr one might see a feature in the models in that range. The extent to which such a feature would be evident depends on the initial (power law) distribution as well as the later evolution, as emphasized in my talk. Also – at $a \lesssim 0.01$ pc binary evolution is predominantly due to successive penetrations of clouds. In particular, there will be some sensitivity to the adopted subclump radius for these calculations. We have not done simulations that focus on this range of semi-major axes.

VALLS-GABAUD – I have not understood your first conclusion. Are your results independent of your 1 pc galactic cut-off? Your diffusion calculation must be dependent. You could use a simple galactic potential (Y. Brbanis-Woltjer) to see the validity of the 1 pc value.

WASSERMAN – At large a, the results do of course depend on the tidal cut-off. But in the solar neighbourhood we can use the characteristic radius of the zero-velocity surface, for example, for the motion of a particle in the combined potential due to the galaxy as well as an attracting $a_T \approx$ central mass. This gives about 1 pc for binaries with $M \approx 2\,M_\odot$ (and $a_T \propto M^{1/3}$ roughly), using measured values of the Oort constants.

HOW WE CAN LEARN FROM STUDIES OF WIDE BINARY STATISTICS

M. D. WEINBERG

Center for Radiophysics and Space Research Cornell University, Ithaca, New York, U.S.A.

Summary. We present a versatile technique for comparing wide binary observations with theoretical semi-major axis distributions.

To illustrate this technique, the theoretical implications of the recent Latham *et al.* (1984) data on wide binaries are examined. We conclude that power law models for the underlying distribution are consistent with observations, implying no sharp cut-offs in the semi-major axis distribution. We argue that the detection of 6 wide binaries by Latham *et al.* would require an uncomfortably large local density of wide binaries (semi-major axes $a \gtrsim 10^{-3}$ pc) for a broad range of theoretical models.

Motivated by the apparent high number density of wide binaries implied by the Latham *et al.* sample, we define a criterion for selecting candidate wide binaries form the *Yale Bright Star Catalog*. The candidate list is used to determine an upper bound for wide binary density of $n_{WB} = 4 \times 10^{-3}$ pc^{-3}. Although the actual value is somewhat uncertain, there is no difficulty in principle in reconciling this result with the local stellar number density.

1. Introduction

In this paper, I will outline a general method developed along with Ira Wasserman for analyzing binary star catalogs. Our approach is somewhat heretical in the context of the work presented at this meeting; our aim is to exploit the statistics of binary detections in a given catalog to probe the galactic environment. As Dr Wasserman has described, a wide binary with initial semi-major axis $a \gtrsim 0.01$ pc will evolve significantly in the age of the Galaxy due to encounters with field objects such as other stars and giant molecular clouds. Assuming that we know the birthrate and initial distribution of binary separations, we can predict the present-day distribution. For example, by comparing the theoretical and observed distributions we can, in principle, put limits on the mass of the unseen particles (Bahcall *et al.*, 1985; Wasserman and Weinberg, 1987). Conversely, we may place be able to place limits on the initial distribution of binary pairs by a similar comparison.

After describing the technique in Section 2, I will use the method to obtain the number density of wide binaries in the Weistrop (1972) and the *Yale Bright Star Catalog* (Hoffleit, 1982) in Section 3. I conclude with Section 4.

2. Method

We begin with a set of wide binary detections whose selection statistics are well understood. We want to compare these with a physical theory which may have undetermined parameters. In general, this can not be done directly since some of the quantities m, $\mathbf{r}$, $\mathbf{v}$ are unknown for each star even in the best of circumstances. For example, for many objects the distance is poorly determined so only two of the three of the spatial

Astrophysics and Space Science **142** (1988) 277–287.

coordinates are known. If proper motions are available, the tangential velocity is only known modulo the distance. Rather than attempt to reconstruct the physical distribution from incomplete data, we choose to project the theoretical distributions on the observational plane using the catalog selection criteria. A statistical comparison between the distributions may then be used to place limits on unknown parameters. I emphasize that the success of this program depends on our ability to precisely quantify the inherent observational selection in the catalog's membership. Our list of candidate pairs is selected to best discriminate between the parameters of the theoretical models, not to find the most bound pairs. Also note that the process is iterative. An initially crude or idealized theory may yield only a rough determination of the desired parameters. But, the statistical comparison may suggest both a refinement to the theory and a range over which the observations could be augmented to better determine the unknown quantities.

To summarize, the method consists of the following steps:

(1) The selection criteria by which a catalog of binary stars were in a given region of the sky must be precisely quantified.

(2) A theoretical model or perhaps a family of models with unknown parameters is projected against the sky to simulate distributions that would be found if the theoretical model described physical reality.

(3) The theoretical and observed distributions are compared statistically to ascertain the probability that the observed distribution was drawn from the theoretical one or to determine the most probable values of the unknown parameters.

3. Examples: the Binary Number Density

As an example of our method, I will attempt to place limits on the binary number density using the Weistrop (1972) and the *Yale Bright Star Catalog* (Hoffleit, 1982). The following discussion is meant to be illustrative and by no means complete. For more details on the results of these analyses, refer to Wasserman and Weinberg (1987), Weinberg and Wasserman (1988) and the review by Wasserman in this volume

3.1. Physical model

To begin, we define a simple physical model which we will use for the analysis of both catalogs in the following sections. We assume that the semi-major axes of wide binaries are as $\propto a^{-\lambda}$ at birth. Although we have no cosmogonical support for this assumption, we find rough corroboration in the distribution of separations of doubles for $a \gtrsim 0.001$ pc presented by Abt (1983). In addition, since most star formation takes place in molecular clouds, we expect the distribution of subclumps to play a role in the initial separations of bound pairs. Blitz (1987) has found that the masses and sizes of subclumps have power-law distributions over many orders of magnitude. Thus, we might speculate that the distribution of subclump separations might also be described by a power-law.

Furthermore, as Wasserman has discussed (this volume), if one begins with power-law distributions in semi-major axes, the evolved distributions show little evidence for breaks except for the most extreme power-laws ($\alpha \approx 0$). For $\alpha \approx 1.5$, the distribution appears to evolve self-similarly. We will thus assume that the intrinsic distribution of wide binaries at the current epoch is power-law:

$$N_{WB}(a) = n_{WB} a^{-\lambda} . \tag{1}$$

Clearly, we need to specify upper and lower limits for the distribution given by Equation (1). A natural upper limit for the distribution is provided by the galactic tide. This tidal limit occurs roughly at $a_T = (M_b/M_\odot)^{1/3}$ pc for a pair with total mass M_b. There is no such physical constraint of the lower limit. However, we have mentioned that for $a \lesssim 0.01$ pc there is very little evolution in the age of the Galaxy. Thus, we want to choose the lower limit at the point at which we no longer believe that the current distribution may be represented by a power-law. We again refer to Abt's (1983) distribution which has a break at 0.001 pc which Abt attributes to a division between the close and wide binary populations. We will thus take our lower limit to be $a_{\rm min} = 0.001$ pc.

To summarize, we will assume that *wide binary semi-major axes have a power-law distribution at the present time for* $0.001 \text{ pc} \le a \le (M_b/M_\odot)^{1/3}$ pc. (For a discussion of possible problems with this assumption, see Wasserman and Weinberg, 1987.)

3.2. Weistrop Catalog

As a first example of our technique, we will use the Weistrop (1972) catalog to find n_{WB}. Weistrop's catalog covers an area of 13.5° square in the direction of the North Galactic Pole and includes all stars with

$$9 \le m_v \le 12 . \tag{2}$$

The first search for binaries in this region was made by Bahcall and Soneira (1981) who searched for nearest neighbours with angular separation

$$10'' \le \Delta\theta \le 2' . \tag{3}$$

Based on the radial velocity measurement of their 19 candidate pairs by Latham *et al.* (1984), six of the candidate pairs are probably bound (Wasserman and Weinberg, 1987). (See the article by T. Mazeh in this volume for a report on the latest observations of this region.)

We now need to project the theoretical model defined in Section 3.1 onto the observer plane according to the above selection criteria. To begin, we project the distribution of semi-major axes in volume along the line-of-sight. We assume that the orbital elements of binaries with a given semi-major axis are distributed uniformly in phase space; that is, all orientations are equally likely and eccentricities e are distributed as e. The projected distribution then becomes

$$\mathcal{N}(s)\,{\rm d}s = {\rm d}s \int_{s/2}^{\infty} \frac{{\rm d}a}{a}\, N_{WB}(a) F(s/a) , \tag{4}$$

where $F(x)$ takes into account the eccentricity and angular averaging for the pairs. (The function $F(x)$ vanishes at $x = 0$ and $x = 2$ with a peak at $x \approx 1$.) By use of Equation (1), Equation (4) becomes

$$\mathcal{N}(s)\,\mathrm{d}s = s^{-\lambda}\,\mathrm{d}s \int_{\max(1/2,\, a_T/s)}^{\max(1/2,\, a_{\min}/s)} \frac{\mathrm{d}x}{x}\, x^{-\lambda} F(1/x) \sim s^{-\lambda}\,\mathrm{d}s\,, \qquad (5)$$

where the final relation obtains for s far from $a_{\min}$ or a_T. At a given point in space, Equation (5) gives the volume density of binaries with separations projected along the line-of-sight.

Clearly a binary pair at a particular physical position **r** will only be detected if both members have apparent magnitude with in the catalog's range (Equation (2)). The probability that a pair can be observed, then, depends upon the absolute magnitude of the members and its position in the Galaxy since disk scale heights of the stellar populations and obscuration are direction specific. To take these effects into account, we assume the distribution of wide binary members is the same as the single star distribution given by Bahcall and Soneira (1980) with scale heights given by the youngest member. This last assumption embodies the prejudice that each member of a wide pair has roughly the same age. We then define the 'effective volume' $V(s)$ as the true space volume in which a binary with projected separation s could be registered in the catalog consistent with Equations (2) and (3). (For more complicated selection criteria or binary

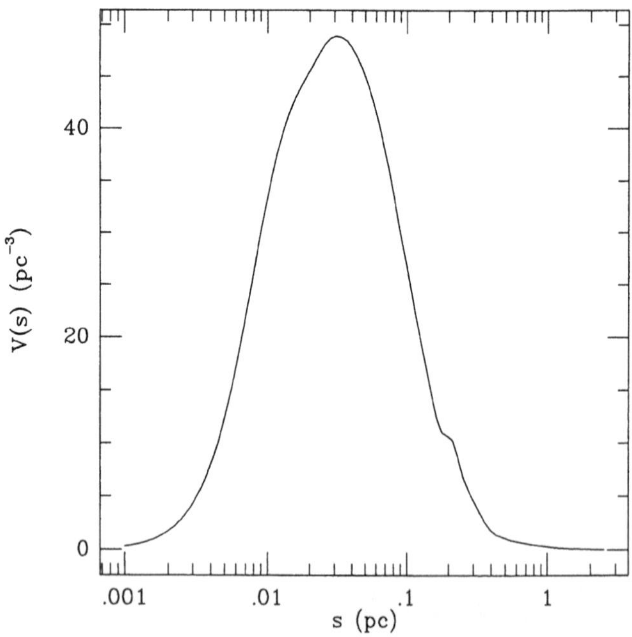

Fig. 1. Effective volume $V(s)$ as a function of projected separaton s is shown for the Bahcall and Soneira (1981) wide binary search of the Weistrop catalog.

member distributions it may not be possible to define the effective volume. See Wasserman and Weinberg (1987) for a more general treatment.) The effective volume for the catalog is shown in Figure 1. Note that the search is only sensitive to binaries with separations 0.01 pc $\lesssim s \lesssim$ 0.1 pc.

From the projected number density $\mathcal{N}(s)$ and the effective volume $V(s)$, the expected number of binaries in the catalog with a given separation s becomes

$$\psi(s)\, ds = \mathcal{N}(s) V(s)\, ds\,. \tag{6}$$

From Equation (6), we can immediately form two interesting quantities. First, the overall number of expected pairs is given by the integral of ψ over all s, $N_{\text{total}} = \int_0^\infty ds\; V(s)\mathcal{N}(s)$. From N_{total} and the actual number of binaries detected in the catalog, we easily determine the binary number density n_{WB}. Information on the shape of the distribution is provided by the cumulative distribution of ψ, $\chi(s) = \int_0^s ds\; V(s)\mathcal{N}(s)/N_{\text{total}}$. We use the Kolmogorov–Smirnov (KS) test to estimate the likelihood that observed distribution is given by the theoretical model. The computed binary number density and KS significance for our family of models is shown in Figure 2. Since $V(s)$ is sharply peaked

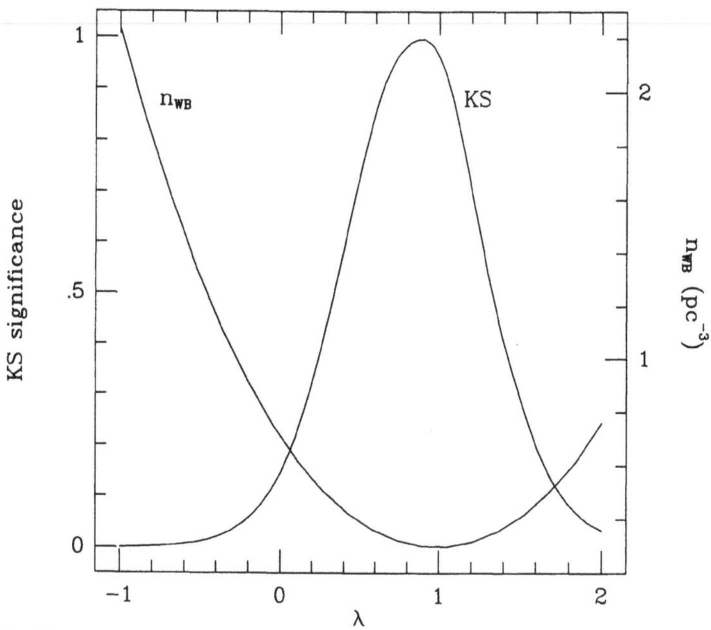

Fig. 2. Wide binary number density n_{WB} and Kolmogorov–Smirnov significance for power-law model shown as a function of power-law index λ implied by the 6 binaries found by Latham *et al.* (1984).

it is no surprise that n_{WB} increases for both large and small λ. For small (large) λ one is invoking the existence of binaries at large (small) separations that cannot be observed. The best fit model has $\lambda \approx 0.9$ (Figure 3) and $n_{WB} \approx 0.3$ pc^{-3}. Such a huge binary

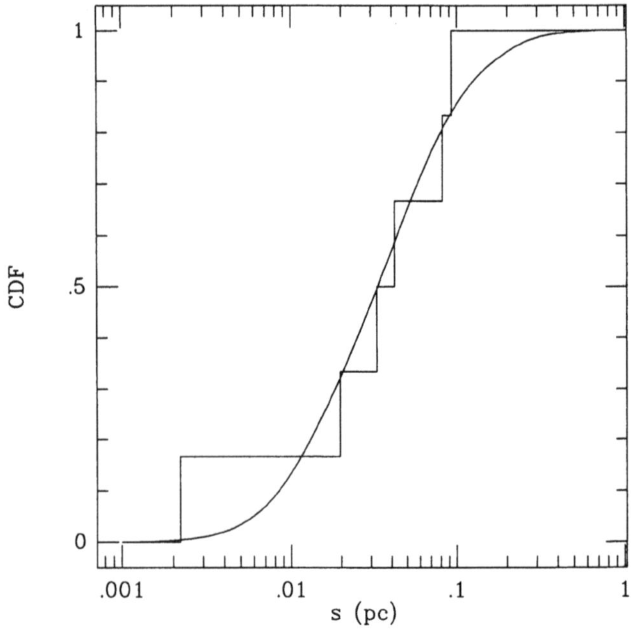

Fig. 3. The best fit power-law model ($\lambda = 0.9$) is shown along with the CDF for the 6 detected pairs.

number density, however, would require a single star number density 4 times the observed value of $n_* \simeq 0.15 \text{ pc}^{-3}$!

There are two possible explanations for this perplexing result: there is something wrong with the model or there is something pecular about the Weistrop sample. For example, we have assumed that wide binary members show no bias for a particular spectral type. However, if we restrict binary membership to spectral types F5–K5, we find that the binary number density is marginally consistent with the single star number density. This is a suspicious explanation since nearly all of the pairs observed have members with spectral types F5–K5. The second possibility – Weistrop's sample is peculiar – can only be checked by analyzing a different catalog which brings us to the next example.

3.3. Yale Bright Star Catalog

The *Yale Bright Star Catalog* (YBS, Hoffleit, 1982) is a full sky catalog that is complete to $m_V = 6.5$ including 8401 stars. We explicitly exclude O and B stars to avoid contaminating the list of candidate binaries with widely separated, multiply paired stars that are members of relatively distant OB associations. When O and B stars are removed, 6691 stars with $m_V \leq 6.5$ remain. The most common spectral classes in this subset of YBS are K and A, with substantial (but smaller) numbers of F and G stars as well. The large A star membership distinguishes YBS from the Weistrop catalog. To

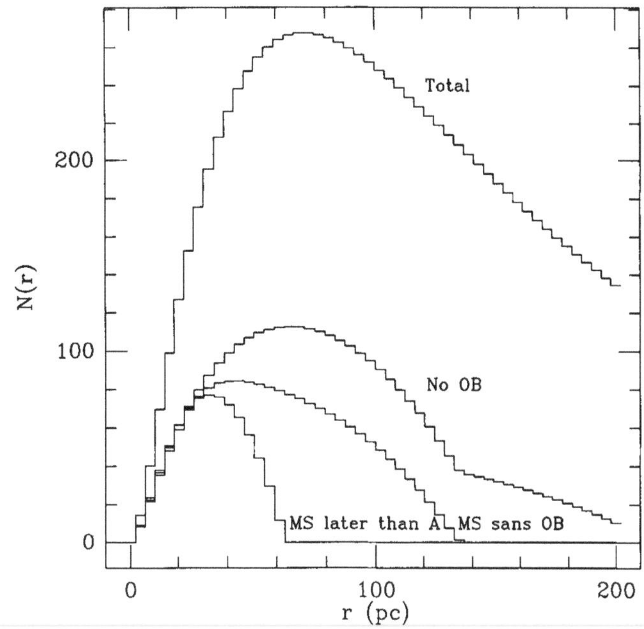

Fig. 4. The expected number of stars in YBS at a given distance from the observer $N(r)$ is shown as a binned distribution. The various curves show the effects of restrictions in spectral range.

get some idea of the spatial distribution of stars in the catalog, I histogrammed the expected number of stars as a function of distance (Figure 4) for various spectral ranges. For the sample without O and B stars, we see that the distribution is well peaked and the most likely distance for an arbitrary star D_{eff} is ≈ 80 pc (median value).

Angular positions, proper motions and spectral types are available for essentially all catalog members, and radial velocities for more than 90% of the entries. Parallaxes – and, hence, accurate distances – are only available for roughly one-third of the YBS stars. Furthermore, the radial velocities are of low quality, certainly not sufficient for reliably discriminating between chance superpositions and bound pairs. Because distances and radial velocities are relatively poorly known, we base our binary search criteria only on angular positions and proper motions, which are accurately determined, distance independent quantities tabulated in YBS. Since we expect wide binaries to have small angular separations and relative velocities, we identify as candidate binaries all pairs satisfying the simultaneous criteria

$$\Delta\theta \leq \theta_0 , \qquad \Delta\mu \leq \mu_0 , \tag{7}$$

where θ_0 and μ_0 are judiciously chosen cut-offs. Because a pair of stars at a distance D separated by an angle $\Delta\theta$ has a projected separation

$$s \dot{\approx} 4 \times 10^{-4}\,\text{pc} \left(\frac{\Delta\theta}{1\ \text{arc sec}}\right)\left(\frac{D}{80\ \text{pc}}\right), \tag{8}$$

we choose $\theta_0 = 1$ deg, which is comparable to the mean angular separation of YBS stars. At $D_{\rm eff} \approx 80$ pc, we are sensitive to binary separations 4×10^{-4} pc $\lesssim s \lesssim 1.4$ pc for $\theta_0 = 1$ deg. For context, the galactic tidal limit for binaries consisting of two A stars is ≈ 1.3 pc. Of course, this choice of θ_0 precludes identification of very nearby wide binaries, at $D \ll D_{\rm eff}$, with large separations, $\Delta\theta \lesssim 1$ deg. Choosing a larger value of θ_0, however, would necessarily pollute the candidate list with unwanted noise due to random superpositions of bright catalog members at large distances. Similarly, a pair of stars at distance D with relative proper motion $\Delta\mu$ has a projected velocity difference

$$v_\perp \approx 1.9 \text{ km s}^{-1} \left(\frac{\Delta\mu}{0.005 \text{ arc sec yr}^{-1}}\right)\left(\frac{D}{80 \text{ pc}}\right), \tag{9}$$

so we want to choose μ_0 to be as small as possible. However, any value of $\mu_0 < \sqrt{2}\,\sigma_\mu$ is less than the random differences due to measurements uncertainties and, therefore, of dubious significance. Consequently, we choose $\mu_0 = 0.01$ arc sec yr^{-1} in our binary search. The relative proper motions of the candidate pairs then satisfy $\Delta\mu \leq (1 - 1.4)\sigma_\mu$. We thus have compromised our candidate list by suppressing the chance proper motion coincidences at the risk of rejecting up to half of the pairs that truly have $\Delta\mu \leq \mu_0$.

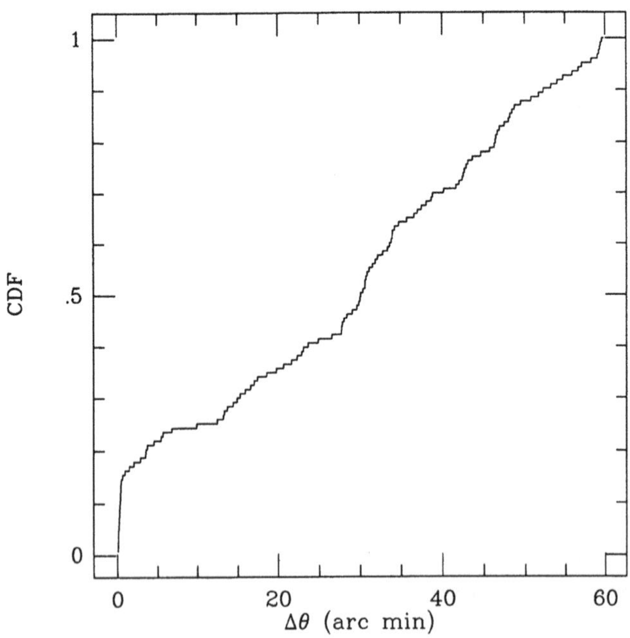

Fig. 5. CDF in separation for the YBS candidate list.

The cumulative distribution of the candidate list angular separations is shown in Figure 5. The candidates appear to be the sum of two populations: (1) a component whose semi-major axis distribution is a steeply decreasing function of angular separation for $\Delta\theta \leq 10'$; and (2) a component whose distribution is roughly flat in separation for $\Delta\theta > 10'$. The second population does not appear to be a binary population for the following reasons. First, the extrema separations of this population correspond to ≈ 0.2 pc and 1.4 pc at D_{eff} which includes the tidal radius. If this population were due to binaries, we would expect a drop off at large angular separations corresponding to physical separations near the tidal radius which is not seen. Secondly, we know that part of the tail is due to random superpositions of known members of clusters and associations. We have examined the single star density about pairs in this flat tail which are not known cluster members, and found that it, too, is significantly above average. For the time being then, we ascribe this second population to large scale clustering and concentrate on the population with $\Delta\theta \leq 10'$.

Similar to the analysis of the Weistrop catalog in the previous section, we use Equations (1) and (7) to predict the observed distribution in angular separation, $\psi(\Delta\theta)$. Because of the proper motion criterion, the calculation of ψ is lengthy and we omit the details here. We show the derived n_{WB} and the KS significance for the model fits in

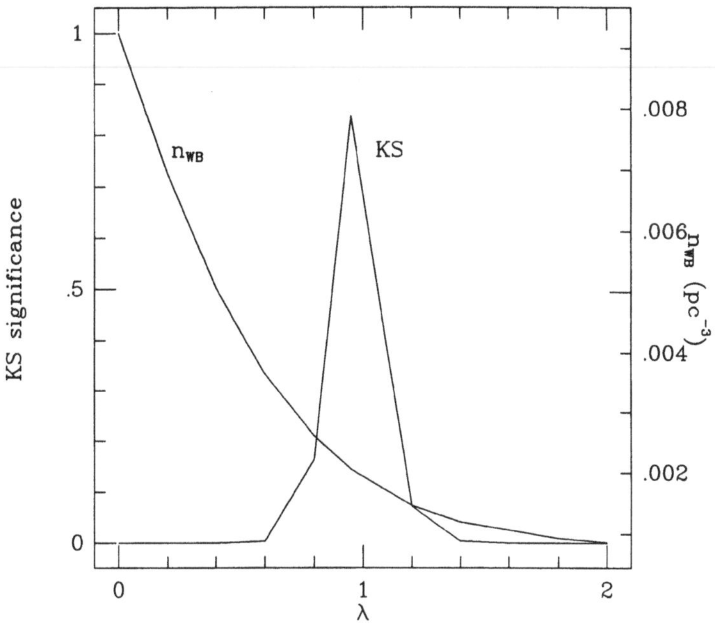

Fig. 6. Wide binary number density n_{WB} and Kolmogorov–Smirnov significance for power-law model shown as a function of power-law index λ for the candidate list derived from YBS with $\Delta\theta \leq 10'$.

Figure 6. The best fit power-law obtains for $\lambda \approx 1$ with a number density of $n_{WB} \approx 0.002$ pc^{-3}. This implies that only 3% of the single stars make up wide binaries. Even after admitting that the candidate list may be incomplete by perhaps as much as a factor of 2, there is no normalization conundrum in the analysis of the *Yale Bright Star*

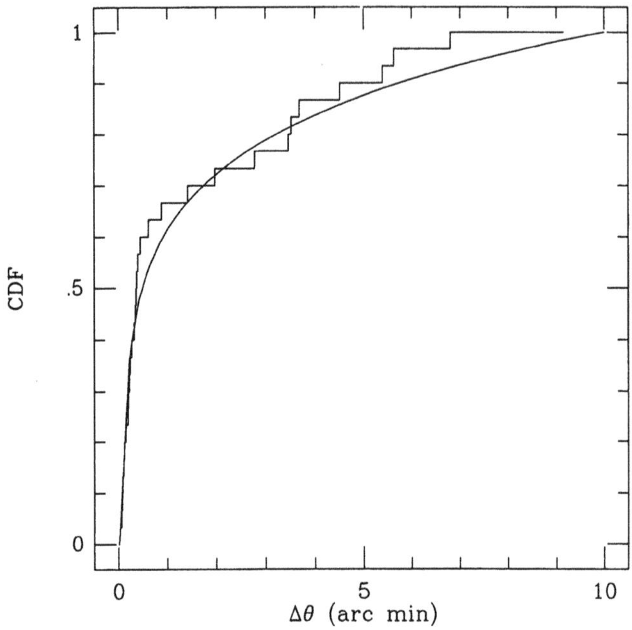

Fig. 7. The best fit power-law model ($\lambda = 1$) is shown along with the CDF for the YBS candidate list with $\Delta\theta \leq 10'$.

Catalog. Of course, as I previously emphasized, this analysis is sensitive to a different spectral population of stars than the previous analysis, so the wide binary enigma in the Weistrop catalog is yet to be explained.

4. Conclusions

Although wide binaries systems are interesting in and of themselves, I hope I have convinced you that the study of wide binary statistics may tell an interesting story about galactic evolution. In the first part of this paper, I outlined a general approach for comparing physical hypotheses with observational results. I again emphasize that in order to use the great wealth of data that has been accumulated over the years on wide biunary system, the selection criteria need to be known precisely. Poveda *et al.* (1982) have shown how this might be done *a posteriori*.

In the latter part of this paper, I used the method to place limits on the wide binary number density n_{WB} using two catalogs. In the first example, I considered the 6 wide binary detections of Latham *et al.* (1984) in the Weistrop catalog (1972). Although we found that the simple evolutionary models that we have discussed previously (Weinberg *et al.*, 1987) cannot be ruled out, the value for n_{WB} is impossibly large. We have no convincing explanation for this enigmatic result. To further investigate n_{WB}, we developed a search criterion for sifting the *Yale Bright Star Catalog* for candidate binary pairs. The candidate list appears to consist of two distinct populations: a population at $\Delta\theta > 10'$ which we ascribe to unidentified clusters and associations, and a population with $\Delta\theta \leq 10'$. On the basis of the number and distribution of the candidate pairs, we derive the upper limit of $n_{WB} \lesssim 0.002\ \mathrm{pc}^{-3}$. Thus, we have found the binary number density is consistent with the single star number density for pairs with bright (early-type) members.

These preliminary investigations have already yielded some thought provoking results. As more accurate and extensive astrometry becomes available, for example from the Hipparcos experiment, we may realize our goal of putting limits on the mass of unseen disk particles and investigating the nascent conditions of wide pairs.

Acknowledgements

I would like to thank the International Astronomical Union and American Astronomical Society for travel grants. This research was supported in part by National Science Foundation Grant No. AST 84–15162 to Cornell University.

References

Abt, H. A.: 1983, *Ann. Rev. Astron. Astrophys.* **21**, 343.
Bahcall, J. N.: 1981, *Astrophys. J.* **264**, 122.
Bahcall, J. N., Hut, P., and Tremaine, S.: 1985, *Astrophys. J.* **290**, 15.
Bahcall, J. N. and Soneira, R. M.: 1980, *Astrophys. J. Suppl.* **44**, 73.
Blitz, L.: 1987, Preprint.
Hoffleit, D.: 1982, *The Bright Star Catalog*, Yale University Press, New Haven.
Latham, D. W., Tonry, J., Bahcall, J. N., Soneira, R. M., and Schechter, P. S.: 1984, *Astrophys. J.* **281**, L41.
Poveda, A., Allen, C., and Parrao, L.: 1982, *Astrophys. J.* **258**, 589.
Wasserman, I. and Weinberg, M. D.: 1987, *Astrophys. J.* **312**, 390.
Weinberg, M. D. and Wasserman, I.: 1987, *Astrophys. J.* (in press).
Weistrop, D.: 1972, *Astron. J.* **77**, 366.

POVEDA – I am surprised that, in your analysis, you have not considered the rich material that you can find. In the IDS (or FIDS) catalogue or in the binaries contained in Gliese's catalogue, after all they are much more numerous than the Weistrop–Bahcall/Soneira binaries.

WEINBERG – The reasons that we have only looked at the Weistrop catalog and the *Yale Bright Star Catalog* are purely historical: these data were more immediately available in machine-readable form. Analyzing Gliese's catalog would be interesting because the analysis would be more sensitive to rather late type stars, unlike the two analyses we have performed. An analysis of FIDS would be interesting but only possible if we can understand the selection criteria in separations.

EVIDENCE FOR MASS LOSS IN VISUAL BINARY STARS

D. VALLS-GABAUD

Institut d'Astrophysique de Paris, France

Summary. The present status of the observational evidence about mass loss in visual binary and multiple systems is reviewed. Particular emphasis is placed on the importance of mass-loss rates measures in these stars, as a systematic study would provide useful constraints in two fields. First, as mass-outflows processes are not modified by the interaction of the two stars, and distances, masses, and evolutionary stages are usually fairly good known, an absolute stellar scale for $\dot{\mathcal{M}}$ can more easily and accurately be derived for these systems. Second, $\dot{\mathcal{M}}$ rates are essential to study the detailed orbital evolution of wide pairs, providing clues to their origin. Detailed numerical simulations including mass loss have been carried out. It is suggested that mass loss may explain in part the recent statistics on separation of wide visual binaries as a function of spectral type and age.

1. Introduction

"[...] we shall have to examine whether this hitherto unconsidered possibility is capable of solving the mystery of the observed distribution of the orbits of binary stars."

J. H. JEANS (1924)

The observed frequencies of wide binaries are good tracers of the galactic history and stellar dynamics. The distribution of separations contains clues on the initial formation, evolution and destruction processes of these pairs. Therefore, the study of wide visual or common-proper-motion (c.p.m.) systems may be fundamental to galactic dynamics and stellar formation theories, but it has been neglected for a long time (Dommanget, 1984). The first results of new investigations on this subject are encouraging (e.g., Brosche and Hoffman, 1979; Latham *et al.*, 1984) even if most efforts continue to deal with closer pairs.

Unfortunately, statistics of visual (VBs) and spectroscopic (SBs) binaries have often lead to controversial conclusions on the distributions of periods, semi-major axes (see, e.g., Heintz, 1969; Abt and Levy, 1976; Abt, 1979, 1983; Van Dessel and Dommanget, 1980; Halbwachs, 1983) and mass ratios (e.g., Trimble, 1978; Gieseking, 1982; Halbwachs, 1987, Giannuzzi, 1987).

When selection effects and biases are properly taken into account, part of these contradictory results may dissappear but many problems remain, in particular the shape of initial distributions, the formation of wide pairs, and their maximum separations as a function of age.

Here it is suggested that mass loss could provide an astrophysically plausible mechanism to produce wide binaries from close systems. It may explain their formation and present separations, and provides some clues to derive their initial distributions. This mechanism was introduced by Jeans (1924) and developed by Smart (1925), Dommanget (1963) and Hadjidemetriou (1963). Dommanget has advocated this evolutionary scenario since 1963, as it may explain the period-eccentricity relation

Astrophysics and Space Science **142** (1988) 289–304.

(Dommanget, 1963; Hadjidemetriou, 1966b; see also Fulconis and Bücher, 1978), but no analysis has been made so far taking into account astrophysical constraints (Jorissen, 1984). Detailed numerical simulations that include mass loss as a driving factor for the evolution of VBs have been performed for the first time.

Section 2 reviews current determinations of mass-loss rates across the HR diagram as a function of age. The dynamical effects of mass loss are applied to the evolution of binary systems in Section 3, where it is shown how orbital amplification modify significantly the initial distributions. Section 4 discusses the evidence for mass loss in VBs that can be derived from present data.

2. Mass-Loss Rates Across the HR Diagram

"It is now widely recognized that many stars have an emergent mechanical flux as well as a radiative flux."

S. Kwok (1984)

The first detection of mass loss was made by Beals in 1924, who identified the so-called P Cygni profiles, which are due to an expanding envelope, on Balmer lines in a sample of Wolf–Rayet stars. The discovery of stationary, blueshifted absorption lines in α Her (Deutsch, 1956) was interpreted as due to a circumstellar (CS) envelope formed by the ejection of matter from the star. The binarity of α Her (in fact, a multiple system as α Her B is a spectroscopic double) permitted the first evaluation of the mass-loss rate in a red supergiant (RSG) star, found to be around 3×10^{-8} solar masses per year. As mass-loss rates are usually expressed in solar masses per year, we define m as $-\log(-\dot{\mathcal{M}}(M_\odot\, \mathrm{yr}^{-1}))$.

Nowadays, mass-loss rates are derived from several observables. These include:

(1) Radio measurements (OH, SiO masers, free-free emission) which are quite accurate and lead to m values that are not very sensitive to the assumed mass-loss model (see, e.g., Knapp and Morris, 1985).

(2) IR observations of CS envelopes (Barlow and Cohen, 1977).

(3) Visible and UV spectra, which can be used to infer m from CS lines (Ca II, Mg II), asymmmetries in chromospheric resonance lines, P Cygni profiles, and emission lines.

The large number of these determinations lead to the well-established fact that strong stellar winds associated with intense mass loss develop in stars mostly at the red and blue edges on the HR diagram. Several regions have been identified in the blue edge: (i) hypergiants or luminous blue variables (LBV) as P Cyg, η Car, and S Dor, which have extreme mass-loss rates around $m \sim 4$, (ii) Wolf–Rayet stars ($m \sim 4$ to 5), (iii) O stars ($m_{\mathrm{O\,I}} \sim 5$–6, $m_{\mathrm{O\,V}} \sim 5$–8), (iv) B stars ($m_{\mathrm{B\,I}} \sim 5$–7.5, $m_{\mathrm{B\,V}} \sim 7$–10), and (v) nuclei of planetary nebulae ($m_{\mathrm{NPN}} \sim 7$). At the other edge, red giants (RG) have $m_{\mathrm{RG}} \sim 7$–9, whereas their supergiant counterparts have $m_{\mathrm{RSG}} \sim 5$–7. All these figures may be uncertain by a factor of about 10 in $\dot{\mathcal{M}}$ (i.e., ± 1 in m) as different observations may yield different global mass-loss rates. Detailed reviews on mass loss in the different parts of the HR diagram are given by Kwok (1984), Lamers (1981), Conti (1981), Cassinelli and Lamers (1987) for hot stars, and by Reimers (1983), Dupree (1986), and Dupree and Reimers (1987) for cool stars.

Mass-loss rates depend, therefore, on the evolutionary stage. For instance, pre-Main-Sequence (PMS) stars are believed to loose mass at a rate of $m_{\text{PMS}} \gtrsim 7.5$ in some 5×10^7 yr (Hartmann *et al.*, 1982). However, during the bipolar jet phase ($\Delta t \approx 10^6$ yr), m may attain values as high as 6. Many binaries are suspected to exist among T Tauri stars (Herbig and Rao, 1972; Cohen and Kuhi, 1979), even if only one spectroscopic binary has been discovered so far (Mundt *et al.*, 1985).

Except for OB stars, mass loss on the MS is believed to be very small. The solar wind provides the value $m_\odot \sim 13.7$. However, observations are exceedingly difficult and it has been recently suggested that m_{MS} could be in fact as high as 9 (Wilson *et al.*, 1987). See Section 4 for a discussion on the implications of this hypothesis. Strong winds start to develop once the star leaves the MS and exhausts first H and then He. In the RGB, the mass lost by a typical 0.8 $\mathcal{M}_\odot$ star is in the range 0.01 to 0.2 solar masses. Once in the AGB phase, m could be very high, but the main contribution comes from the short superwind phase ($m \leq 5$) that leads to PN formation. In a time scale of some 10^3 yr, m may attain values of 3 to 5. For more details on the evolution of massive pairs the reader is referred to the review by Chiosi and Maeder (1986).

Different parametrizations have been formulated, either theoretically (radiation-driven winds, fluctuation theory) and empirically for OB and RSG stars. It seems clear that OB rates are proportional to the luminosity (Garmany and Conti, 1984; Migozzi and Lafon, 1985), whereas for RSG Reimers (1977) formulation (m proportional to $\log(L/gR)$, where g is the gravity and R the stellar radius) is more appropriate. These empirical fits have been criticized by Vardya (1984, 1987) who claims that these mere least-squares fits have no physical meaning.

No connection appears between hot and cool stars according to these standard parametrizations, as the mechanisms supposed to be responsible for the stellar winds are totally different. However, Waldron (1985) found empirical correlations in a sample of 272 stars representing all spectral types throughout the HR diagram. These correlations relate the wind luminosity (defined as $L_w = \frac{1}{2}\dot{\mathcal{M}} v_\infty^2$, where v_∞ is the wind terminal velocity) to the bolometric luminosity, and the terminal velocity to the effective temperature. From his results, one can find the relation,

$$\log(-\dot{\mathcal{M}}) = -1.01 + 1.96 \log(L/L_\odot) - 3.54 \log T_{\text{eff}}, \tag{1}$$

that holds in the sampled region $3.2 \leq \log T_{\text{eff}} \leq 4.8$ and $2.0 \leq \log(L/L_\odot) \leq 6.5$. The geometrical location of the points with the same m (iso-mass-loss loci) are straight lines with slope -1.8 in a ($\log T_{\text{eff}}$, $\log(L/L_\odot)$) diagram.

De Jager *et al.* (1986) fitted the mass loss rates of a sample of 189 stars to the expression

$$m = \sum_{i=0}^{N} \sum_{j=0}^{N} a_{ij} (\log T_{\text{eff}} - b)^i (\log L/L_\odot - c)^j . \tag{2}$$

The corresponding iso-mass-loss loci are illustrated in Figure 1. Note that they differ strongly from straight lines. WR and C stars do not fit the relation derived from the rest

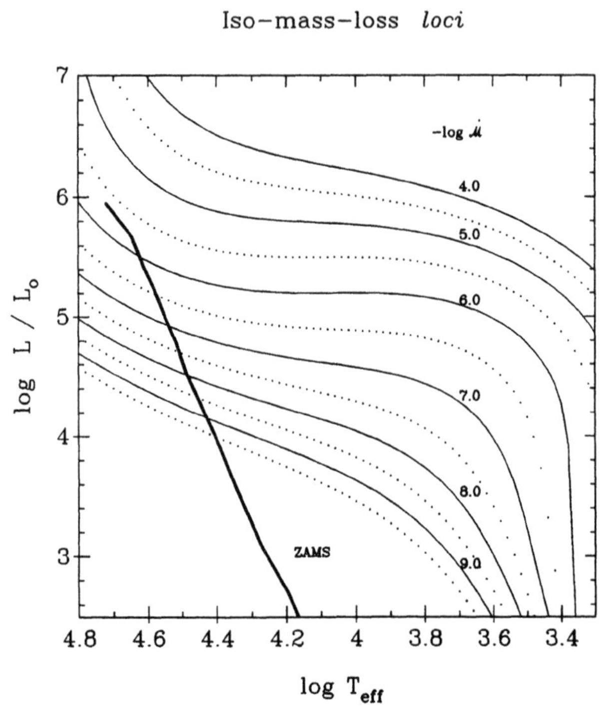

Fig. 1. Iso-mass-loss loci computed with Equation (2) from De Jager *et al.* (1986). Lines blueward the ZAMS are an artifact as very few stars contributed to the fit. The lower loci are very uncertain, due to observational difficulties. Full lines are labelled with the corresponding value of *m*. Dotted lines have half-integer *m*-values, ranging from 4.5 to 9.5.

of the sample. A recent improvement (De Jager *et al.*, 1987) includes a larger sample (264 O to M stars, 17 Be, 3 O subdwarfs, 11 C, 4 S, 32 WR and 15 nuclei of planetary nebulae) and the mass-loss rate is developed into a series of orthonormal Chebyschev polynomia of the first kind T_k and their cross-products

$$m = \sum_{n=0}^{N} \sum_{i=0}^{n} a_{i,n-i} T_i(\log T_{\text{eff}}) T_{n-i}(\log L/L_\odot)\,. \tag{3}$$

The fit is quite good, even if a slight dependence on other parameters is not ruled out. All these fits do not support, however, a physical connection between hot and cool mass loss mechanisms. Recent evolutionary models (Maeder and Meynet, 1987) take into account these global fits, instead of local parametrizations. The enormous advantage is that at least the mass loss rates have no spurious discontinuities and are in fair agreement with observations.

3. Dynamical Effects of Mass Loss and Orbital Evolution

"Large-scale mass loss must affect, at some stage, the components of wide binary systems."

Z. KOPAL (1984)

It seems clear, from the preceding section, that a non-negligible mass is lost on a nuclear time scale, inducing *a priori* significant dynamical effects on the orbit, depending on the amount of mass lost. The first theoretical description of these effects was made by Jeans (1924), who used the method of adiabatic invariants to derive, under the assumption of isotropic mass loss, the integral

$$\mathcal{M}a = \text{constant}\,, \tag{4}$$

where $\mathcal{M}$ is the total mass of the system and a the semi-major axis. In fact, it can be easily shown (Dommanget, 1963; Hadjidemetriou, 1963) that the true integral is the product of the mass and the semi-latus rectum p of the orbit (i.e., $\mathcal{M}p = \mathcal{M}a\sqrt{1-e^2}$) and, in consequence, the angular momentum per unit reduced mass is conserved.

The perturbing force $\overline{R}$ due to an isotropic variation of mass is shown to be tangential to the orbit, so that only the planar orbital elements (i.e., a, e, and ω) will be perturbed. It is quite straightforward to derive the variations of these elements (see, e.g., Dommanget, 1963; Kopal, 1978):

$$\frac{\mathrm{d}\log a}{\mathrm{d}t} = -\frac{aV^2}{G\mathcal{M}}\frac{\mathrm{d}\log\mathcal{M}}{\mathrm{d}t} = -\frac{1+e\cos E}{1-e\cos E}\frac{\mathrm{d}\log\mathcal{M}}{\mathrm{d}t}\,, \tag{5}$$

$$\frac{\mathrm{d}e}{\mathrm{d}t} = -(e+\cos v)\frac{\mathrm{d}\log\mathcal{M}}{\mathrm{d}t} = -\frac{(1-e^2)\cos E}{1-e\cos E}\frac{\mathrm{d}\log\mathcal{M}}{\mathrm{d}t}\,, \tag{6}$$

$$\frac{\mathrm{d}\omega}{\mathrm{d}t} = -\frac{\sin v}{e}\frac{\mathrm{d}\log\mathcal{M}}{\mathrm{d}t} = -\frac{\sqrt{1-e^2}\sin E}{e(1-e\cos E)}\frac{\mathrm{d}\log\mathcal{M}}{\mathrm{d}t}\,, \tag{7}$$

where V is the orbital velocity around the centre of mass, E the eccentric anomaly, and v the true anomaly of the osculating orbit. Care must be paid when expressing these equations in terms of the eccentric anomaly, as Kepler's equation is no longer valid (Dommanget, 1963) and has to be replaced by

$$\frac{\mathrm{d}E}{\mathrm{d}t} = \left(\frac{2\pi}{P} + \frac{\mathrm{d}\log\mathcal{M}}{\mathrm{d}t}\frac{\sin E}{e}\right)(1-e\cos E)^{-1}\,, \tag{8}$$

which reduces to the classical $E - e\sin E = 2\pi(t - T_{\text{per}})/P$ when $\dot{\mathcal{M}} = 0$.

Most studies on this field have looked for finding analytical solutions to the above system [Equations (5), (6), (7)] of differential equations, adopting simple expressions for $\dot{\mathcal{M}}$ (see Dommanget,1963,for a review of the different attempts, and Hadjidemetriou, 1966a, 1969). For instance, Jeans (1924) took $\dot{\mathcal{M}} = -\alpha\mathcal{M}^n$, with $n = 3$, found by setting a mass-luminosity relation $L \sim M^3$ on the mass-loss law assumed to be $L = -c^2\dot{\mathcal{M}}$. That is, the mass loss is due to the radiative loss, an idea suggested earlier by Laplace (1976). Martin (1962) adopted a rather arbitrary position-dependent mass-loss rate ($\dot{M} = k(t)r^2$) that increases the mass loss at periastron. The β Arietis system might show such an effect (Dommanget, 1969). The recent discovery of close companions to α Ori (Karovska *et al.*, 1986) has been invoked to explain the changes of the line profiles of Mg II, which are indicators of mass loss. It seems that when the epoch of periastron coincides with the epoch of maximum expansion velocity of the photosphere, mass ejection is triggered by the induced instabilities (Karovska *et al.*, 1986). Whether this 'periastral effect' is astrophysically possible in VBs or SBs and provides an explanation to the period-eccentricity relation is not clear (Dommanget, 1981).

As shown in Section 2, mass loss is a complicated function of time and position on the HR diagram and, therefore, it seems difficult to apply any of these studies to real situations and a quite general result, independent of the assumed mass-loss law, must be used. For instance, the secular variation per orbital revolution in the semi-major axis, eccentricity, and argument of periastron are readily computed integrating over one revolution Equations (5), (6), and (7). Neglecting the small term in $\sin E/e$ in Equation (8), one obtains

$$\Delta a = -\frac{2\pi c^3}{G^2 \mathcal{M}^2} \frac{\bar{a}}{\sqrt{1-e^2}} \frac{\mathrm{d}\log \mathcal{M}}{\mathrm{d}t}, \tag{9}$$

$$\Delta e = 0, \tag{10}$$

$$\Delta \omega = 0, \tag{11}$$

where $\bar{a}$ is the mean semi-major axis during the revolution and c the areal constant. Thus, the orbital evolution under the assumption of isotropic mass loss is very simple (to first order): the semi-major axis increases secularly, while the eccentricity and argument of periastron remain secularly constant, even if they suffer periodic oscillations with variable amplitude within each revolution.

Equation (10) was first derived by Jeans (1925) and was proved to be valid, assuming Jeans's law, up to second order in α by Dommanget (1963). The small third-order secular term found by Hadjidemetriou (1966a) is negligible even in ~10 nuclear time-scales. Equation (11) was first deduced by MacMillan (1925), and holds up to first order only in α (Hadjidemetriou, 1966a). If we assume that the mean $\bar{a}$ is given by the arithmetic mean of the a values in two consecutive revolutions (i.e., $2\bar{a} = a_{n+1} + a_n$, where n is the revolution index) the difference equation formed with Equation (9) admits of the solution

$$a_{n+1} = a_0 \left(\frac{1+k}{1-k}\right)^n, \tag{12}$$

with k given by $\Delta a/2\bar{a}$ (Hadjidemetriou, 1966a). That is,

$$k = \frac{\Delta a}{2\bar{a}} = -\frac{\pi c^3}{G^2 \mathcal{M}^2} \frac{1}{\sqrt{1-e^2}} \frac{\mathrm{d}\log \mathcal{M}}{\mathrm{d}t}. \tag{13}$$

Simple simulations for arbitrarily chosen pairs have been performed to estimate the orbital amplification, defined as $a(t)/a_{\text{initial}}$. Each component evolves according to the stellar evolution models supplied by Maeder and Meynet (1987), which include the HR-position dependence of mass loss according to De Jager *et al.* (1986). Figure 2 gives the mass evolution of individual components in these models, where the last points indicate the central exhaustion of helium. It is worth to note that massive stars (e.g., $M \geq 30\, M_\odot$) loose more than half their mass in a few *million* years, whereas the less massive ($1 \leq M \leq 30$) will do it in a time-scale smaller than a few hundred million years. If Equation (4) holds, Figure 2 gives the reciprocal of the orbital amplification $a_0/a(t) = M(t)/M(0)$ for systems with very small mass ratios.

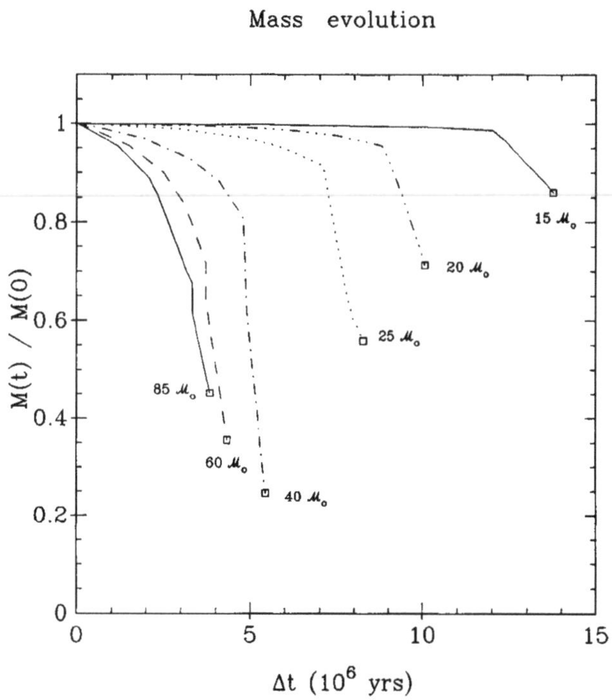

Fig. 2. Mass evolution of massive stars, according to Maeder and Meynet (1987). These models have an overshooting distance of 0.25 pressure scale heights and include the mass-loss rates of Figure 1. Squares indicate the central exhaustion of He. Initial masses $M(0)$ are shown along each curve.

If we assume that Equation (12) holds for each time interval Δt_i where the mass loss $\dot{\mathcal{M}}_i$ occurs, and taking n_i as $\Delta t_i/P_{n_i}$, P_{n_i} being the orbital period at n_i, the condition of steady, isotropic mass loss is given by $n_i \gg 1$. If this condition is not fulfilled at some stage (i.e., explosive mass loss occurs) then, assuming instantaneous ejection, the new orbital elements are,

$$a = a_0 \frac{Q}{1 - 2(a/r)(1 - Q)}, \tag{14}$$

$$e = \sqrt{1 - (1 - e_0^2)\frac{a_0}{a}\frac{1}{Q}}, \tag{15}$$

where

$$Q = \frac{M_1(t) + M_2(t)}{M_1(0) + M_2(0)}. \tag{16}$$

See Mitalas (1980), Noerdlinger (1982), Hills (1983), and references therein for more details. Note that the new elements depend on the location (given by r) within the orbit where the components lay at the moment of the explosion. The condition for disruption is, from Equation (14), $Q = 1 - (r/2a)$.

As mass loss may occur either steadily or quasi-explosively, Equations (5) to (7) must be solved numerically, as Equations (9)–(10) and (14)–(15) are mere asymptotic expressions to the actual situations. For each time step the model computes the mass of each component, the corresponding separation, and the new orbital elements. A sample simulation is illustrated in Figure 3, where the secular orbital amplification is

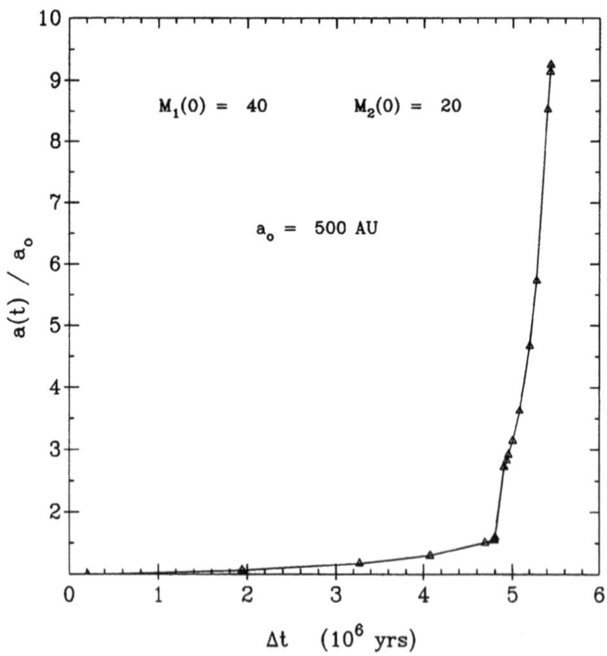

Fig. 3. Secular orbital amplification in a massive binary of initial mass 40 + 20 $\mathcal{M}_\odot$. A factor ~8 is obtained in 5×10^6 years due to the intense mass loss during the final stages (see Figure 2). Numerical integration is necessary for long period binaries.

represented versus time. In this particular case, the amplification reaches a factor 10 in 6×10^6 years, whereas other pairs may attain factors of 1.3 to nearly 20, almost independently of the initial semi-major axis a_0. Full details on the influence of each parameter will be given elsewhere. A Monte Carlo simulation in progress will allow the comparisons between the computed distributions and the actual ones, once biases are removed. Note, however, that the distributions of separation s and semi-major axis a are not statistically identical. Van Albada (1968a) has shown that

$$\langle \log s \rangle - \langle \log a \rangle = \langle \log(1 + \sqrt{1 - e^2}) - 0.4343 \sqrt{1 - e^2} \rangle , \qquad (17)$$

which is approximately -0.11 for the whole sample of VBs that have quasi-definitive orbital elements.

It seems difficult to avoid the conclusion that mass loss can modify significantly the distribution of semi-major axis in massive pairs in a quite short time, as well as in less massive systems in a somewhat larger time-scale.

4. Evidence for Mass Loss in VBs

"L'existence même des étoiles doubles, surtout celle des étoiles doubles visuelles, serait donc un indice d'une perte de masse séculaire des étoiles."

J. Dommanget (1969)

Even if direct evidence seems difficult to obtain if a dedicated, systematic program is not forwarded, there are some clues that may lead to derive constraints on the orbital amplification.

Indirect evidence is provided by the existence of visual binaries with white dwarf (WD) components (e.g., Sirius and Procyon). Typical MS + WD systems had initial masses $0.5 + 5\ M_\odot$, whereas present values are around $0.5 + 0.7\ M_\odot$. Current stellar evolution models (Iben, 1987) indicate that white dwarfs are the end products of 1 to 8 $M_\odot$ stars, which lose mass severely during the AGB phase and eject the remainder of their envelope later on as OH/IR sources and planetary nebulae.

White dwarf masses are found to be distributed in the range 0.6 to 1.1 $M_\odot$. Therefore, these WD + MS systems probably lose an average of $\sim 80\%$ of their mass. A photometrically-selected sample of 68 of such pairs has a median separation of $\log s$ (AU) = 3.1 (Greenstein, 1986a), but there is an obvious selection effect as photometry can only been performed in wide systems ($\rho \geq 5''$). The high mass ratios deduced seem to differ from those predicted by the simple, mass-conservative analysis made by Halbwachs (1985). Further MS + WD systems will be found in the catalog compiled by Agayev *et al.* (1982).

Even if a large proportion of these systems remains to be detected, it seems difficult not to invocate mass loss as a primary mechanism to produce such high-s distribution from the observed MS + MS distribution, which peaks around $\log s \sim 1.3$. In fact it has been claimed (Greenstein, 1986a) that the statistically significant different distribution is a proof of the prolifical total mass loss during the final evolutionary stages. A factor

~ 10 may be gained in the giant phase, while the remaining amplification may be due to the PN phase. A detailed analysis will be made in a forthcoming publication.

Double-degenerate pairs (DD) provide a useful tool to check this hypothesis. The deduced $M_{\text{initial}}/M_{\text{final}}$ values lie between 1.7 and 7.3 and, therefore, the orbital amplification could be very high. If mass loss is responsible for the larger separations, one expects the s-distribution of a DD sample to be even more shifted towards large s than the MS + WD sample. Apparently, this is not the case. The statistics of seven well-studied DD systems (Greenstein, 1986b) give a mean $\log s = 2.34$, with $\sigma(\log s) = 0.84$.

However, one must be cautious when interpreting DD data as mass loss might be quasi-explosive, as shown in the preceding section. Sudden explosions (i.e., $t_{\text{exp}}/P_{\text{orb}} \ll 1$) are *not* always a very efficient mechanism for orbital amplification, even if they can disrupt the system as in the case of supernovae explosions (Hills, 1983). Beech (1987) has shown that for very short periods ($P \lesssim 10$ days) the eccentricity of the surviving systems is essentially random, while the period distribution *peaks* at the initial period, with a large tail that extends up to ~ 1000 times the initial period. Not-so-sudden explosions in wider pairs are under detailed investigation with our code.

Direct evidence is tiny. There are some images of α Ori obtained from interferograms in the pupil frame (Roddier *et al.*, 1985) and the well-known ζ Aur/VV Cep systems. These close systems are composed of a cool SG with an extended atmosphere and a hot B star companion. Typical periods are 1000 days and mass loss rates are $m \approx 7.5$ (see, e.g., Che *et al.*, 1983). They could provide a test for orbital amplification if a long series of observations is accumulated, but the interaction with the wind and the chromosphere is difficult to model (Hadjidemetriou, 1969), due, particularly, to the variable mass *accretion* rate on the secondary. The small separation of α Ori B (some 4.7 AU) may indicate a mass transfer stage in the recent past.

This is a clear case for non-isotropic mass loss. As shown in the interferometric image reconstruction of Roddier *et al.* (1985), there existed an asymmetric CS structure at some 1.3 stellar radii from the surface of the RSG. According to Karovska *et al.* (1986), this may be due to an outburst from the primary during the periastron passage of the companion. Mass loss may explain why the orbit has not been circularized and why the companion has not spiralled down into α Ori A. In a general way, the effects of anisotropic mass loss on wide binaries are presumed to be negligible as the solid angle subtended by the secondary from the primary is very small.

Another tool to detect the effects of mass loss in VBs is the apparent age difference between their components. The work of Oblak and Chareton (1980) has shown that the isochrone-fitting procedure gives in some cases a substantially different age for primaries and secondaries. Even if a statistical bias can explain in part this discrepancy and confirmation must wait the derivation of more reliable masses, the very same effect is found in other evolved systems (Olsen, 1982) and even in pairs which are not very far from the MS (Duncan, 1983). It is especially important to check these binaries as in some cases the mass-loss rates needed may attain $m \sim 9$, which is nearly the same value postulated by Willson *et al.* (1987). The same phenomenon may have been observed in close binaries as well (Corbally and Garrison, 1980; Corbally, 1984a, b). The proposed spaceborne stellar interferometer of Massa and Endal (1987a, b) should provide an ideally-suited instrument to achieve this goal, as well as HIPPARCOS.

Further support to the idea of mass loss as a driving evolutionary mechanism to form wide binaries from close ones may come from the recent study of the distributions of separation as a function of spectral types and ages (Abt, 1986, 1988). In his analysis of 11 Trapezium systems, Abt (1986) found that the maximum component separation from the primary depends on the spectral type. Upper limits are found to be around 2 to 4×10^4 AU among early-type pairs whereas the limit for late-type systems is 3×10^3 AU. However, Trapezium data are difficult to interpret as numerous encounters between their members bias the statistics. For instance, 50% of members will have one or more encounters with an impact parameter of 100 AU in some 12×10^6 years (Herbig and Terndrup, 1986). These are poor statistics (11 systems) but if the trend is confirmed, orbital amplification seems a very simple explanation indeed.

Abt (1988) has studied the maximum separations of VBs in the *Yale Bright Star Catalogue* as well. Two striking results are derived. First, B stars appear to be more separated than other MS stars (peaks at 5100 and 40 AU, respectively, see his Figure 2). This result could be due to selection effects, even if cluster members have been eliminated. The fact to use a sample of bright and therefore nearby stars may bias the distribution. In the case of nearby OB stars, it is well-known that they lie within Gould's belt. In their recent study, Bahcall *et al.* (1987) have shown that 46% of the early-type stars in this catalog are associated with this density enhancement. Thus the quoted B star distribution may just represent (in part) the *local* distribution in the belt. The second unexpected result is that GM giant stars are also more separated than their MS progenitors (see Figure 3 of Abt, 1988). Again, a selection bias might seem to be, as in the case of OB stars, the limited angular resolution that makes it easier to detect bright wide systems at larger distances. The same result, namely, that evolved systems or MS + evolved companion are more separated than MS pairs, can be observed in the study of Wasserman and Weinberg (1987, Figure 2).

The strong dependence of separations on spectral type was also found by Trimble and Cheung (1976) in a sample of spectroscopic binaries. Again, evolved systems appear to be more separated (mean around 5 AU), whereas the 'unevolved' ones have smaller separations (peak around 0.5 AU). This result has been confirmed recently with late-type stars by Griffin (1985). Once more, there is an obvious selection effect due to the broadness of OB spectral lines which prevents the detection of systems with velocity amplitudes less than ~ 15 km s^{-1}.

A simple account for selection effects may be done estimating in the whole sample the biasing function $\mathscr{F}_B$ of each system, defined as $1 - \mathscr{B}$, where $\mathscr{B}$ is the probability of detection. A very careful analysis made by Halbwachs (1983, 1986) has shown that when the biasing function is properly calculated, no statistically significant discrepancy is found between OB and late-type sub-samples. So, once again, no firm conclusion can be drawn until selection effects will be fully understood.

If we assume that biases are small, these new results are quite intriguing. Mass loss and orbital amplification seem a natural explanation, even if further study is clearly needed. Abt (1988) explains his results as due to the disruption of open clusters, as suggested earlier in a different way by Van Albada (1968b). It is well known that due to the low star formation efficiency, open clusters have a stellar density too low to remain stable against galactic tides, suffer relaxation, and their members must expand once the residual gas is removed. If the binary binding energy is low enough, the energy provided by the violent relaxation of the system may disrupt the binary. The time-scale for this scenario is of the order of 15×10^6 years (Lada, 1986). Therefore, very wide binaries are disrupted and the remaining pairs are much closer, thus the decrease in the distribution with age is explained. Note that this mechanism is too slow to account for the observed distribution. Terlevich (1987) has shown that the inclusion of mass loss in N-body simulations of open clusters enhances the depletion of light pairs, and very few late-type MS systems must be found. Further, it is difficult to explain why binaries are preferentially formed in wide pairs, as in principle no orbital periods or separations are to be preferred, from the limit imposed by the coalescence of the pair to the range given by the binding energy and the galactic tidal field. For instance, Mouschovias (1977) has proposed a semi-quantitative scenario to account for the distributions of periods. If the collapse of the GMC core and star formation occurs at constant angular momentum, then the densities at which the magnetic field decouples from the matter of the cloud determines univoquely the periods of the systems. If the density is sufficiently high and the angular momentum small, single stars may form as well but very wide pairs are difficult to form.

The statistics on mass ratios will be reviewed in detail elsewhere, as the claimed bimodal distribution (peaks at $q = 1$ and around 0.3) is probably a pure selection effect (see Gieseking, 1982; Trimble and Walker, 1986; Halbwachs, 1987; Giannuzzi, 1987). If real, the spike at $q = 1$ in the distribution of mass-ratios may be explained (Lucy and Ricco, 1979) if the formation of close binaries ($P \leq 25$ days) is due to fragmentation (formation by the break up of a rotating protostar after the dynamical collapse phase) whereas fission (i.e., bifurcation of a rotating protostar during the Kelvin–Helmholtz contraction to the ZAMS) would produce systems with q near 0.3. However, as shown by Cohen and Kuhi (1979), the Kelvin–Helmholtz phase starts at small radii, limitating the orbital periods of binaries formed by fission. The formation of $q = 1$ systems is also elusive even by fragmentation, and it seems likely that this feature is not an intrinsic property of the distribution. Anyway, whether fission or fragmentation really occurs is still a matter of debate (see Lucy, 1981; Zinnecker, 1984).

Very close binaries may suffer mass transfer and, therefore, this analysis can not be applied as the evolutionary scenarios are very different. On the other hand, close binaries are difficult to form if orbital amplification is very efficient. The different modes of orbital evolution have been reviewed by Eggleton (1986), who suggested that, under some conditions, low mass systems ($M_{\text{tot}} \leq 8\ M_{\odot}$) may evolve into close pairs. However, it seems likely that the mass-loss rate is enhanced by the presence of a close enough companion (say, when the stellar radius is greater than about 0.4 times the Roche lobe radius).

The maximum separation found by Abt (1986, 1988) is about 35000 AU (0.17 pc) in agreement with the study of Bahcall and Soneira (1981) and the detections of Latham *et al.* (1984). This limit has been explained as due to encounters with field stars and molecular clouds (Retterer and King, 1982). However, the Monte Carlo simulations of Weinberg *et al.* (1987) show that the cut-off in the distribution of semi-major axes is not dictated by dynamical arguments. In fact, the existence of such a cut-off as been criticized by Eggen (1986). In any case, a mechanism is needed to form these binaries with such low binding energies. The diffusion process due to the cumulative effects of encounters at large impact parameters proposed by Weinberg *et al.* (1987) just modifies the initial distribution, but does not create wide binaries, as orbital amplification does. Late-type MS wide pairs such as those detected by Latham *et al.* (1984) are not explained by this mechanism, unless the initial distributions give a non-negligible number of these stars at large s or pre-MS and MS mass loss is substantial.

If mass loss is triggered in the pulsational instability strip (Willson and Bowen, 1984), then MS stars in this stage can suffer mass-loss rates as high as $m \sim 9$ (Willson *et al.*, 1987) and, therefore, evolve downwards the MS. For instance, the recent solar model of Guzik *et al.* (1987) assumes that *one* solar mass is lost in some 4×10^9 years, that is, the Sun was initially twice as massive as today. Even if this controversial hypothesis is not clearly verified, such an enhanced rate on the MS will shift the distribution of separations in solar-type stars (in fact, from 1 to 2.5 solar masses) accordingly. An observational test of this hypothesis on the statistics of solar-type stars would easily be done.

Another point is that some wide pairs are not evolved systems, and some close, non-interacting binaries are old. This may be explained as due to the fact that in the mass-loss phase there are clear indications that quite different amounts of mass have been lost by initially similar stars (i.e., from the spread on the HB in globular clusters; see Eggleton, 1986).

A final remark concerns very low mass binaries. Low mass stars do not evolve significantly in a Hubble time, and no *intrinsic* evolutionary scenario can be applied. If further low mass wide binaries are detected, as predicted by Wasserman and Weinberg (1987), mass loss can not provide a simple explanation except if pre-MS mass loss is important.

5. Conclusions

"It may be remarked that the theory of secular decrease in mass has certain potentialities."

W. M. SMART (1925)

It has been shown that orbital amplification due to mass loss is a plausible mechanism to form wide binaries from close systems. The properties of this mechanism are in good agreement with the observed (biased?) statistics, although a more detailed study, particularly of selection effects, would be desirable.

This evolutionary scenario predicts a clear trend in the distribution of separations: samples of evolved (i.e., mass-losing) systems must be more separated that non-evolved pairs. To check this prediction, a study of early-type systems (Lindroos, 1985a, b) is going to be made. Young OB binaries must be less separated than old OB pairs. However, in this case nearly the same effect is expected from the disintegration of open clusters but on a larger time-scale. A better test to discriminate these hypothesis may come from the fact that the Wasserman–Weinberg (1987) model predicts that wide binaries are predominantly made of F5–K5 stars and that a major fraction of all these stars are in wide systems, whereas the mass loss mechanism suggests that most wide pairs are composed of at least one evolved, mass-losing companion.

The time-scale for the evolutionary scenario of Weinberg *et al.* (1987) is about 1 Gyr, comparable to the low ($m \sim 9$) mass-loss time-scale. Orbital amplification acts on a much shorter time-scale (say, 10^{6-8} years), populating the high-s tail, while advection and diffusion tends to deplete this tail on time-scales of about 10^{8-9} years. Thus, only a very careful analysis of separation versus age may disentangle these effects that act together.

Finally, it may be remarked that as mass-outflows processes are not modified by the interaction of the two stars, and distances, masses, and evolutionary stages are usually fairly good known, an absolute stellar scale for mass loss can more easily be derived. This would help to study in more detail the correct (if any) parametrization of mass-loss rates throughout the whole HR diagram.

Acknowledgements

This work is dedicated to the memory of the late Harry J. Frank (1961–1987). It is a pleasure to thank J. Dommanget for encouragement and useful discussions. I am indebted to Z. Kopal for comments on non-isotropic mass loss, and to F. Van 't Veer and M. Friedjung for a critical reading of the manuscript. A travel grant from the IAU is gratefully acknowledged.

References

Abt, H. A.: 1979, *Astron. J.* **84**, 1591.
Abt, H. A.: 1983, *Ann. Rev. Astron. Astrophys.* **21**, 343.
Abt, H. A.: 1986, *Astrophys. J.* **304**, 688.
Abt, H. A.: 1988, *Astrophys. Space Sci.* **142**, 111 (this issue).
Abt, H. A. and Levy, S. G.: 1976, *Astrophys. J. Suppl. Ser.* **30**, 273.
Agayev, A. G., Guseinov, U. H., and Novruzova, H. I.: 1982, *Astrophys. Space Sci.* **81**, 5.
Bahcall, J. N. and Soneira, R. M.: 1981, *Astrophys. J.* **246**, 122.
Bahcall, J. N., Casertano, S., and Ratnatunga, K. U.: 1987, *Astrophys. J.* **320**, 515.
Barlow, M. J. and Cohen, M.: 1977, *Astrophys. J.* **213**, 737.
Beech, J. N.: 1987, *Astrophys. Space Sci.* **132**, 269.
Brosche, P. and Hoffman, M.: 1979, *Astrophys. Space Sci.* **63**, 467.
Che, A., Hempe, K., and Reimers, D.: 1983, *Astron. Astrophys.* **126**, 225.
Casinelli, J. P., and Lamers, H. J. G. L. M.: 1987, in Y. Kondo (ed.), *The Scientific Accomplishments of the IUE*, D. Reidel Publ. Co., Dordrecht, Holland, pp. 139–155.
Chiosi, C. and Maeder, A.: 1986, *Ann. Rev. Astron. Astrophys.* **24**, 329.
Cohen, M. and Kuhi, L. V.: 1979, *Astrophys. J. Suppl. Ser.* **41**, 743.
Conti, P. S.: 1981, 'Effects of Mass Loss in Stellar Evolution' *IAU Colloq.* **52**, 1.

Corbally, C. J.: 1984a, *Astrophys. J.* **285**, 195.
Corbally, C. J.: 1984b, *Astron. J.* **89**, 1887.
Corbally, C. J. and Garrison, R. F.: 1980, *Publ. Astron. Soc. Pacific* **92**, 493.
De Jager, C., Nieuwenhuijzen, H. and Van der Hucht, K. A.: 1986, 'Luminous Stars and Associations in Galaxies', in C. W. H. De Loore *et al.* (eds.), *IAU Symp.* **116**, 109.
De Jager, C., Nieuwenhuijzen, H., and Van der Hucht, K. A.: 1987, in H. J. G. L. M. Lamers and C. W. H. De Loore (eds.), *Instabilities in Luminous Early Type Stars*, D. Reidel Publ. Co., Dordrecht, Holland, pp. 205–207.
Deutsch, A. J.: 1956, *Astrophys. J.* **123**, 210.
Dommanget, J.: 1963, *Ann. Obs. Roy. Belgique*, 3ème Série, **IX** (5), 216.
Dommanget, J.: 1969, in M. Hack (ed.), *Mass Loss from Stars*, D. Reidel Publ. Co., Dordrecht, Holland, pp. 274–277.
Dommanget, J.: 1981, 'Effects of Mass Loss on Stellar Evolution', in C. Chiosi and R. Stalio (eds.), *IAU Colloq.* **59**, 507.
Dommanget, J.: 1984, *Astrophys. Space Sci.* **99**, 23.
Duncan, D. K.: 1984, *Astron. J.* **89**, 515.
Dupree, A. K.: 1986, *Ann. Rev. Astron. Astrophys.* **24**, 377.
Dupree, A. K. and Reimers, D.: 1987, in Y. Kondo (ed.) *The Scientific Accomplishments of the IUE*, D. Reidel Publ. Co., Dordrecht, Holland, pp. 321–353.
Eggen, O. J.: 1986, *Astron. J.* **92**, 125.
Eggleton, P. P.: 1986, in J. Truemper *et al.* (eds.), *The Evolution of Galactic X-Ray Binaries*, D. Reidel Publ. Co., Dordrecht, Holland, pp. 87–105.
Fulconis, M. and Bücher, A.: 1978, *Astron. Astrophys.* **62**, 121.
Garmany, C. D. and Conti, P. S.: 1984, *Astrophys. J.* **284**, 705.
Giannuzzi, M. A.: 1987, *Astrophys. Space Sci.* **135**, 245.
Gieseking, F.: 1982, *Mitt. Astron. Ges.* **57**, 143.
Greenstein, J. L.: 1986a, *Astron. J.* **92**, 859.
Greenstein, J. L.: 1986b, *Astron. J.* **92**, 867.
Griffin, R. F.: 1985, in P. P. Eggleton and J. E. Pringle (eds.), *Interacting Binaries*, D. Reidel Publ. Co., Dordrecht, Holland, pp. 1–12.
Guzik, J. A., Willson, L. A., and Brunish, W. M.: 1987, *Astrophys. J.* **319**, 957.
Hadjidemetriou, J. D.: 1963, *Icarus* **2**, 440.
Hadjidemetriou, J. D.: 1966a, *Icarus* **5**, 34.
Hadjidemetriou, J. D.: 1966b, 'The Theory of Orbits in the Solar System and in Stellar Systems', in G. Contopoulos (ed.), *IAU Symp.* **25**, 129.
Hadjidemetriou, J. D.: 1969, *Adv. Astron. Astrophys.* **5**, 131.
Halbwachs, J. L.: 1983, *Astron. Astrophys.* **128**, 399.
Halbwachs, J. L.: 1985, *Astrophys. Space Sci.* **110**, 149.
Halbwachs, J. L.: 1986, *Astron. Astrophys.* **168**, 161.
Halbwachs, J. L.: 1987, *Astron. Astrophys.* **183**, 234.
Hartmann, L., Edwards, S., and Avrett, E.: 1982, *Astrophys. J.* **261**, 279.
Heintz, W. D.: 1969, *R.A.S.C. J.* **63**, 275.
Herbig, G. H. and Rao, N. K.: 1972, *Astrophys. J.* **174**, 401.
Herbig, G. H. and Terndrup, D. M.: 1986, *Astrophys. J.* **307**, 609.
Hills, J. G.: 1983, *Astrophys. J.* **267**, 322.
Iben, I.: 1987, in G. Gilmore and R. Carswell (eds.), *The Galaxy*, D. Reidel Publ. Co., Dordrecht, Holland, pp. 365–374.
Jeans, J. H.: 1924, *Monthly Notices Roy. Astron. Soc.* **85**, 2.
Jeans, J. H.: 1925, *Monthly Notices Roy. Astron. Soc.* **85**, 912.
Jorissen, A.: 1984, *Bull. Astron. Obs. Roy. Belgique* **IX**(6), 325.
Karovska, M., Nisenson, P., and Noyes, R.: 1986, *Astrophys. J.* **308**, 260.
Knapp, G. R., and Morris, M.: 1985, *Astrophys. J.* **292**, 640.
Kopal, Z.: 1978, *Dynamics of Close Binary Systems*, D. Reidel Publ. Co., Dordrecht, Holland, pp. 264–276.
Kopal, Z.: 1984, *Astrophys. Space Sci.* **99**, 3.
Kwok, S.: 1984, in R. Stalio and J. B. Zirker (eds.), *Relations between Chromospheric-Coronal Heating and Mass Loss in Stars*, Trieste Workshop, pp. 187–210.

Lada, C. J.: 1986, 'Star Forming Regions', in M. Peimbert and J. Jugaku (eds.), *IAU Symp.* **115**, 1.
Lamers, H. J. G. L. M.: 1981, *Astrophys. J.* **245**, 593.
Laplace, P. S.: 1796, *Exposition du système du monde*, Vol. 4, Ch. 17, Paris, p. 401.
Latham, D. W., Tonry, J., Bahcall, J. N., Soneira, R. M., and Schechter, P.: 1984, *Astrophys. J.* **281**, L41.
Lindroos, K. P.: 1985a, *Astron. Astrophys. Suppl. Ser.* **60**, 183.
Lindroos, K. P.: 1985b, *Astron. Astrophys.* **156**, 223.
Lucy, L. B.: 1981, 'Fundamental Problems in the Theory of Stellar Evolution', *IAU Symp.* **93**, 75.
Lucy, L. B. and Ricco, E.: 1979, *Astron. J.* **84**, 401.
MacMillan, W. D.: 1925, *Monthly Notices Roy. Astron. Soc.* **85**, 904.
Maeder, A. and Meynet, G.: 1987, *Astron. Astrophys.* **182**, 243.
Martin, E. L.: 1962, *Mem. Soc. Astron. Ital.* **33**(4), 1.
Massa, D. and Endal, A. S.: 1987a, *Astron. J.* **93**, 579.
Massa, D. and Endal, A. S.: 1987b, *Astron. J.* **93**, 761.
Migozzi, M. C. and Lafon, J.-P. J.: 1985, in A. B. Underhill and A. G. Michalitsianos (eds.), *The Origin of Non-Radiative Heating/Momentum in Hot Stars*, NASA CP-2358, 236–239.
Mouschovias, T. C.: 1977, *Astrophys. J.* **211**, 147.
Mitalas, R.: 1980, *Am. J. Phys.* **48**, 226.
Mundt, R., Walker, F. M., Feigelson, E. D., Finkenzeller, U., Herbig, G. H., and Odell, A. P.: 1983, *Astrophys. J.* **269**, 229.
Noerdlinger, P. D.: 1982, *Am. J. Phys.* **50**, 733.
Oblak, E. and Chareton, M.: 1980, *Astron. Astrophys. Suppl. Ser.* **41**, 255.
Olsen, E. H.: 1982, *Astron. Astrophys.* **110**, 215.
Reimers, D.: 1977, *Astron. Astrophys.* **61**, 217.
Reimers, D.: 1983, in R. Stalio and R. N. Thomas (eds.), *Effects of Variable Mass Loss in the Local Stellar Environment*, Trieste Workshop Series, pp. 149–162.
Retterer, J. M. and King, I. R.: 1982, *Astrophys. J.* **254**, 214.
Roddier, F., Roddier, C., and Karovska, M.: 1985, in M. Morris and B. Zuckerman (eds.), *Mass Loss from Red Giants*, D. Reidel Publ. Co., Dordrecht, Holland, p. 63–64.
Smart, W. M.: 1925, *Monthly Notices Roy. Astron. Soc.* **85**, 423.
Terlevich, E.: 1987, *Monthly Notices Roy. Astron. Soc.* **224**, 193.
Trimble, V.: 1978, *Observatory* **98**, 163.
Trimble, V. and Cheung, C.: 1976, in P. P. Eggleton, S. Mitton, and J. Whelan (eds.), 'Structure and Evolution of Close Binary Systems', *IAU Symp.* **73**, 369.
Trimble, V. and Walker, D.: 1986, *Astrophys. Space Sci.* **126**, 243.
Van Albada, T. S.: 1968a, *Bull. Astron. Inst. Neth.* **20**, 47.
Van Albada, T. S.: 1968b, *Bull. Astron. Inst. Neth.* **20**, 57.
Van Dessel, E. L. and Dommanget, J.: 1980, *Irish Astron. J.* **14**, 226.
Vardya, M. S.: 1984, *Astrophys. Space Sci.* **107**, 141.
Vardya, M. S.: 1987, *Astrophys. J.* **312**, 812.
Waldron, W. L.: 1985, in A. B. Underhill and A. G. Michalitsianos (eds.), *The Origin of Non-Radiative Heating/Momentum in Hot Stars*, NASA CP-2358, pp. 95–98.
Wasserman, I. and Weinberg, M. D.: 1987, *Astrophys. J.* **312**, 390.
Weinberg, M. D., Shapiro, S. L., and Wasserman, I.: 1987, *Astrophys. J.* **312**, 367.
Willson, L. A. and Bowen, G. H.: 1984, *Nature* **312**, 429.
Willson, L. A., Bowen, G. H., and Struck-Marcell, C.: 1987, *Comm. Astrophys.* **12**, 17.
Zinnecker, H. A.: 1984, *Astrophys. Space Sci.* **99**, 41.

CONCLUDING REMARKS

ZDENĚK KOPAL
Department of Astronomy, University of Manchester, England

In conclusion of the colloquium which brought us all to Brussels this week, the first pleasant duty of the last speaker should be to express – on behalf of us all – our very sincere thanks to the Scientific and Local Organizing Committees of the occasion – and, in particular, to their chairmen, Dr Jean Dommanget and Dr E. van Dessel of the Observatoire Royal de Belgique – as well as to all other sponsoring organizations for the excellent arrangements which they made to run the entire programme so effectively and pleasantly for all participants.

Perhaps the most fruitful idea underlying the entire colloquium was its subject itself, dedicated to a discussion of the physical state of *wide* components in double and multiple stellar systems, and to problerms arising with their interpretation. This subject has indeed been much neglected in recent decades; with the spotlight turned on 'close' binary systems – those 'trouble doubles' which claimed so much attention at the centre of the stage of double-star astronomy in the past generation – witness the ratio of 65 to 260 of current membership of the relevant IAU Commissions 26 and 42 by the end of 1985. While it is perhaps not for me to deprecate the significance of close binaries for the general framework of stellar astronomy, it is a fact (and one, perhaps, not as generally realized as it should be) that the larger number of the 'degrees of freedom' possessed by close pairs is also apt to make an interpretation of their observations more ambiguous than is generally realized; and this ambiguity (if not properly kept in mind) can easily mislead their students into entirely wrong conclusions.

In contrast – and precisely because of their greater separations – wide binaries can provide observational evidence much more specific than close pairs for an interpretation of their observations. Therefore, it is only natural that a major part of our programme has been devoted to a survey of the observational facts now in our possession on wide binary systems – ranging from binaries with components separated by a few astronomical units to wide pairs whose components are characterized only by common proper motions. 'Observe, observe and observe' was the clarion call of the late Robert Grant Aitken (1864–1951) – that grand master of double-star astronomy in the first half of this century – but this call carried only one-half of the relevant message: for while necessary, observations alone are not sufficient to disclose the full wealth of their message without appropriate analysis, necessary to establish the proper place of our subject in contemporary stellar astronomy.

What is the principal significance of binary or multiple stars – close or wide – for stellar astronomy? It is the fact that (whatever may be the more specific way of their formation) they represent physical associations of objects originating at essentially the same time and from material of essentially the same chemical composition; and if so,

Astrophysics and Space Science **142** (1988) 305–310.

their subsequent evolutionary tracks can differ only as a consequence of an initial difference in mass. The same is true, of course, of multiple systems of different orders of magnitude generally referred to as clusters. But the masses of cluster stars cannot be inferred otherwise than from the luminosity by an appeal to specific models of internal structure. Multiple stars represent, on the other hand, rudimentary 'clusters' in which the masses of the individual components can be determined directly from the observations – without any appeal to their luminosity. If, however, their initial masses are different, such systems represent a veritable royal road for studies of *differential evolution* of the stars, caused by possible difference in mass – either from the time of their origin, or acquired since in the course of their subsequent evolution.

Moreover – and this is essential for our subject – the observable effects of such a differential evolution are documented by wide pairs more convincingly than by the close ones in which interaction phenomena may complicate subsequent evolution. A biologist might, therefore, say that – in contrast to close pairs – wide binaries represent in a much more real sense a 'pure culture' of stellar population with a more limited number of degrees of freedom, which permits us to study their evolution from cradle to grave in simpler terms than would be the case with more complicated cultures; while retaining the validity of their basic characteristics – i.e., of equal age and initial composition – up to mutual separations of the order of 10^4 AU (cf. Section I-1 of Kopal, 1978).

As far as the *origin* of wide binaries is concerned, an important feature of the problem was brought out at these meetings by Dommanget (see pp. 23–30 of this volume), demonstrating that virtually all wide binaries in our stellar neighbourhood with known visual orbits constitute a relatively small number of groups characterized by *parallelism* of their orbital planes – reminiscent no doubt of the dynamical characteristics of pre-existing vortices in interstellar matter by a disintegration of which the binary groups originated.

That the percentage of stars in the sky constituting physical double or multiple systems is very high has been known for a long time. Already Kuiper (1942) estimated that their percentage in our galactic spiral arm may exceed 80%; and subsequent work has tended to push this estimate (limited always by observational selection) ever closer to 100%. Dommanget's recent work has shown that most pre-existing vortices of interstellar gas collapse into binary or multiple systems; do all stars regarded by us as single have their origin in the dissolution of such systems? A positive answer to this question is not yet known. However, new material presented at this Colloquium suggests with ever increasing verisimilitude that this may indeed be the case; and that the percentage of stars now appearing as single reflects only the rate at which gravitational perturbation keeps dissolving wide multiple systems in the course of time.

But is this true also of our Sun? And if so, was the Sun with its planetary system initially born as the component of a binary (or more complicated system) dissolved in the past 4–5 × 10^9 yr? A convincing answer is again not yet known; but the possibility that it may turn out to be in the affirmative is on the increase. If this was indeed the case, the divorced partner of a former union should be sought somewhere within a few degrees of the ecliptic as a star exhibiting no secular proper motion, but an annual parallax well in excess of 1 arcsecond.

Direct terrestrial (geological) evidence that this may have been the case is conspicuous by its absence; for next to nothing has been preserved in the Earth's crust that is older than 4×10^9 yr. However, on the Moon it is a very different story; and the micro-chemistry of the lunar soil brought back to the Earth by successive Apollo missions between 1969–1972, and of radiogenic ages exceeding 4×10^9 yr, disclosed that it possesses a nuclear composition so complicated that – in order to account for it by the action of nearby supernovae (a favourite hypothesis entertained at that time) – not less than 17 such discrete events would have to be postulated to account for the outcome! This, while not impossible, is very unlikely; and its unlikelihood should only encourage us to consider alternative ways to account for the observed facts. Could, perchance, there will be a possibility (nay, probability) that the early Sun was attended by another star, whose radiation complicated the chemical imprint preserved in the oldest strata of the solar system accessible to our analysis? Again only the future can, in due course, pronounce an appropriate verdict.

But let us not leave our subject at this purely hypothetical stage, and call instead to our witness stand certain *individual* wide binary systems whose components are found to differ widely in their evolutionary stages – i.e., pairs of stars in which (in extreme cases) a Main-Sequence star may be attended by a white dwarf. Such systems are by no means unknown (or even rare): in the close neighbourhood of the Sun, Sirius (α CMa) or Procyon (α CMi) represent typical examples of such a situation; and turning further afield, Agayev *et al.* (1982) identified not less than 67 such systems among wide binaries listed in their catalogue. White dwarfs detected in such systems are as a rule by many magnitudes fainter than their Main-Sequence mates and as, accordingly, observational selection operates strongly against their discovery, the actual percentage of such queer pairs in the sky is likely to be considerably higher.

What is so significant about such pairs? It is the fact that (in every case which could be subjected to an observational test) white-dwarf components turn out to be *less* of the same age and started their life with the same composition, how could a star now less massive have overtaken its more massive mate in their respective evolutionary tracks by so wide a margin?

Within the framework of our present knowledge of nuclear evolution of the stars, the only physically sensible way out of this evolutionary paradox is to admit that the present white dwarf must indeed have embarked on its evolution with a larger mass than it possesses today, but *lost* much of it since its present stage was attained. Such a conclusion should, perhaps, no longer surprise us; for the situations in which the components of more advanced evolutionary types (subgiants!) turned out to be less massive than their early-type mates were also encountered in *close* binary systems (the Algol paradox!) a generation before (cf. Kopal, 1955, 1956) and processes put forward to account for this anomaly, connected with geometrical properties of the Roche Model. Whatever merit such processes may, however, possess for close binary system*, these

* The conjuring word for the latter has been 'mass transfer' (a process in which the masses of the components can change, but their sum remains the same) – in contrast with 'mass loss' of one or both stars (a process by which the mass of the system as a whole will undergo a secular decrease). The first process

are clearly irrelevant for wide binaries like Sirius or Procyon – pairs so wide that none of their components could approach (let alone attain) their respective Roche limits even if the initial mass of one (or both) components had been several times as large as it is at the present time. Yet the 'evolutionary paradox' in all binaries appears to be of the same type (more extreme, in fact, in wide binaries than in the close ones), and probably due to the same reason. What can it be?

In an effort to localize it, let us depart from certain relevant facts established by the observations. According to van de Kamp (1971) the present most probable masses $m_A + m_B$ are 2.3 + 0.98 ⊙ for Sirius; 1.74 + 0.63 ⊙ for Procyon; and their current mean separations of the components are 20.1 and 15.8 astronomical units. If, moreover, the mass of Sirius A has remained constant and equal to 2.3 ⊙ since the time of its formation, the initial mass of Sirius B could not have been much less than 5 ⊙ to attain its present state; rendering the combined mass of the initial mass of the Sirius system close to 7 ⊙ – rather than 3 ⊙ as it is at the present time.

What would have been the size of its relative orbit? At the mutual separation (now, or in the past) of the components their physical interaction (other than gravitational attraction) is insignificant, the only conceivable way of mass removal from the system would be by *stellar wind*, acting *isotropically* at a velocity of several hundred km s^{-1}. If so, it follows from a well-known integral of the two-body problem with variable mass (cf. Hadjidemetriou, 1963) that a product of the semi-latus rectum times the total mass of the system must be conserved. If this theorem is applied to Sirius, it follows that a diminution of its total mass from 7 ⊙ to 3 ⊙ should have increased the size of its relative orbit 7/3 times, implying that if its present size is equal to 20.1 AU, its initial dimensions should have been only 8.6 AU or 1850 radii of the Sun. This latter figure would still render the Roche limit of the (then) more massive component very much larger than a star of mass 5 ⊙ could ever hope to attain in the course of its evolution; and, therefore, large-scale mass removal could have been carried out only by stellar wind.

When did this happen, and where is the lost mass now? At the beginning of this century – and long before the story of evolution was placed on a sound physical basis, that colourful American astronomer T. J. J. See pointed out that, in certain astronomical sources of the ancient world the star Sirius had been consistently described as 'red' (cf. See, 1926); should it actually have been so, this would have dated the transition of Sirius B from a red giant to a white dwarf to a relatively recent past. This argument was not accepted without criticism (cf., e.g., Dittrich, 1927; Meissner, 1927; Stentzel, 1927); but revived very recently by Schlosser and Bergmann (1985) or Bruhweiler *et al.* (1986) with reference to a 5th-century chronicle by Gregory of Tours; though their

---

can be described as 'cannibalism' perpetrated by one component on its mate; ut it can operate effectively only at low velocities of mass ejection – unproven by the observations, and postulated mainly by hypothesis. The second process (not requiring the mass of the system to remain invariant) can, on the other hand; be described (by its consequences) as 'fratricide' – by which an uneven stellar wind brings about an increase in size of the system, leading to its eventual dissolution. In such a terminology, the present evolutionary trends in wide binary systems can thus be succinctly described as 'fratricide, but no cannibalism'.

identification was again vigorously contested by van Gent (1986a, b) and also by Tang (1986). At the present time the case may still be regarded as open, and the timing of the event uncertain. 'Thus, it may have been the incomprehending eye of a Pithecanthropus, rather than that of an Egyptian priest or Greek sage, which may have beheld this heavenly wonder; but the event was long forgotten before writing was invented' (cf. Kopal, 1959, p. 543); or are we still uncertain about the rate of thermonuclear runaways? Bruhweiler *et al.* (1986) seem to think so; but it may be wise to await further developments on both astronomical and philological fronts before drawing more definitive conclusions.

The patent uncertainty about the time-scale of the phenomenon should not, of course, lessen our curiosity on the whereabouts of the mass which must have been lost by Sirius some time in the past to reduce its originally more massive component to its state of the present white dwarf. An amount of it equal to several solar masses must have been dispersed to accomplish this purpose; what has happened to it and where is it now?

If the late Sir Arthur Eddington (1882–1944) were still with us, and preparing a new edition of his delightful book on *Stars and Atoms* (Eddington, 1927), as an avid reader of detective stories he would have no doubt augmented his 'Story of the Companion of Sirius' (*op. cit.*, pp. 48–53) by an addendum entitled 'The Mystery of the Missing Mass'; for the recent discovery of an X-ray source in the proximity of Sirius (cf. Chlebowski *et al.*, 1981), or of a peculiar flare star near Sirius (cf. Sharma *et al.*, 1983) has given the subject of our inquiry an entirely new twist.

Both these enigmatic objects – separated from Sirius A by 6–9 arc min (but, incidentally, much less from each other) – cannot be any nearer to it than several thousand astronomical units; and according to their observers, they are very probably background objects separated from Sirius by many parsecs in space. This may well be the case; but if so, the whereabouts of the 'missing mass' continues to confront us with an open problem. As the minimum distance between these new objects and Sirius AB system cannot be less than 1000 AU, the orbital periods of such hypothetical 'third bodies' would be much too long to disclose indications of any orbital motions still for many centuries to come. The only practicable way to establish their possible physical association with Sirius would be by a commonality of their apparent proper motions. It is to be hoped some attention may be paid to these in the near future; for should they – against all expectations – prove to be the same as that of the visible pair, the whole problem of the mass loss accompanying the formation of a white dwarf in our close stellar neighbourhood could still take an unexpected turn.

References

Agayev, A. G., Guseinov, O. H., and Novruzova, H. I.: 1982, *Astrophys. Space Sci.* **81**, 5.
Bruhweiler, F. C., Kondo, Y., and Sion, E. M.: 1986, *Nature* **324**, 235.
Chlebowski, T., Halpern, J. P., and Steiner, J. E.: 1981, *Astrophys. J.* **247**, L35.
Dittrich, A.: 1927, *Astron. Nachr.* **231**, 385.
Eddington, A. S.: 1927, *Stars and Atoms*, Clarendon Press, Oxford.
Hadjidemetriou, J.: 963, *Icarus* **2**, 440.
Kopal, Z.: 1955, *Ann. Astrophys.* **18**, 379.

Kopal, Z.: 1956, *Ann. Astrophys.* **19**, 298.
Kopal, Z.: 1959, *Close Binary Systems*, Chapman-Hall and John Wiley, London and New York.
Kopal, Z.: 1978, *Dynamics of Close Binary Systems*, D. Reidel Publ. Co., Dordrecht, Holland.
Kuiper, G. P.: 1942, *Astrophys. J.* **95**, 201.
Meissner, O.: 1927, *Astron. Nachr.* **231**, 391.
Schlosser, W. and Bergmann, W.: 1985, *Nature* **318**, 45.
See, T. J. J.: 1926, *Astron. Nachr.* **229**, 245.
Sharma, D. P., Marar, T. M. K., Seetha, S., Shylaja, K. S., Kasturirangan, K., Rao, U. R., Bhattacharyya, Ju. C., and Rozario, J. M.: 1983, *Astrophys. Space Sci.* **91**, 467.
Stentzel, A.: 1927, *Astron. Nachr.* **231**, 387.
Tang, T. B.: 1986, *Nature* **311**, 532.
Van de Kamp, P.: 1971, *Ann. Rev. Astron. Astrophys.* **9**, 103.
Van Gent, R. H.: 1986a, *Sky Telesc.* **72**, 556.
Van Gent, R. H.: 1986b, *Zenit* **13**, 212.

GPSR Compliance
The European Union's (EU) General Product Safety Regulation (GPSR) is a set of rules that requires consumer products to be safe and our obligations to ensure this.

If you have any concerns about our products, you can contact us on

ProductSafety@springernature.com

In case Publisher is established outside the EU, the EU authorized representative is:

Springer Nature Customer Service Center GmbH
Europaplatz 3
69115 Heidelberg, Germany

www.ingramcontent.com/pod-product-compliance
Ingram Content Group UK Ltd.
Pitfield, Milton Keynes, MK11 3LW, UK
UKHW020239290926
15470UKWH00002B/625
* 9 7 8 9 0 2 7 7 2 7 3 7 4 *